Soil Erosion Measurement Techniques and Field Experiments

Soil Erosion Measurement Techniques and Field Experiments

Editors

Vito Ferro
Alessio Nicosia

MDPI

Basel • Beijing • Wuhan • Barcelona • Belgrade • Novi Sad • Cluj • Manchester

Editors
Vito Ferro
Department of Agricultural, Food and Forest Sciences
University of Palermo
Palermo
Italy

Alessio Nicosia
Department of Agricultural, Food and Forest Sciences
University of Palermo
Palermo
Italy

Editorial Office
MDPI
St. Alban-Anlage 66
4052 Basel, Switzerland

This is a reprint of articles from the Special Issue published online in the open access journal *Water* (ISSN 2073-4441) (available at: www.mdpi.com/journal/water/special_issues/soil_erosion_field_experiments).

For citation purposes, cite each article independently as indicated on the article page online and as indicated below:

Lastname, A.A.; Lastname, B.B. Article Title. *Journal Name* **Year**, *Volume Number*, Page Range.

ISBN 978-3-7258-0938-7 (Hbk)
ISBN 978-3-7258-0937-0 (PDF)
doi.org/10.3390/books978-3-7258-0937-0

Cover image courtesy of Alessio Nicosia

Contents

About the Editors

Vito Ferro

Prof. Vito Ferro is a full professor of hydraulics and hydrology, soil erosion and conservation, and environmental restoration at the University of Palermo. He has been involved in many projects, such as the National Biodiversity Future Center (NBFC). In 2007, he won the "Karl Emil Hilgard Hydraulic Prize".

Alessio Nicosia

Dr. Alessio Nicosia is an assistant professor at the University of Palermo. His research deals with hydraulics and hydrology, soil erosion and conservation, and environmental restoration. The main subject of his research is the study of flow resistance in rills.

water

MDPI

Editorial

Soil Erosion Measurement Techniques and Field Experiments

Vito Ferro [1,2,*] and Alessio Nicosia [1]

1 Department of Agricultural, Food and Forestry Sciences, University of Palermo, Viale delle Scienze, Building 4, 90128 Palermo, Italy; alessio.nicosia@unipa.it
2 National Biodiversity Future Center (NBFC), 90133 Palermo, Italy
* Correspondence: vito.ferro@unipa.it

Soil erosion is a process in which soil particles are first detached from the soil surface and then transported by erosive agents such as rainfall, overland flow and channelized flows in rills, ephemeral gullies and gullies [1–3]. Accelerated soil erosion affects both natural and anthropogenic environments. It is also responsible for land productivity decrease due to the removal of soil organic matter and plant nutrients [4,5]. The negative effects of soil erosion include in-site effects, such as degradation of soil structure, loss of organic matter and nutrient content, and reduction in the cultivable soil layer [6]. Erosion also determines off-site damages due to soil particles entering the water system, such as sedimentation into channels, loss of reservoir storage, eutrophication of waterways, and contamination due to fertilizer and chemical pesticides [7].

Accurate and repeatable measurements of erosion processes are required both for understanding and realizing correct modeling. Experiments provide an opportunity to investigate to what extent the concepts used in models are a truly valid descriptions of the erosion processes occurring.

The main aim of this Special Issue was to collect papers dealing with (1) experimental sites for measuring soil erosion at different spatial (plot, hillslope, basin) scales, (2) field experiments that aim to study the soil erosion processes (interrill, rill, and gully erosion), and (3) new methods and procedures for measuring soil erosion processes (e.g., three-dimensional photo-reconstruction techniques, measurement of erosion features using aerial and terrestrial acquisition platforms, and tracers).

Eleven articles (ten research articles and one review) are published in this Special Issue. They cover different aspects of erosion processes, investigate cause–effect relationships, and develop new models for predicting soil erosion. The effects of rainfall, cropping system, etc., on soil erosion have been discussed, and different types of erosion (i.e., rills, gullies) have been studied.

Among the research articles, the papers by de Oliveira et al. [8] and Nicosia et al. [9] deal with rill erosion. In particular, the first [8] presents the development and verification of an improved and cost-effective flume apparatus and corresponding testing methodology. The authors tested both the apparatus and methodology using statically compacted specimens of a latosol from the central region of Brazil. They produced erosion curves with repeatability that were superior with respect to their initial linear and transition portions. The second [9], instead, assessed the influence of the rill profile shape on flow resistance law. The analysis demonstrated that the component of the Darcy–Weisbach friction factor due to the profile shape varies in the range 0.68–14.6% of the overall friction factor for the concave profile, and from 3.4 to 26.9% for the convex profile. The authors also proved that the concave profile leads to an eroded rill volume lower than those detected for the uniform and convex profiles (reduction of 57.9% compared to the uniform profile).

Only the paper by Nkonge et al. [10] deals with gully erosion. They studied the susceptibility of the Kakia–Esamburmbur catchment in Narok, Kenya, to this kind of erosion. The authors obtained that land use/cover, distance to road, sediment transport index, and topographic wetness index significantly influence gully occurrence in the catchment.

Citation: Ferro, V.; Nicosia, A. Soil Erosion Measurement Techniques and Field Experiments. *Water* **2023**, *15*, 2846. https://doi.org/10.3390/w15152846

Received: 28 July 2023
Revised: 31 July 2023
Accepted: 1 August 2023
Published: 7 August 2023

Many papers investigated new solutions for preventing soil erosion phenomena or proposed their relationship with the examined variables. Merlo et al. [11], in turn, evaluated the relationship between soil properties (microbial biomass carbon, basal soil respiration, and metabolic quotient) and erosion in areas managed with different cropping system practices under no-tillage in the Brazilian Cerrado. They demonstrated that not only physical and chemical, but also biological properties are deeply affected by erosion. Regarding the application of agricultural practices, Vianney Nsabiyumva et al. [12] evaluated, in three provinces of Morocco, their effects, combined with olive tree plantations, on the hydrological response (final infiltration, imbibition of rainwater, runoff coefficient, and soil detachment). The authors demonstrated that vegetation has an important role in moisture conservation in surficial depths in all sites and reduces runoff. Banu and Attom [13] proposed the use of quicklime to stabilize cohesionless soils, demonstrating its efficiency for both poorly graded and well-graded soils. Its use significantly improves the strength, critical shear stress, and erosion rate index of the soil. Bombino et al. [14] presented an interesting case study, developed in Calabria, using burned felled logs for controlling soil erosion and favoring forest self-regeneration in post-fire conditions, obtaining encouraging results.

Todisco et al. [15] developed a runoff correction factor for the USLE using rainfall and satellite antecedent soil moisture data. They also validated the obtained estimates of runoff and soil loss using plot-scale measurements obtained at SERLAB (Soil Erosion Laboratory) of the University of Perugia. The authors found that the event rainfall depth added to the antecedent soil moisture is a suitable predictor of the runoff.

The article by De Girolamo et al. [16] deals with the measurements of the suspended sediment load in two mountainous river basins in Apulia. The authors developed sediment rating curves to address gaps in the suspended sediment concentration time series. They also obtained that the majority of the suspended sediment load was transported during high-flow conditions, accounting for over 80% of the total load, while, for low-flow conditions, it constituted less than 1% of the total load.

Vergni and Todisco [17] used the datasets comprising many years of soil loss observations at the plot-scale experimental site SERLAB, and applied the random forest machine learning model. This model achieved a global accuracy of 84.8% in recognizing erosive and non-erosive events, demonstrating slightly better performances than non-machine learning methodologies.

Finally, the review paper by Pampalone et al. [18] provides an insight into the scientific activity conducted from the 2000s using the data collected in the plots installed at the Sparacia experimental area for soil erosion measurement in Sicily (South Italy). The authors also described methods and procedures for quantify soil erosion processes (sediment sampling and water level reading in the storage tanks for total erosion measurements; profilometer, and structure from motion technique for rill erosion measurements).

The research's new and relevant findings will help to study soil erosion and understand the dynamics of some processes regarding this topic. The chosen articles advance our knowledge on how water soil erosion can be controlled and reduced, and could give further ideas about what can be investigated in future and what the main challenges are in this field of study.

Data Availability Statement: No new data were created or analyzed in this study. Data sharing is not applicable to this article.

Conflicts of Interest: The authors declare no conflict of interest.

References

1. Di Stefano, C.; Nicosia, A.; Palmeri, V.; Pampalone, V.; Ferro, V. Estimating flow resistance in steep slope rills. *Hydrol. Process.* **2021**, *35*, e14296. [CrossRef]
2. Di Stefano, C.; Nicosia, A.; Palmeri, V.; Pampalone, V.; Ferro, V. Rill flow velocity and resistance law: A review. *Earth Sci. Rev.* **2022**, *231*, 104092. [CrossRef]

3. Di Stefano, C.; Ferro, V.; Pampalone, V.; Sanzone, F. Field investigation of rill and ephemeral gully erosion in the Sparacia experimental area, South Italy. *Catena* **2013**, *101*, 226–234. [CrossRef]
4. Panagos, P.; Borrelli, P.; Ballabio, C.; Lugato, E.; Meusburger, K.; Montanarella, L.; Alewell, C. The new assessment of soil loss by water erosion in Europe. *Environ. Sci. Pol.* **2015**, *54*, 438–447. [CrossRef]
5. Di Stefano, C.; Nicosia, A.; Pampalone, V.; Ferro, V. Soil loss tolerance in the context of the European Green Deal. *Heliyon* **2023**, *9*, e12869. [CrossRef] [PubMed]
6. Food and Agricultural Organization (FAO). *Soil Erosion: The Greatest Challenge to Sustainable Soil Management*; FAO: Rome, Italy, 2019; 100p, Available online: https://www.fao.org/documents/card/en/c/ca4395en (accessed on 26 July 2023).
7. Carollo, F.G.; Di Stefano, C.; Nicosia, A.; Palmeri, V.; Pampalone, V.; Ferro, V. A new strategy to assure compliance with soil loss tolerance at a regional scale. *Catena* **2023**, *223*, 106945. [CrossRef]
8. de Oliveira, V.N.; Gitirana, G.d.F.N., Jr.; dos Anjos Mascarenha, M.M.; Sales, M.M.; Varrone, L.F.R.; da Luz, M.P. An Enhanced Flume Testing Procedure for the Study of Rill Erosion. *Water* **2021**, *13*, 2956. [CrossRef]
9. Nicosia, A.; Di Stefano, C.; Palmeri, V.; Pampalone, V.; Ferro, V. Evaluating the Effects of the Rill Longitudinal Profile on Flow Resistance Law. *Water* **2022**, *14*, 326. [CrossRef]
10. Nkonge, L.K.; Gathenya, J.M.; Kiptala, J.K.; Cheruiyot, C.K.; Petroselli, A. An Ensemble of Weight of Evidence and Logistic Regression for Gully Erosion Susceptibility Mapping in the Kakia-Esamburmbur Catchment, Kenya. *Water* **2023**, *15*, 1292. [CrossRef]
11. Merlo, M.N.; Avanzi, J.C.; Silva, L.d.C.M.d.; Aragão, O.O.d.S.; Borghi, E.; Moreira, F.M.d.S.; Thebaldi, M.S.; Resende, Á.V.d.; Silva, M.L.N.; Silva, B.M. Microbiological Properties in Cropping Systems and Their Relationship with Water Erosion in the Brazilian Cerrado. *Water* **2022**, *14*, 614. [CrossRef]
12. Vianney Nsabiyumva, J.M.; Apollonio, C.; Castelli, G.; Petroselli, A.; Sabir, M.; Preti, F. Agricultural Practices for Hillslope Erosion Mitigation: A Case Study in Morocco. *Water* **2023**, *15*, 2120. [CrossRef]
13. Banu, S.; Attom, M. Internal Erosion Stabilization of Cohesionless Soil Using Lime. *Water* **2023**, *15*, 1992. [CrossRef]
14. Bombino, G.; Barbaro, G.; Pérez-Cutillas, P.; D'Agostino, D.; Denisi, P.; Foti, G.; Zimbone, S.M. Use of Logs Downed by Wildfires as Erosion Barriers to Encourage Forest Auto-Regeneration: A Case Study in Calabria, Italy. *Water* **2023**, *15*, 2378. [CrossRef]
15. Todisco, F.; Vergni, L.; Ortenzi, S.; Di Matteo, L. Soil Loss Estimation Coupling a Modified USLE Model with a Runoff Correction Factor Based on Rainfall and Satellite Soil Moisture Data. *Water* **2022**, *14*, 2081. [CrossRef]
16. De Girolamo, A.M.; Ricci, G.F.; Abdelwahab, O.M.M.; Lo Porto, A.; Milillo, F.; Netti, A.M.; Gentile, F. Suspended sediment transport in Mediterranean streams: Monitoring and load estimation. *Water* **2023**, *15*, 2715. [CrossRef]
17. Vergni, L.; Todisco, F. A Random Forest Machine Learning Approach for the Identification and Quantification of Erosive Events. *Water* **2023**, *15*, 2225. [CrossRef]
18. Pampalone, V.; Carollo, F.G.; Nicosia, A.; Palmeri, V.; Di Stefano, C.; Bagarello, V.; Ferro, V. Measurement of Water Soil Erosion at Sparacia Experimental Area (Southern Italy): A Summary of More than Twenty Years of Scientific Activity. *Water* **2022**, *14*, 1881. [CrossRef]

water

MDPI

Article

Suspended Sediment Transport in Mediterranean Streams: Monitoring and Load Estimation

Anna Maria De Girolamo [1], Giovanni Francesco Ricci [2,*], Ossama M. M. Abdelwahab [2], Antonio Lo Porto [1], Fabio Milillo [2], Addolorata Maria Netti [2] and Francesco Gentile [2]

[1] Water Research Institute, National Research Council, 70132 Bari, Italy; annamaria.degirolamo@ba.irsa.cnr.it (A.M.D.G.); antonio.loporto@ba.irsa.cnr.it (A.L.P.)
[2] Department of Soil, Plant and Food Sciences, University of Bari Aldo Moro, 70126 Bari, Italy; ossama.abdelwahab@uniba.it (O.M.M.A.); fabio.milillo@uniba.it (F.M.); a.netti21@studenti.uniba.it (A.M.N.); francesco.gentile@uniba.it (F.G.)
* Correspondence: giovanni.ricci@uniba.it; Tel.: +39-080-5442958

Abstract: The suspended sediment (SS) load provides valuable insights into soil loss magnitude, requiring comprehensive monitoring of streamflow (Q) and SS concentrations (SSC) across various hydrological conditions. The primary aim of this paper was to quantify SS loads in two mountainous river basins: the Carapelle (506 km^2) and the Celone (72 km^2) located in Apulia (Southeast Italy) where different monitoring strategies were adopted (i.e., continuous and discrete). The specific objective was to develop sediment rating curves to address gaps in the SSC time series. An optical probe was used to continuously monitor the SSC during 2007 to 2011 in a river section of the Carapelle river, while Q was measured with the ultrasonic method. A comprehensive dataset comprising continuous Q measurements and discrete SSC measurements was systematically acquired for the Celone river over the period of 2010 to 2011. Distinct sediment rating curves were formulated for three specific subsets of data delineated by discernible hydrological conditions (i.e., high, normal, and low flow) and SSCs were computed for the missing daily records. The annual specific sediment load exhibited a range of 2.4 to 6.06 t ha^{-1} yr^{-1} for the Celone river, while the Carapelle river displayed a range of 0.9 to 7.45 t ha^{-1} yr^{-1}. A significant majority of the SS load was transported during high-flow conditions, accounting for over 80% of the total load. In contrast, during low-flow conditions, the SS load constituted less than 1% of the total load. The findings of this study highlight the significance of the hydrological regime as a critical factor influencing sediment transport in mountainous Mediterranean rivers. Furthermore, it demonstrates that the duration of the sampling period, along with its specific characteristics, such as dry or wet conditions, can have a substantial impact on the accurate quantification of the sediment load.

Keywords: monitoring; suspended sediment concentrations; streamflow; sediment rating curves; specific sediment load; temporary rivers

Citation: De Girolamo, A.M.; Ricci, G.F.; Abdelwahab, O.M.M.; Lo Porto, A.; Milillo, F.; Netti, A.M.; Gentile, F. Suspended Sediment Transport in Mediterranean Streams: Monitoring and Load Estimation. *Water* **2023**, *15*, 2715. https://doi.org/10.3390/w15152715

Academic Editors: Vito Ferro and Alessio Nicosia

Received: 21 June 2023
Revised: 24 July 2023
Accepted: 25 July 2023
Published: 27 July 2023

1. Introduction

Soil erosion and land degradation are prevalent issues observed in the majority of river basins across the Mediterranean region [1]. In these basins, the geomorphologic and climatic characteristics and the agricultural practices adopted (e.g., tillage) contribute to soil loss [2]. Local, national, European, and international policymakers have shown a growing interest in erosion and its effects on soil and surface waterways in the last few decades. Extensive research efforts have been dedicated to investigating the factors influencing the transport of suspended sediment (SS) and developing models for predicting sediment yield and load [3–5].

Soil erosion may greatly affect soil and surface ecosystems. SS transport may be of concern because it plays a significant role in the transport of contaminants to surface waters resulting from anthropogenic activities (such as farming and breeding) [2,6]. Furthermore, sediment transfer may contribute to lake eutrophication and reservoir siltation [2,7].

The hydrological regime of Mediterranean rivers exerts a substantial influence on the processes of soil erosion and sediment delivery [8,9]. As a result of the pronounced heterogeneity observed in rainfall patterns across space and time, Mediterranean rivers commonly exhibit abrupt fluctuations in flow and a period without flow. These hydrological dynamics give rise to the notable transport of suspended materials and erosional processes within such rivers [10]. These peculiarities make it more challenging to take precise and ongoing measurements of suspended sediment concentrations (SSC) [11], and it also suggests that computing the SS load is highly challenging [12].

Several techniques were used to estimate loads by using measurements of SSC and streamflow (Q) for medium and large catchments [13,14]. The selection of which method is going to be adopted mainly depends on the availability of data, catchment size, and streamflow regime even though each of these techniques has its own drawbacks [12]. All the methods require measurements of Q and SSC that must cover all of the hydrograph characteristics, especially flood events. Also, approaches based on modelling soil losses were developed to quantify sediment loads at different times and spatial scales [15–18]. However, models rely on extensive measurements of Q and SSC, which serve as essential components for validating the model's prognostications [19].

Continuous monitoring of SS transport involves employing infrared optical probes to measure turbidity, while laboratory analysis of water samples is used to determine SSC. These methodologies are commonly adopted for monitoring the transport of SS. Continuous measurements allow us to determine the sediment load during flood events with sufficient accuracy. Since optical technology is regarded as a reliable tool for measuring suspended sediment, its usage for continuous river monitoring has greatly expanded in recent years [18]. These optical instruments demonstrate sensitivity toward the particle size distribution flowing in the water. This trait could be of less significance in small catchments (where rainfall regimes and erodibility are uniform) while it is of great impact in large catchments [20].

In situations where continuous measurements are obtained for both Q and SSC, the sediment load transported through a specific river section within a given time interval is determined by calculating the integral of the product of SSC and Q over that time interval. This equation cannot be applied when SSC are discrete measurements [12]. In this case, sediment load computation requires the estimation of the SSC for the days when no records are available. A common approach for estimating SSC is based on a regression equation, the so-called "sediment rating curve" (SRC), that relates SSC to Q at the time of sampling.

SRCs are also used for filling the gaps in the SSC time series in the case of the failure of continuous measurements, since interruptions due to the malfunction noise of the instruments are quite common. However, according to Horowitz [21], due to the significant degree of dispersion of the SSC data, the SRC leads to an underestimation of loads. Hydrologists have invested a lot of time and energy into developing reliable approaches for reducing bias [22,23]. In-depth knowledge of the flow regime, seasonality, and sediment properties may be useful to adopt a strategy to improve SRC [12].

The present paper aimed to (i) quantify SS loads in two river basins of the Daunia Mountains: the Carapelle river and the Celone river (Apulia Region, Southeast Italy), and (ii) develop different SRCs based on the flow regime (i.e., high flow, normal flow, and low flow conditions) to fill the gaps in the SSC time series. The results are expected to advance the knowledge of SS load quantification in Mediterranean rivers and to provide water resource decision-makers with a rapid and affordable tool for SSC and load estimations.

2. Study Area: The Carapelle and the Celone River Basins

The experimental site for monitoring SSC and Q is situated at the Ordona bridge (Figure 1) along the main course of the Carapelle River, a prominent watercourse in northern Apulia. The stream rises in the Apennine mountains, runs across the Tavoliere floodplain, and discharges into the Adriatic Sea. The catchment area (considering the outlet at the gauging station) is 506 km^2, and the elevation ranges from 120 to 1075 m a.s.l., with an average slope of 8.2%. The mountainous hilly parts of the catchment experience considerable erosion mainly due to agricultural activities [2]. They are made up of flysch formations, whilst clay–sand Plio-Pleistocene sediments define the alluvial plain [16]. The majority of the soils in the region are fine clay loam textured Entisols with low levels of organic matter.

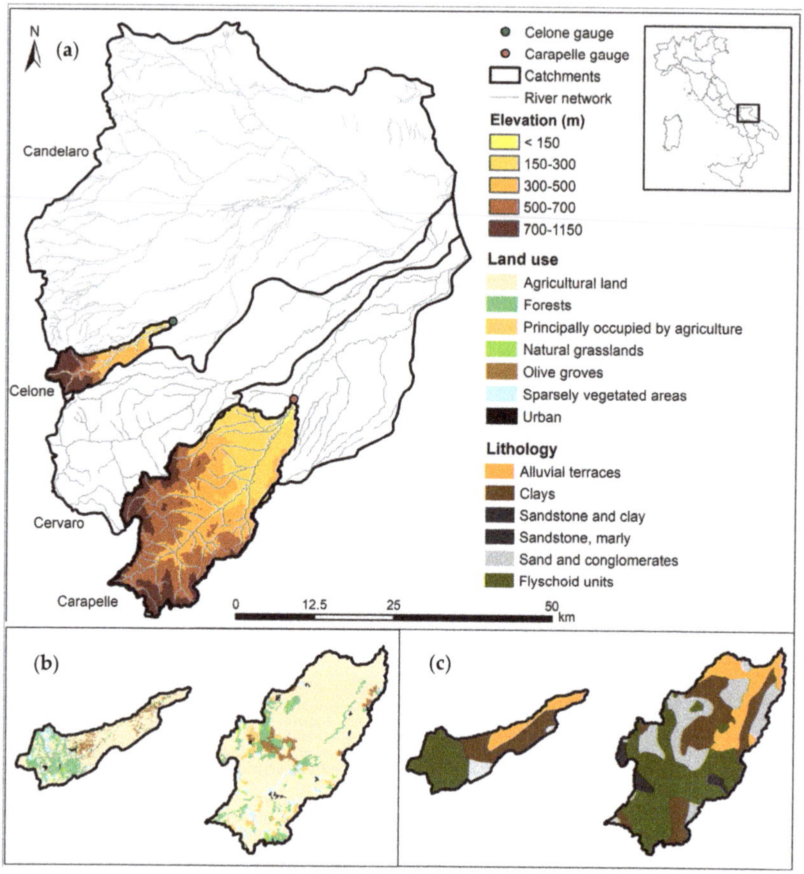

Figure 1. Study area: Celone catchment and Carapelle catchment (Apulia, Italy). (**a**) Elevation; (**b**) Land Use; and (**c**) Lithology. Green and red dots represent the position of the gauging stations of the Celone and of the Carapelle, respectively.

The primary economic activity is agriculture, with deciduous oaks and hardwoods (*Quercus pubescens* s.l. and *Quercus cerris* L.) covering the higher hillslopes along with conifers, pastures, and meadows [24]. Flat areas and low-lying plains are primarily covered by wheat cultivation, and to a lesser extent, olive orchards and other agricultural crops. The catchment has a Mediterranean climate with wet autumn/winter and dry spring/summer seasons with average temperatures that vary from 10 °C to 16 °C. Average annual rainfall ranges from 779 mm (1921–2012), at Bisaccia station, to 531 mm (1921–2012) at Castelluccio dei Sauri station according to the Department of Civil Protection. August (6.4 mm) is the driest month, while March (94.9 mm) and November (81.4 mm) represent the rainiest [2,25]. The hydrological regime is characterized by considerable spatial and temporal variability, with exceptionally low flow conditions or absence of flow in some areas of the river network during the summer (June to August).

The Celone river basin, spanning 72 km^2, encompasses steep slopes ranging from 150 m to 1150 m above sea level (a.s.l.). The river channel exhibits incision in the upper part of the basin and transitions into a braided course within the initial alluvial plain, which serves as a deposition zone for significant quantities of both suspended and bed-load material. In the lowland areas of the basin, the soils predominantly consist of deep clay loam and sandy clay loam textures, reaching depths of 1.5 to 2 m. In the hill and mountainous regions, the soils are moderately deep. The Celone river basin is primarily an agricultural area, with major cultivations including winter and durum wheat (45%), sunflower (9%), pasture (6%), and olive groves (8%). Vegetables and vineyards are minor land uses (2%). Forests (29%) are widely spread in mountainous areas. Urbanized regions account for only 1% of the total area. The basin experiences a climate characterized by a wet season extending from November to May, followed by a dry season. Historical records from 1960 to 2001 indicate an average annual rainfall of 792 mm in the upper part of the basin and 623 mm in the lowland areas. The Celone river shows an intermittent nature with a long dry season interrupted by flash flood events that may occur leading to high SS material transportation. Both catchments experience soil erosion caused by water due to the implementation of conventional tillage practices. Indeed, in autumn and winter, most of the agricultural fields (seeded and ploughed fields for spring crops) are unprotected by the vegetation from erosion. Both soil erosion (i.e., sheet and rill) and river bank erosion contribute to SS transport.

3. Materials and Methods

3.1. Monitoring SSC and Q

At the Carapelle Ordona bridge gauge Q and SSC were continuously monitored (2007–2011). However, due to the lack of data caused by the maintenance of the gauging station, 2009 was not included in this study [8]. The Hach–Lange Solitax hs-line sc probe was installed for measuring SSC (Figure 2) as it is able to measure high values of SSC (SSC > 15 g L^{-1}) [26]. For fouling prevention, the probe has a screen wiper with a customizable time interval. By merging signals, the Hach–Lange Solitax sc100 controller transforms them into SSC [27] (Figure 2c). This approach considers the stray light effects, organic matter, and the colour of the watery medium while allowing for the examination of a wide range of concentrations (0.001–4000 Nephelometric Turbidity Units—NTU—for turbidity and 0–150 g L^{-1} for suspended particles).

The probe is securely housed in a drilled tube, providing protection against the impact of coarse materials in the stream and preventing potential measurement errors caused by stray radiant energy entering the infrared field. At the same time, the holes drilled on the tube (Figure 2d) allow an efficient exchange of the flowing suspension, as observed through several simultaneous manual samplings, which gave the same SSC values inside and outside the shelter tube.

The mechanical system consists of a pulley (Figure 2c), float (Figure 2d), and counter-weight group which allow the probe to be always submerged 20 cm below the free water surface. By employing a data acquisition framework and a transmission infrastructure, the data are sent to a server in the Department. An ultrasonic stage recorder, managed by the Centro Funzionale Decentrato (CFD) of the Sezione Protezione Civile della Regione Puglia, is operated for Q measurements. CFD developed the rating curves and provided Q data.

Laboratory experiments evaluated the functionality of the Solitax sc probe in relation to river sediments, offering versatile applications across various water conditions. Operating in the turbidity-nephelometric mode for low turbid water and suspended sediment-ratio detection system mode for highly concentrated sediments, the instrument's two-detector optical system enabled accurate measurements of high Suspended Sediment Concentrations (SSCs) while compensating for colour variations, light fluctuations, and stray light. The study involved tests on suspensions with fixed granulometric mixtures and varying ratios of sandy and fine fractions, derived from sediment samples collected from the Carapelle stream's riverbed. Gravimetric analysis revealed predominant sandy content (87.5%) with minor proportions of silt (8.8%) and clay (3.7%). The suspensions were prepared in wet, oven-dried, and separated forms to examine their impact on measurements. The experiments yielded valuable insights into the probe's performance, the influence of sediment composition, and the relationship between optical data and gravimetric SSCs (Figures A1–A3 in Appendix A). Following that, the equipment was field-calibrated considering the most significant flood events, and the SSC was plotted against discharge with a 30 min time step. Detailed information about instrument calibration is reported by Gentile et al. [28]. Over the study period, several gaps in continuous measurements of SSC were recorded due to problems with the powering. The percentage of SSC missing data was 21.7% (20 gaps in 2007; 165 in 2008; 21 in 2010; and 112 in 2011).

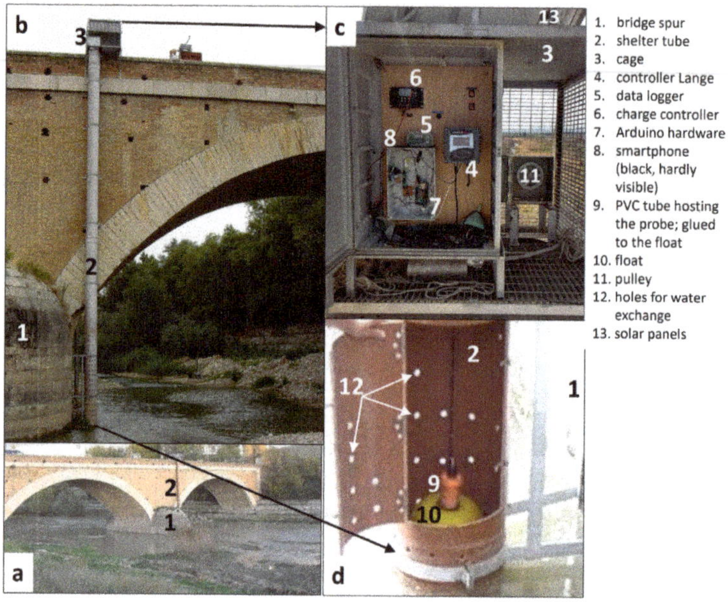

Figure 2. Gauging station at the Ordona bridge (view from upstream), Carapelle river (Apulia Region, Italy). (**a**) View from upstream of the gauging station, (**b**) shelter tube and cage of data acquisition (from upstream), (**c**) data acquisition instruments, and (**d**) side view of the probe sheltered within the tube. The sensor is submerged at 20 cm.

SSC and Q were monitored at the Celone river at the M. Pirro gauge (Figure 3) for a year (July 2010 to June 2011). Discrete samplings for SSC determination were carried out by using an automatic sampler (ISCO model 6712FS; with 24 bottles; pumped volume 1 L). The instrument was connected to a Flow Module (ISCO 750 Area Velocity) for continuous measurements of Q. The sensor provided continuous (5 min) measurements of flow velocity and stream water stage. These values were converted to Q by using a stage-discharge rating curve based on the river cross-section, which had a regular and permanent shape. Therefore, by using a template (Excel worksheet), the cross-sectional area corresponding to each water stage was calculated. Q was determined by multiplying flow velocity by the corresponding cross-sectional area. Several tests were completed to verify the sensor measurements.

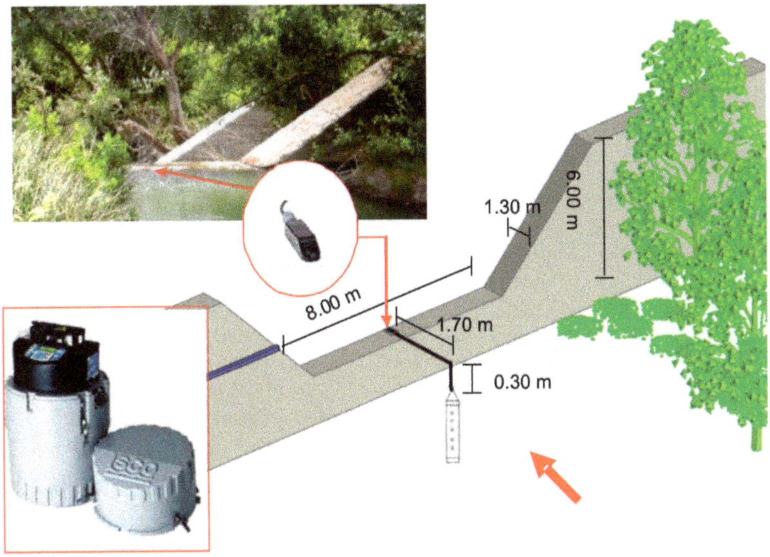

Figure 3. Gauging station and data acquisition system at the Celone M. Pirro, Celone river (Apulia Region, Italy). The red arrow shows the flow direction.

The automatic sampler was set for two different programs. A regular protocol was established for timed samplings to ensure periodic sample collection at biweekly or monthly intervals during the arid season (from July to October). Subsequently, as the season shifted from November to June, the sampling frequency escalated to once or twice a week. The second program was oriented to monitor flood events. Specifically, it was based on fluctuations in water level during the hydrograph rising limb and flow velocity variations during the recession phase. The sampling intervals within the rising limb ranged from 15 min to 2 h, while in the recession limb, they varied from 2 h to 1 day. Over the course of the study, a comprehensive total of 216 samples were meticulously collected, encompassing all hydrological conditions. The IRSA-CNR laboratory conducted the analysis of SSCs using the esteemed APAT-IRSA analytical standard method. More details about instruments and monitoring plans are reported in [3,29].

3.2. Developing Sediment Rating Curves

SRCs were created to fill the gaps in the SSC recorded dataset at the Carapelle monitoring station, and for quantifying the SSC for those days during which no measurements were available at the Celone gauge.

The SRC is a regression equation determined by using a log-transformed least squares regression (Equation (1)), where the Q is the independent variable and SSC' is the dependent variable.

$$Log\ SSC' = \log a + \mathrm{b}\ \log Q \quad (1)$$

The result of Equation (1) is needed in the original units. The back-transformation to the original scale by exponentiation introduces a bias that is an underestimation of the mean value on the original scale [30,31]. Several attempts have been carried out to enhance the SSC–Q relationships by introducing a bias correction factor [30,32–34]. In the present work, the bias correction factor proposed by Duan [33], the so-called Smearing Estimator (CF), was adopted (Equation (2)). The CF does not require any assumptions about the distribution of residuals.

$$CF = \frac{\sum_1^N 10^{ei}}{N} \quad (2)$$

where N is the number of measurements, εi is the residual (Equation (3)).

$$\varepsilon i = log\,(SSC_i) - log(SSC_i') \tag{3}$$

where SSC_i is the observed concentration and $SSC_i{}'$ is the predicted values obtained through regression.

To improve the estimation of SSC′, several authors proposed to develop SRC for subsets of data defined on a seasonal basis, or on classes of Q rather than the entire available data set [23,35]. De Girolamo et al. [12] highlighted that for intermittent rivers, using subsets of data based on the flow regime provides higher accuracy than seasonal data stratification when estimating SSC′ and load with SRCs. In this research, three distinct datasets were identified based on classes of Q: high flow, normal flow, and low flow. These datasets were utilized to construct logarithmically transformed linear functions. To identify the sub-sets, the Flow Duration Curve (FDC) was used. FDC represents graphically the percentage of time (*X*-axis) in which a specific value of flow (*Y*-axis) is equalled or exceeded. All the flow conditions are represented in FDC, with lower percentages corresponding to high flows (i.e., floods) and higher percentages corresponding to low discharges (i.e., low flows) [36]. This methodology allows representation of the hydrologic response of a river basin and, therefore the Q variability [37,38].

The following three data stratifications were adopted: R1 (0–5%; 5–70%; 70–100%), R2 (0–10%; 10–70%; 70–100%), and R3 (0–20%; 20–70%; 70–100%), by using different thresholds for high-flow conditions. Specifically, the flex point observed at 70% of FDC was used to set the low flow class, which is the same for all the stratifications (70–100%). This range is associated with low-flow conditions [39]. For high flow instead, three different ranges were considered 0–5, 0–10%, and 0–20%. The values of 5%, 10%, and 20% are associated with flood and high-flow conditions [8,11].

The mean error (*E*) (Equation (4)) was calculated for each dataset as a percentage of the variances between SSC_i and $SSC_i{}'$ [21],

$$E\,(\%) = \left[\frac{\Sigma_1^N\left(\frac{(SSC_i'-SSC_i)}{SSC_i}\right)}{N}\right] \times 100 \tag{4}$$

after evaluating the *SSC′*, specific SS loads (t ha^{-1} yr^{-1}) were computed both at the Carapelle gauge and at the Celone gauge at daily, and yearly time scales by the integration of the product of *SSC* and *Q*.

4. Results

4.1. Monitoring Streamflow and SSC

The transport of SS is greatly influenced by the flow regime of the rivers (Figures 4 and 5). It is characterized by a long period (summer) of extremely low flow and a wet period (winter–spring) with continuous flow and floods. The pattern of SSC strictly follows the hydrograph for both rivers.

At the Carapelle gauge, the highest Q value (94.4 m^3 s^{-1}) was registered on 10 November 2010 when also the highest SSC value (18.92 g L^{-1}) was recorded. In the wet season, the largest flood can generate a high value of SSC due to the whole basin contributing to sediment generation [8]. Low values of Q (0.005 m^3 s^{-1}) and SSC (<0.05 g L^{-1}) are generally recorded in late summer. In the study period, the absence of flow was not recorded at the Ordona bridge. However, as reported in the historical time series of Q there are several years with zero flow conditions and the river is classified as a temporary river [8]. Different SSC magnitudes were recorded for equal discharge peaks, for instance, the floods recorded on 3 November 2010 (22.19 m^3 s^{-1}; 1.53 g L^{-1}), and 10 March 2010 (22.03 m^3 s^{-1}; 0.52 g L^{-1}). The observed behaviour can be attributed to various factors, including the impact of vegetation, spatial variability of precipitation, and variations in antecedent soil moisture content. Indeed, lands with the main crop production (wheat) in the area are unprotected in November, meanwhile, they are completely covered by vegetation in March.

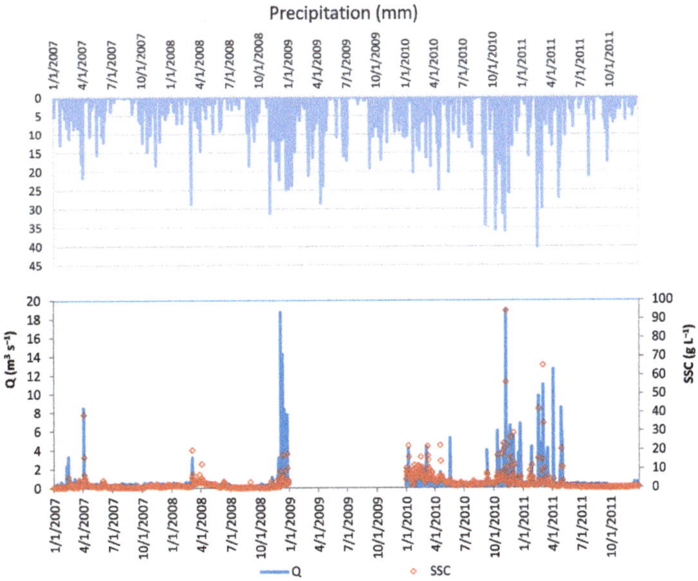

Figure 4. Observed flow (Q), observed suspended sediment concentrations (SSC), and observed precipitation at the Carapelle Ordona bridge gauge. The year 2009 was not included in this study due to the lack of data caused by the maintenance of the gauging station.

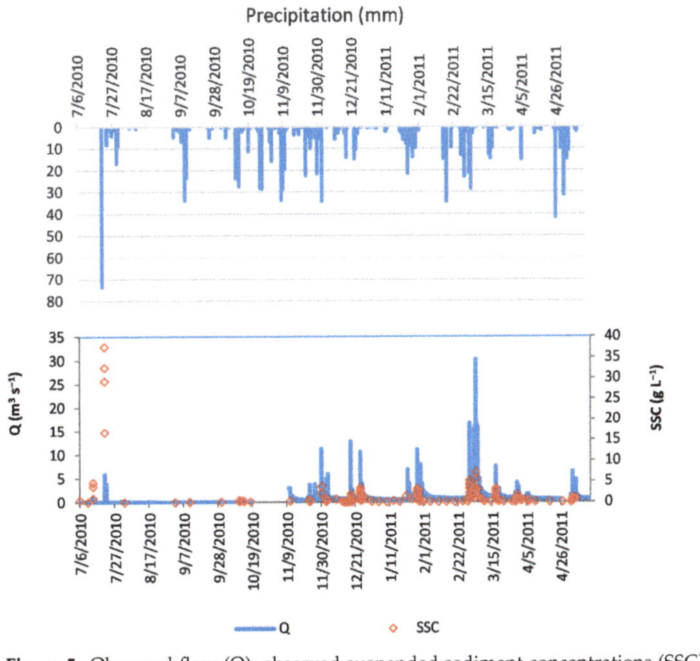

Figure 5. Observed flow (Q), observed suspended sediment concentrations (SSC), and observed precipitation at the Celone M. Pirro gauge. The period between the end of October and the beginning of November 2010 was not considered in this study due to a malfunctioning of the gauge station.

In the case of the Celone river, it is noteworthy that similar peak values of Q resulted in a range of magnitudes of SSC, as depicted in Figure 5. For instance, on 30 November (10.83 m^3 s^{-1}; 3.92 g L^{-1}) and 16 May (10.44 m^3 s^{-1}; 0.81 g L^{-1}), it is evident that the interplay of vegetation, potential variations in rainfall patterns, and variations in antecedent soil moisture content plays a significant role in soil erosion and the transport of SS. The maximum SSC recorded in the wet season was 7.13 g L^{-1}, associated with a streamflow value of 23.5 m^3 s^{-1} (5 March 2011). In summer, flash flood events may be critical for SS transport. Particularly, on 21 July 2010, during a period of intense precipitation (Faeto gauge recorded 73.6 mm h^{-1}), the recorded SSC was 37.60 g L^{-1}, accompanied by a Q value of 5.95 m^3 s^{-1}. This event marked the highest ratio of SSC to Q observed throughout the study period. In summer, flash floods characterized by high SSC despite low Q, are frequent. This is caused by the fact that in summer most of the wheat fields are harvested and, therefore, the soils in the basin are bare [12]. In this case, if an intense rainfall event occurs, infiltration excess is the dominant process that generates the overland flow, therefore, Q increase and decrease quickly and a large amount of SS can be transported into the river including the sediment accumulated on the river bed [11].

As described by De Girolamo et al. [12] flash floods were considered outliers, therefore, they were not included in the dataset for developing the SRC.

Based on the dataset presented in this study, at the Carapelle gauge, SSC and Q are significantly correlated (r = 0.69). The coefficient of determination (R^2 = 0.48) shows that the variance of SSC may be explained for 48% by Q, meanwhile, other factors such as rainfall factors (i.e., intensity, spatial distribution), vegetation cover, and tillage may hold for the 52% of the variance (Figure 6b). At the Celone gauge, after excluding four outliers, Q and SSC data from the monitoring campaign (July 2010 to June 2011) were significantly correlated. The coefficient of determination (R^2 = 0.66) designates that the variance of SSC may be explained for 66% by Q (Figure 6a).

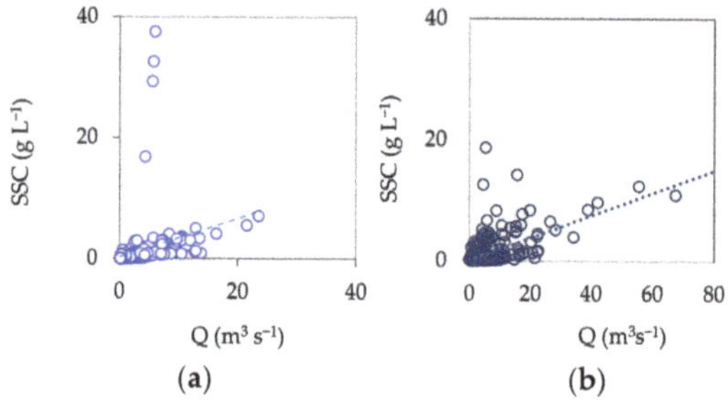

Figure 6. Measured flow (Q) and observed suspended sediment concentrations (SSC) at the Celone M. Pirro gauge (**a**), and at the Carapelle Ordona bridge gauge (**b**).

4.2. Developing Sediment Rating Curves

To identify the subsets of data for developing SRC, the FDC was used. For both rivers, SSC values resulted scattered, especially in high-flow conditions (Figure 7a,b) indicating that for the same value of Q, different values of SSC were recorded. At the Celone gauge, the coefficient of variation (CV) for SSC ranged from 109 (low flow) to 251 (high flow; 0–20%). The data stratification R1 showed the lowest CV in high-flow conditions. At the Carapelle gauge, the CV varied from 28 (70–100%) to 204 (5–70%). R1 showed the lowest CV for the high-flow conditions compared with R2 and R3. Log-transformed linear equations were derived to construct SRCs for the previously mentioned data stratifications (Figures 8 and 9).

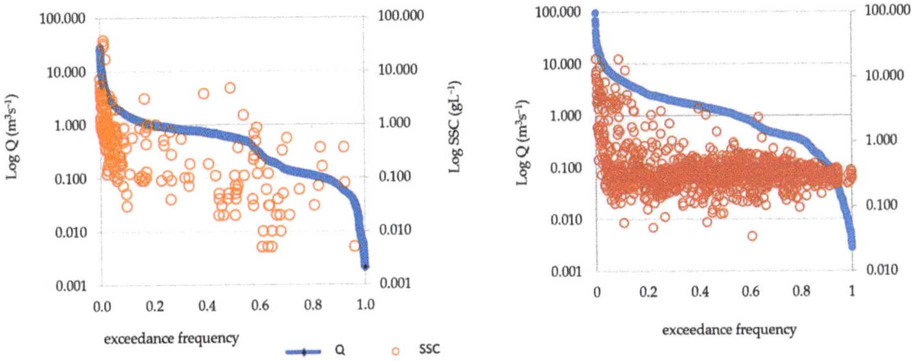

(a) (b)

Figure 7. Flow duration curve (FDC) and measured suspended sediment concentration (SSC, red circles): (**a**) Celone M. Pirro gauge; (**b**) Carapelle Ordona bridge gauge. The X-axes and Y-axes are on a logarithmic scale.

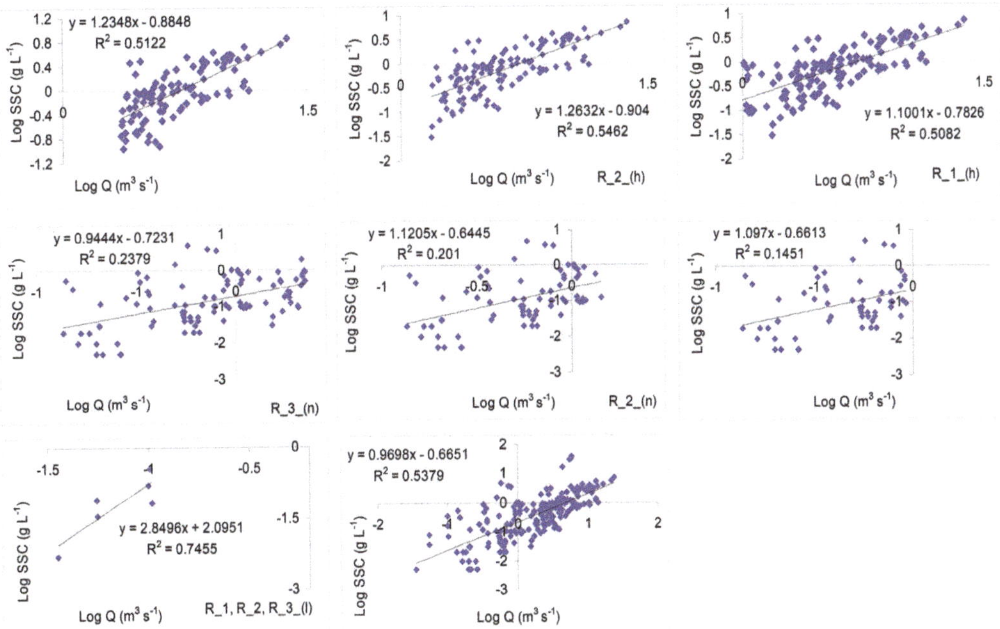

Figure 8. Sediment rating curves (SRC) developed at the Celone M. Pirro gauge. The Y-axes are on a logarithmic scale. Statistical significance: p-value < 0.01.

The CF values computed for high-flow conditions were higher than the values recorded in normal- and low-flow conditions for all the stratification data sets (Table 1) at the Carapelle gauge. This result indicates that the differences between measured and modelled values are higher in high flow than in the other conditions. At the Celone gauge, the highest values of CF were detected for the normal-flow condition clearly indicating very high residuals in this flow condition. The number of observations and their distribution among the flow conditions may have influenced the CF values. The latter hypotheses could also explain the fact that at the Celone gauge the CF values were higher than those recorded at the Carapelle gauge. High variability was also detected in the values assumed by E (%) among the datasets R1, R2, and R3 for both rivers (Table 2). High values of E (%) were detected for the Celone river, especially for normal- and low-flow conditions. This result is due to the different monitoring strategies adopted in the two catchments, discrete and continuous.

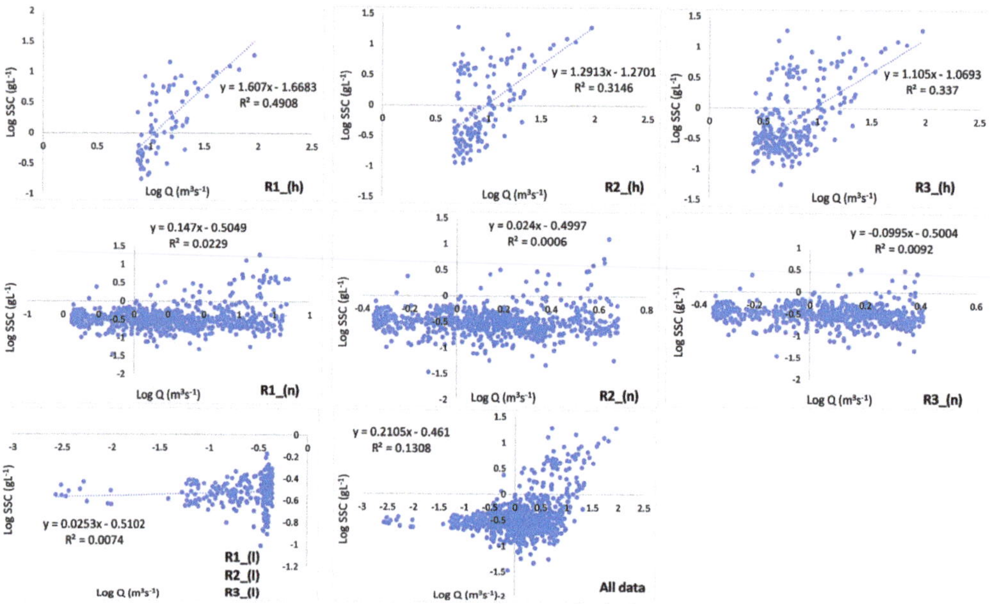

Figure 9. Model fitting of sediment rating curves (SCR) for the Carapelle river. The Y-axes are on a logarithmic scale. Statistical significance: p-value < 0.01.

Table 1. Duan's Correction Factor (CF) computed for the classes of flow (Q) at the Celone gauge and Carapelle gauge. R1 is for the data stratification 0–5%, 5–70%, 70–100%; R2 is for the data stratification 0–10%, 10–70%, 70–100%; R3 is for the data stratification 0–20%, 20–70%, 70–100%.

	Celone River			Carapelle River		
	R1	R2	R3	R1	R2	R3
High flow	2.04	2.12	2.20	1.64	2.06	1.87
Normal flow	4.69	3.62	3.13	1.41	1.25	1.16
Low flow	1.24	1.24	1.24	1.04	1.04	1.04

Table 2. Percentage error, E (%), computed for the classes of flow (Q) at the Celone gauge and Carapelle gauge. R1 is for the data stratification 0–5%, 5–70%, 70–100%; R2 is for the data stratification 0–10%, 10–70%, 70–100%; R3 is for the data stratification 0–20%, 20–70%, 70–100%.

	Celone River			Carapelle River		
	R1	R2	R3	R1	R2	R3
High flow	19.10	29.61	35.52	36.64	59.98	40.08
Normal flow	98.94	191.49	126.39	19.98	15.53	13.15
Low flow	99.97	99.97	99.97	4.67	4.67	4.67

4.3. Sediment Loads

The annual specific SS load estimated for the Carapelle catchment showed a high inter-annual variability (Table 3), from 0.878 (2007) to 7.452 t ha^{-1} yr^{-1} (2010). Specifically, its mean annual value ranged from 3.30 to 3.34 t ha^{-1} yr^{-1}, computed without and with CF, respectively. Slight differences among the three data stratifications (R1, R2, and R3) resulted in specific SS loads (Table 2). In addition, upon analysis, slight variations were observed in the results when considering the back-transformation correction CF, compared to the results obtained without its consideration (i.e., 1.479 t ha^{-1} yr^{-1} and 1.542 t ha^{-1} yr^{-1}, respectively) (Table 3).

Table 3. Annual specific suspended sediment (SS) loads (t ha^{-1} yr^{-1}) evaluated by applying three different data stratifications (R1; R2; and R3) and with the back-transformation correction (R1_CF; R2_CF; and R3_CF) at the Carapelle gauge and Celone gauge.

	R1	R1_CF	R2	R2_CF	R3	R3_CF
Carapelle						
2007	0.878	0.881	0.878	0.880	0.878	0.879
2008	1.479	1.542	1.475	1.543	1.474	1.565
2010	7.426	7.437	7.430	7.452	7.431	7.449
2011	3.420	3.439	3.424	3.459	3.429	3.465
Celone						
2010–2011	2.610	6.060	2.550	5.780	2.400	5.200

As previously described, the whole data set of the Celone river was limited to one year (from July 2010 to June 2011), hence, no investigations were carried out regarding the inter-annual variability. On a yearly basis, the specific SS load, computed by using the three data stratifications spanned in the interval 2.400 to 2.610 t ha^{-1} yr^{-1} (Table 3). A wide difference in specific SS loads was detected when using the CF (5.200 to 6.060 t ha^{-1} yr^{-1}). Therefore, the actual value of specific SS load could vary in the interval 2.610–6.060 t ha^{-1} yr^{-1} showing a large uncertainty in the estimation.

SS load was mainly delivered during high flow in both catchments. Specifically, the suspended material transported during the low flow accounts for less than 0.1% of the total suspended material with slight differences between the basins. More than 80% is transported in high-flow conditions. Indeed, approximately 94% and 93% were delivered in the high-flow conditions (>20% of exceedance) in the Carapelle and Celone rivers, respectively (Table 4).

Table 4. Average specific suspended (SS) sediment loads (t ha^{-1} yr^{-1}) evaluated for high, normal, and low flow by using three different data stratification (R1; R2; and R3) and with the back-transformation correction (R1_CF; R2_CF; and R3_CF) at the Carapelle gauge and Celone gauge.

	R1 (t ha^{-1} yr^{-1})	%	R1_CF (t ha^{-1} yr^{-1})	%	R2 (t ha^{-1} yr^{-1})	%	R2_CF (t ha^{-1} yr^{-1})	%	R3 (t ha^{-1} yr^{-1})	%	R3_CF (t ha^{-1} yr^{-1})	%
Carapelle												
High Flow	2.74	83.2	2.75	82.8	3.00	91.0	3.02	90.8	3.14	95.2	3.17	95.0
Normal Flow	0.54	16.5	0.56	16.9	0.29	8.7	0.30	8.9	0.15	4.5	0.15	4.6
Low Flow	0.01	0.4	0.01	0.4	0.01	0.4	0.01	0.4	0.01	0.4	0.01	0.4
Celone												
High Flow	2.23	85.6	4.91	81.1	2.27	89.2	4.83	83.5	2.27	94.3	4.62	88.8
Normal Flow	0.36	13.7	1.12	18.5	0.26	10.1	0.93	16.1	0.26	4.9	0.56	10.7
Low Flow	0.02	0.7	0.02	0.4	0.02	0.7	0.02	0.4	0.02	0.7	0.02	0.4

5. Discussion

This study showed that among the factors influencing the SS load determination are the duration of the sampled period and its characteristics (i.e., dry, wet conditions). Indeed, substantial year-to-year fluctuations were recorded for the Carapelle catchment, where the specific SS load ranged from 0.9 to 7.45 t ha^{-1} yr^{-1} (2007–2011). Also, monitoring strategies such as the frequency, the type of instrument (i.e., automated water sampling, turbidity monitoring), and the methodological approach adopted for estimating loads proved to influence the SS load computation. At the Celone M. Pirro, the specific SS load ranged from 2.40 to 6.06 t ha^{-1} yr^{-1} (from July 2010 to June 2011) depending on the computation method.

The present study identified the flow regime as the primary factor influencing SS delivery in mountainous catchments. Therefore, it is essential to carefully design monitoring plans that consider the streamflow behaviour.

Regarding discrete SSC samplings, it is essential to carefully determine the samplings timing and frequency. This ensures that the temporal variations in SSC are adequately captured, allowing for a representation of the complete range of Q conditions [40]. Both steady and sporadic measurements of Q and SSC offer distinct advantages and disadvantages. Steady continuous measurement of SSC offers the advantage of closely tracking the variations in Q, thereby facilitating the determination of sediment loads. However, this approach can be expensive and pose challenges, particularly when conducting measurements over extended periods of time. Infrared-based turbidity probes offer the advantage of providing continuous measurements of SSC. However, the use of these probes presents certain challenges. Some of the identified issues include the sensitivity of the probe's output to environmental conditions and the variability in particle size, shape, and colour. To improve measurements, particle size could be further analysed using laser diffraction [41]. However, a cross-calibration with laboratory determinations is required. Furthermore, it should be noted that these turbidity probes have a limited operational range. Consequently, during situations with extreme variations in particle composition, there is a possibility of underestimating the SSC, particularly in conditions with very high values of SSC [42]. Automatic samplers offer the advantage of collecting regular samples and the ability to operate also in flood events. However, it is important to note that their usage can be costly. A limitation in their use is that the number of bottles available (typically 24) may not be sufficient to adequately cover certain floods, especially those of extended duration. In addition, in the case of discrete measurements of SSC, the SS load estimation may be affected by a high uncertainty level depending on the method used for its computation.

In this study, SRCs were formulated and employed to estimate SSC. This methodology was selected considering its ease of use, and its possibility to be applied in semi-arid regions characterized by data scarcity [11,43,44]. Indeed, other complex hydrological and morphodynamics models may be used, but they require a large input dataset to be implemented and a tough training phase [2,11,45,46]. Prior research has pointed out that SRCs underestimate high concentrations and overestimate low concentrations of SSC [9,11,12,21,23,31,47]. Hence, to enhance the accuracy of the SRCs, in this study, the complete dataset was split into subsets, based on the flow regime using FDC. This methodology allows us to generally assess the characteristics of Q and of a specific river basin, through a comparison between the magnitude and frequency [37]. On the other hand, one of the limits associated with this methodology is that the seasonality of the flow cannot be considered. Indeed, since all the data were plotted excluding the date of occurrence, the time-dependent information is lost [48]. Consequently, when FDC is associated with rating curves for the SS load estimation the accuracy can be low due to the non-linear relationship between Q and SSC [37]. This issue could be accentuated in Mediterranean rivers, in which, due to several factors (i.e., rainfall regime, the shape of the basin, and the location of the sediment sources), the peak of the SSC may occur before or after the peak of Q [8,49]. Moreover, flash floods in the dry season may be characterized by high SSC and low Q [11]. One solution which can be adopted to improve the accuracy is to develop FDC using sub-daily data [50].

The application of CF was then employed to address the error associated with the back-transformation process. The results in terms of annual specific SS loads are included in a very short interval when using the three data stratifications at the Carapelle gauge. Also, the use of CF showed slight differences in specific SS loads on a yearly basis (<0.3%).

For the Celone river, differences in specific SS load estimations among the three data stratifications were detected (i.e., 2.40 t ha^{-1} yr^{-1} and 2.61 t ha^{-1} yr^{-1}). In contrast with the Carapelle catchment, a large interval was detected in the SS load estimated by using the CF at the Celone catchment (for instance 2.61 t ha^{-1} yr^{-1} and 6.01 t ha^{-1} yr^{-1}, for the R1 data stratification). These results were due to the fact that at the Carapelle gauge the SRCs were used only for filling the gaps, which in some cases were limited to a few dozen (20 data in 2007, and 21 in 2010), therefore the difference among the three data stratification was deemed negligible, as well as the use of the CF. In the case of the Celone catchment, it was observed that the number of measurements and the specific methodology employed for computing the sediment load had a significant impact on the results. Furthermore, these factors contributed to a notable level of uncertainty associated with the findings [12]. Consistent with expectations, a substantial portion of the SS load was observed to be transported during high-flow conditions, accounting for over 80% of the total load in both the Carapelle and the Celone catchments. Similarly, Benselama et al. [51] highlighted that flood can contribute approximately 64%, and that in wet season almost 78% of SS load was recorded. In contrast, during low-flow conditions, the SS load was significantly lower, comprising less than 0.7% in the Celone catchment and approximately 0.4% in the Carapelle catchment. High variability was also detected in the values assumed by E (%) among the datasets R1, R2, and R3 for both rivers. Data stratification R1 showed the lowest E (%) in the high-flow condition compared with R2 and R3. Considering that most of the SS load was transported during the high flow, R1 is the best data stratification strategy for assessing SS load. The values of specific SS load (t ha^{-1} yr^{-1}) observed in the Mediterranean areas are extremely variable [8]. Liquete et al. [52], for different river basins located in Spain, reported specific SS load values ranging from 0.004 to 2.00 t ha^{-1} yr^{-1}. A slightly higher value (2.94 t ha^{-1} yr^{-1}) was indicated for the Wadi El Maleh basin (Algeria) [51]. Estrany et al. [43] reported an SS load ranging from 2.20 to 4.50 t ha^{-1} yr^{-1}. In Italy, within an analysis involving 40 different basins, Van Rompay et al. [53] observed a range between 0.20–19.6 t ha^{-1} yr^{-1}. A higher value (32 t ha^{-1} yr^{-1}) was evidenced for a basin located in central Apennine [54]. The values obtained for the two studied basins were similar to what is reported here for Mediterranean environment and, also in line with the average value of 6.73 t ha^{-1} yr^{-1} reported in Vanmaercke et al. [55].

Several studies that analysed the soil formation rates clearly indicated that soil formation is highly variable [56,57]. Alewell et al. [58] reported an average soil production rate in alpine areas ranging between 0.54 and 1.13 t ha^{-1} yr^{-1} for old soils and between 1.19 and 2.48 t ha^{-1} yr^{-1} for young soils. Similarly, soil loss tolerance (SLT) can assume different rates based on environmental factors (i.e., soil type, soil depth, and climate) [59,60]. Europe's environment assessment [61] reported SLT in the range 1–5 t ha^{-1} yr^{-1} (shallow sandy soils and deep soils, respectively). However, over the long period, it is clear that annual soil loss higher than 1 t ha^{-1} yr^{-1} may produce irreversible soil damage if the soil formation rate is low [62]. The European Commission [63] recommended that the Member States adopt SLT values suitable to guarantee soil functions and sustainable soil use. Currently, SLTs of 0.3–2 t ha^{-1} yr^{-1} are adopted in Europe depending on the soil formation rate. This study highlighted that both in the Carapelle and the Celone catchments the specific SS load is higher than SLT, and confirms other studies carried out in neighbouring areas [47]. We argue that soil erosion values are higher than STL in the entire mountainous area of "Monti della Daunia", hence, in both catchments, a program of mitigation or conservation measures is needed to reduce soil losses.

6. Conclusions

The annual variation in specific suspended sediment (SS) load in a catchment is influenced by factors such as wet or dry conditions and sediment load estimation methods. This study emphasizes the importance of monitoring streamflow and suspended sediment concentration (SSC) to assess soil loss. The findings reveal that both the Carapelle and Celone catchments experience soil losses exceeding the rate of soil formation.

The study demonstrates the effectiveness of sediment rating curves (SRCs) for estimating SSCs and filling data gaps. However, accurate SRCs require a comprehensive dataset covering diverse hydrological conditions, including flood events. The study implemented a data stratification approach based on flow regimes to enhance sediment load estimation by developing and testing multiple sediment rating curves.

These findings are valuable for water resource managers in quantifying suspended sediment loads. Further research should investigate the relationship between agricultural management practices, soil erosion, and suspended sediment loads. Identifying effectively the best management practices to reduce soil losses in the Carapelle and Celone catchments is crucial. These studies can contribute to strategies for mitigating sediment transport and improving water resources management in the area.

Author Contributions: Conceptualization, A.M.D.G.; Data curation, A.M.D.G., G.F.R. and F.M.; Formal analysis, A.M.D.G., G.F.R., F.M. and A.M.N.; Funding acquisition, A.M.D.G., A.L.P. and F.G.; Investigation, A.M.D.G., G.F.R. and F.M.; Methodology, A.M.D.G.; Supervision, A.M.D.G. and F.G.; Validation, A.M.D.G.; Visualization, A.M.D.G., G.F.R. and O.M.M.A.; Writing—original draft, A.M.D.G., G.F.R. and O.M.M.A.; Writing—review and editing, A.M.D.G., A.L.P. and F.G. All authors have read and agreed to the published version of the manuscript.

Funding: The data concerning the Celone river were collected within the EU FP7 MIRAGE Project (Contr. no. 211732) by the Water Research Institute—National Research Council (IRSA-CNR). The data concerning the Carapelle river were collected by the University of Bari (Principal Investigator F. Gentile) within the projects PRIN2010 "National network for monitoring, modeling and sustainable management of erosion processes in agricultural, hilly and mountainous territories" funded by the Italian Ministry of Education, University and Research and "Soil Erosion in Apulia: Monitoring, Modelling and Control Strategies" funded by the River Basin Authority of the Apulia Region. This study was carried out within the AGRARSENSE (Project: 101095835; HORIZON-KDT-JU-2021-1-IA) by IRSA-CNR and the Agritech National Research Center and received funding from the European Union Next-Generation EU (PIANO NAZIONALE DI RIPRESA E RESILIENZA (PNRR)—MISSIONE 4 COMPONENTE 2, INVESTIMENTO 1.4—D.D. 1032 17/06/2022, CN00000022). This manuscript reflects only the authors' views and opinions, neither the European Union nor the European Commission can be considered responsible for them.

Data Availability Statement: The data presented in this study are available on request.

Acknowledgments: The authors gratefully acknowledge Angelantonio Calabrese and Giuseppe Pappagallo from IRSA—CNR and Giovanni Romano from DiSSPA—UNIBA for collaborating in field activities and analytical determinations of SSC. The authors gratefully acknowledge the anonymous reviewers for their valuable comments which contributed to improving the quality of the research article.

Conflicts of Interest: The authors declare no conflict of interest.

Appendix A

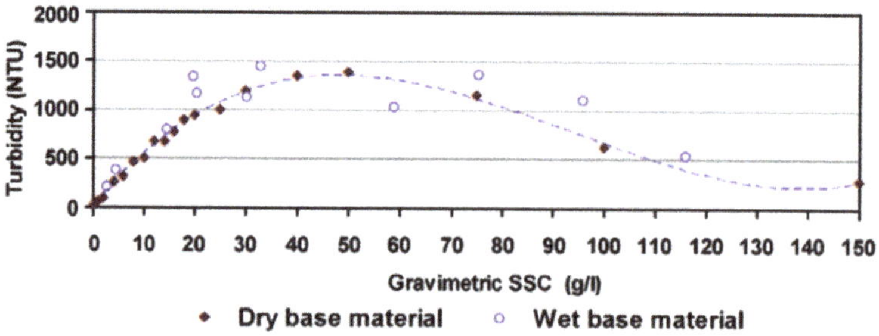

Figure A1. Relation between measured turbidity (NTU) and gravimetric SSC. Reprinted with permission from Gentile et al. [28].

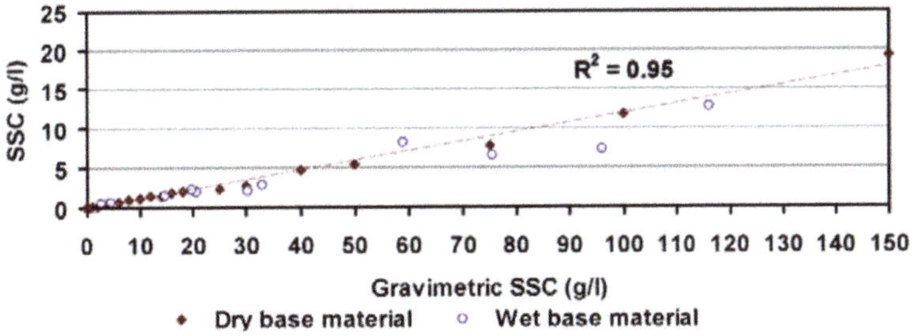

Figure A2. Relation between measured SSC and gravimetric SSC. Reprinted with permission from Gentile et al. [28].

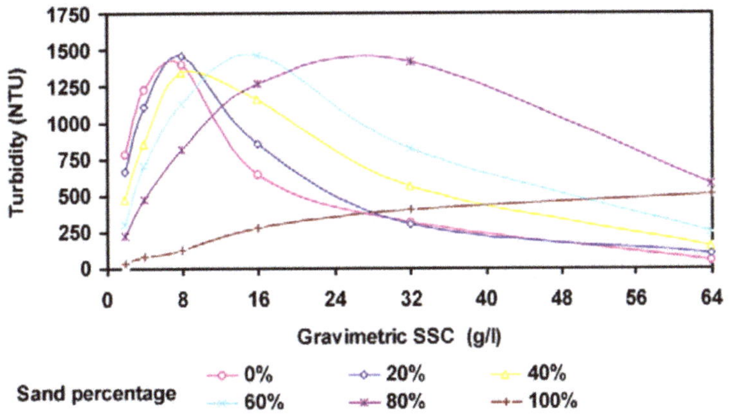

Figure A3. Relation between measured turbidity (NTU) and gravimetric SSC (Original wet and dry batches). Reprinted with permission from Gentile et al. [28].

References

1. Jones, A.; Panagos, P.; Barcelo, S.; Bouraoui, F.; Bosco, C.; Dewitte, O.; Gardi, C.; Erhard, M.; Hervás, J.; Hiederer, R.; et al. *State of Soil in Europe*; European Commission: Brussels, Belgium, 2012. [CrossRef]
2. Ricci, G.F.; D'Ambrosio, E.; De Girolamo, A.M.; Gentile, F. Efficiency and Feasibility of Best Management Practices to Reduce Nutrient Loads in an Agricultural River Basin. *Agric. Water Manag.* **2022**, *259*, 107241. [CrossRef]
3. Di Pillo, R.; De Girolamo, A.M.; Lo Porto, A.; Todisco, M.T. Detecting the Drivers of Suspended Sediment Transport in an Intermittent River: An Event-Based Analysis. *CATENA* **2023**, *222*, 106881. [CrossRef]
4. López-Tarazón, J.A.; Estrany, J. Exploring Suspended Sediment Delivery Dynamics of Two Mediterranean Nested Catchments. *Hydrol. Process* **2017**, *31*, 698–715. [CrossRef]
5. Pagano, S.G.; Rainato, R.; García-Rama, A.; Gentile, F.; Lenzi, M.A. Analysis of Suspended Sediment Dynamics at Event Scale: Comparison between a Mediterranean and an Alpine Basin. *Hydrol. Sci. J.* **2019**, *64*, 948–961. [CrossRef]
6. De Girolamo, A.M.; Spanò, M.; D'Ambrosio, E.; Ricci, G.F.; Gentile, F. Developing a Nitrogen Load Apportionment Tool: Theory and Application. *Agric. Water Manag.* **2020**, *226*, 105806. [CrossRef]
7. Bazzoffi, P.; Abbatista, F.; Vanino, S.; Napoli, R.; Fais, A.; Nino, P. Loss of Water Storage Capacity of Reservoirs in Southern Italy: Economic Implicances of Sedimentation. In Proceedings of the OECD Workshop on Agriculture and Water: Sustainability, Markets and Policies, Adelaide, Australia, 14–18 November 2005.
8. Fortesa, J.; Ricci, G.F.; García-Comendador, J.; Gentile, F.; Estrany, J.; Sauquet, E.; Datry, T.; De Girolamo, A.M. Analysing Hydrological and Sediment Transport Regime in Two Mediterranean Intermittent Rivers. *CATENA* **2021**, *196*, 104865. [CrossRef]
9. Ricci, G.F.; De Girolamo, A.M.; Gentile, F. Monitoring Suspended Sediment Transport in Two Moutainous River Basins: The Carapelle and the Celone (Apulia, Italy). In Proceedings of the 2022 IEEE Workshop on Metrology for Agriculture and Forestry (MetroAgriFor), Perugia, Italy, 3–5 November 2022; pp. 277–281. [CrossRef]
10. Ricci, G.F.; Romano, G.; Leronni, V.; Gentile, F. Effect of Check Dams on Riparian Vegetation Cover: A Multiscale Approach Based on Field Measurements and Satellite Images for Leaf Area Index Assessment. *Sci. Total Environ.* **2019**, *657*, 827–838. [CrossRef]
11. De Girolamo, A.M.; Pappagallo, G.; Lo Porto, A. Temporal Variability of Suspended Sediment Transport and Rating Curves in a Mediterranean River Basin: The Celone (SE Italy). *CATENA* **2015**, *128*, 135–143. [CrossRef]

12. De Girolamo, A.M.; Di Pillo, R.; Lo Porto, A.; Todisco, M.T.; Barca, E. Identifying a Reliable Method for Estimating Suspended Sediment Load in a Temporary River System. *CATENA* **2018**, *165*, 442–453. [CrossRef]
13. Moatar, F.; Meybeck, M. Compared Performances of Different Algorithms for Estimating Annual Nutrient Loads Discharged by the Eutrophic River Loire. *Hydrol. Process* **2005**, *19*, 429–444. [CrossRef]
14. Letcher, R.A.; Jakeman, A.J.; Merritt, W.S.; McKee, L.J.; Eyre, B.D.; Baginska, B. Review of Techniques to Estimate Catchment Exports. *Environ. Prot. Auth.* **1999**. Available online: https://s3-ap-southeast-1.amazonaws.com/ap-st01.ext.exlibrisgroup.com/61SCU_INST/storage/alma/46/23/CF/BC/A4/41/B1/C4/37/90/C5/4E/CF/06/BC/40/fulltext.pdf?response-content-type=application%2Fpdf&X-Amz-Algorithm=AWS4-HMAC-SHA256&X-Amz-Date=20230726T153307Z&X-Amz-SignedHeaders=host&X-Amz-Expires=119&X-Amz-Credential=AKIAJN6NPMNGJALPPWAQ%2F20230726%2Fap-southeast-1%2Fs3%2Faws4_request&X-Amz-Signature=52411230030f2f472e4d1b11f6bc67fb3eea1e357b9987fae64ac71ae0212b15 (accessed on 20 June 2023).
15. Borrelli, P.; Alewell, C.; Alvarez, P.; Anache, J.A.A.; Baartman, J.; Ballabio, C.; Bezak, N.; Biddoccu, M.; Cerdà, A.; Chalise, D.; et al. Soil Erosion Modelling: A Global Review and Statistical Analysis. *Sci. Total Environ.* **2021**, *780*, 146494. [CrossRef] [PubMed]
16. Abdelwahab, O.M.M.; Bingner, R.L.; Milillo, F.; Gentile, F. Evaluation of Alternative Management Practices with the AnnAGNPS Model in the Carapelle Watershed. *Soil Sci.* **2016**, *181*, 293–305. [CrossRef]
17. Ferro, V.; Porto, P. Sediment Delivery Distributed (SEDD) Model. *J. Hydrol. Eng.* **2000**, *5*, 411–422. [CrossRef]
18. Bagarello, V.; Ferro, V.; Pampalone, V.; Porto, P.; Todisco, F.; Vergni, L. Predicting Soil Loss in Central and South Italy with a Single USLE-MM Model. *J. Soils Sediments* **2018**, *18*, 3365–3377. [CrossRef]
19. De Girolamo, A.M.; Cerdan, O.; Grangeon, T.; Ricci, G.F.; Vandromme, R.; Lo Porto, A. Modelling Effects of Forest Fire and Post-Fire Management in a Catchment Prone to Erosion: Impacts on Sediment Yield. *CATENA* **2022**, *212*, 106080. [CrossRef]
20. Lewis, J. Turbidity-Controlled Suspended Sediment Sampling for Runoff-Event Load Estimation. *Water Resour. Res.* **1996**, *32*, 2299–2310. [CrossRef]
21. Horowitz, A.J. An Evaluation of Sediment Rating Curves for Estimating Suspended Sediment Concentrations for Subsequent Flux Calculations. *Hydrol. Process* **2003**, *17*, 3387–3409. [CrossRef]
22. Jansson, M.B. Estimating a Sediment Rating Curve of the Reventazón River at Palomo Using Logged Mean Loads within Discharge Classes. *J. Hydrol.* **1996**, *183*, 227–241. [CrossRef]
23. Asselman, N.E.M. Fitting and Interpretation of Sediment Rating Curves. *J. Hydrol.* **2000**, *234*, 228–248. [CrossRef]
24. Aquilino, M.; Novelli, A.; Tarantino, E.; Iacobellis, V.; Gentile, F. Evaluating the Potential of GeoEye Data in Retrieving LAI at Watershed Scale. *Remote Sens. Agric. Ecosyst. Hydrol. XVI* **2014**, *9239*, 590–600. [CrossRef]
25. Novelli, A.; Tarantino, E.; Fratino, U.; Iacobellis, V.; Romano, G.; Gentile, F. A Data Fusion Algorithm Based on the Kalman Filter to Estimate Leaf Area Index Evolution in Durum Wheat by Using Field Measurements and MODIS Surface Reflectance Data. *Remote Sens. Lett.* **2016**, *7*, 476–484. [CrossRef]
26. Bisantino, T.; Gentile, F.; Milella, P.; Liuzzi, G.T. Effect of Time Scale on the Performance of Different Sediment Transport Formulas in a Semiarid Region. *J. Hydraul. Eng.* **2010**, *136*, 56–61. [CrossRef]
27. Sadar, M.J.; Hach, C.C. *Turbidity Science: Technical Information Series*; Hach Company: Loveland, CO, USA, 1998.
28. Gentile, F.; Bisantino, T.; Corbino, R.; Milillo, F.; Romano, G.; Liuzzi, G.T. Monitoring and Analysis of Suspended Sediment Transport Dynamics in the Carapelle Torrent (Southern Italy). *CATENA* **2010**, *80*, 1–8. [CrossRef]
29. De Girolamo, A.M.; Balestrini, R.; D'Ambrosio, E.; Pappagallo, G.; Soana, E.; Lo Porto, A. Antropogenic Input of Nitrogen and Riverine Export from a Mediterranean Catchment. The Celone, a Temporary River Case Study. *Agric. Water Manag.* **2017**, *187*, 190–199. [CrossRef]
30. Koch, R.W.; Smillie, G.M. Bias in Hydrologic Prediction Using Log-Transformed Regression Models. *J. Am. Water Resour. Assoc.* **1986**, *22*, 717–723. [CrossRef]
31. Ferguson, R.I. River Loads Underestimated by Rating Curves. *Water Resour. Res.* **1986**, *22*, 74–76. [CrossRef]
32. Bradu, D.; Mundlak, Y. Estimation in Lognormal Linear Models. *J. Am. Stat. Assoc.* **1970**, *65*, 198–211. [CrossRef]
33. Duan, N. Smearing Estimate: A Nonparametric Retransformation Method. *J. Am. Stat. Assoc.* **1983**, *78*, 605–610. [CrossRef]
34. Gilroy, E.J.; Hirsch, R.M.; Cohn, T.A. Mean Square Error of Regression-Based Constituent Transport Estimates. *Water Resour. Res.* **1990**, *26*, 2069–2077. [CrossRef]
35. Walling, D.E.; Webb, B.W. Erosion and Sediment Yield: Global and Regional Perspectives. In Proceedings of the International Symposium, Exeter, UK, 15–19 July 1996.
36. Kannan, N.; Anandhi, A.; Jeong, J. Estimation of Stream Health Using Flow-Based Indices. *Hydrology* **2018**, *5*, 20. [CrossRef]
37. Vogel, R.M.; Fennessey, N.M. Flow Duration Curves Ll: A Review of Applications in Water Resources Planning. *Water Resour. Bull.* **1995**, *31*, 1029–1039. [CrossRef]
38. Ganora, D.; Claps, P.; Laio, F.; Viglione, A. An Approach to Estimate Nonparametric Flow Duration Curves in Ungauged Basins. *Water Resour. Res.* **2009**, *45*. [CrossRef]
39. Smakhtin, V.U. Low Flow Hydrology: A Review. *J. Hydrol.* **2001**, *240*, 147–186. [CrossRef]
40. Harmel, R.D.; Cooper, R.J.; Slade, R.M.; Haney, R.L.; Arnold, J.G. Cumulative Uncertainty in Measured Streamflow and Water Quality Data for Small Watersheds. *Trans. ASABE* **2006**, *49*, 689–701. [CrossRef]
41. Di Stefano, C.; Ferro, V.; Mirabile, S. Comparison between Grain-Size Analyses Using Laser Diffraction and Sedimentation Methods. *Biosyst. Eng.* **2010**, *106*, 205–215. [CrossRef]

42. Clifford, N.J.; Richards, K.S.; Brown, R.A.; Lane, S.N. Laboratory and Field Assessment of an Infrared Turbidity Probe and Its Response to Particle Size and Variation in Suspended Sediment Concentration. *Hydrol. Sci. J.* **1995**, *40*, 771–791. [CrossRef]
43. Estrany, J.; Garcia, C.; Batalla, R.J. Suspended Sediment Transport in a Small Mediterranean Agricultural Catchment. *Earth Surf. Process Landf.* **2009**, *34*, 929–940. [CrossRef]
44. Rulot, F.; Dewals, B.J.; Erpicum, S.; Archambeau, P.; Pirotton, M. 6.13—Long-Term Sediment Management for Sustainable Hydropower. In *Comprehensive Renewable Energy*; Elsevier: Amsterdam, The Netherlands, 2012; pp. 355–376, ISBN 9780080878737.
45. Wright, S.A.; Topping, D.J.; Rubin, D.M.; Melis, T.S. An Approach for Modeling Sediment Budgets in Supply-Limited Rivers. *Water Resour. Res.* **2010**, *46*. [CrossRef]
46. Harrington, S.T.; Harrington, J.R. An Assessment of the Suspended Sediment Rating Curve Approach for Load Estimation on the Rivers Bandon and Owenabue, Ireland. *Geomorphology* **2013**, *185*, 27–38. [CrossRef]
47. Walling, D.E.; Webb, B.W. The Reliability of Rating Curve Estimates of Suspended Sediment Yield: Some Further Comments. In Proceedings of the Porto Alegre Meeting, Porto Alegre, Brazil, 14–21 December 1988.
48. Guo, X.; Xu, L.; Su, L.; Deng, Y.; Yang, C. Comparing Flow Duration Curves and Discharge Hydrographs to Assess Eco-Flows. *Water Resour. Manag.* **2021**, *35*, 4681–4693. [CrossRef]
49. Walling, D.E. Limitations of the Rating Curve Technique for Estimating Suspended Sediment Loads, With Particular Reference to The Use of Cs-137 and Pb-210 Measurements for the Assessment of Soil Erosion and Sedimentation in the Riva Basin (Istanbul, NW Turkey) View Project. Available online: https://www.researchgate.net/profile/Des-Walling/publication/253921527_Limitations_of_the_Rating_Curve_Technique_for_Estimating_Suspended_Sediment_Loads_With_Particular_Reference_to_British_Rivers/links/53edb5bb0cf23733e80ad89d/Limitations-of-the-Rating-Curve-Technique-for-Estimating-Suspended-Sediment-Loads-With-Particular-Reference-to-British-Rivers.pdf (accessed on 20 June 2023).
50. Chen, L.; Schumer, R.; Knust, A.; Forsee, W. Impact of Temporal Resolution of Flow-Duration Curve on Sediment Load Estimation. *J. Am. Water Resour. Assoc.* **2012**, *48*, 145–155. [CrossRef]
51. Benselama, O.; Mazour, M.; Hasbaia, M.; Djoukbala, O.; Mokhtari, S. Analysis of the Suspended Sediment Yield at Different Time Scales in Mediterranean Watershed, Case of Wadi El Maleh (North-West of Algeria). *J. Mediterr. Earth Sci.* **2019**, *11*, 3–13. [CrossRef]
52. Liquete, C.; Canals, M.; Ludwig, W.; Arnau, P. Sediment Discharge of the Rivers of Catalonia, NE Spain, and the Influence of Human Impacts. *J. Hydrol.* **2009**, *366*, 76–88. [CrossRef]
53. Van Rompaey, A.; Bazzoffi, P.; Jones, R.J.A.; Montanarella, L. Modeling Sediment Yields in Italian Catchments. *Geomorphology* **2005**, *65*, 157–169. [CrossRef]
54. Borrelli, P.; Märker, M.; Panagos, P.; Schütt, B. Modeling Soil Erosion and River Sediment Yield for an Intermountain Drainage Basin of the Central Apennines, Italy. *CATENA* **2014**, *114*, 45–58. [CrossRef]
55. Vanmaercke, M.; Poesen, J.; Verstraeten, G.; de Vente, J.; Ocakoglu, F. Sediment Yield in Europe: Spatial Patterns and Scale Dependency. *Geomorphology* **2011**, *130*, 142–161. [CrossRef]
56. Alexander, E.B. Rates of Soil Formation: Implications for Soil-Loss Tolerance. *Soil Sci.* **1988**, *145*, 37–45. [CrossRef]
57. Verheijen, F.G.A.; Jones, R.J.A.; Rickson, R.J.; Smith, C.J. Tolerable versus Actual Soil Erosion Rates in Europe. *Earth Sci. Rev.* **2009**, *94*, 23–38. [CrossRef]
58. Alewell, C.; Egli, M.; Meusburger, K. An Attempt to Estimate Tolerable Soil Erosion Rates by Matching Soil Formation with Denudation in Alpine Grasslands. *J. Soils Sediments* **2015**, *15*, 1383–1399. [CrossRef]
59. Bazzoffi, P. Soil Erosion Tolerance and Water Runoff Control: Minimum Environmental Standards. *Reg. Environ. Chang.* **2009**, *9*, 169–179. [CrossRef]
60. Di Stefano, C.; Ferro, V. Establishing Soil Loss Tolerance: An Overview. *J. Agric. Eng.* **2016**, *47*, 127–133. [CrossRef]
61. Europe's Environment: The Second Assessment—European Environment Agency. Available online: https://www.eea.europa.eu/publications/92-828-3351-8 (accessed on 31 May 2023).
62. Di Stefano, C.; Nicosia, A.; Pampalone, V.; Ferro, V. Soil Loss Tolerance in the Context of the European Green Deal. *Heliyon* **2023**, *9*, e12869. [CrossRef] [PubMed]
63. Soil and Land. Available online: https://environment.ec.europa.eu/topics/soil-and-land_en (accessed on 31 May 2023).

water MDPI

Article

Use of Logs Downed by Wildfires as Erosion Barriers to Encourage Forest Auto-Regeneration: A Case Study in Calabria, Italy

Giuseppe Bombino [1,*], Giuseppe Barbaro [2], Pedro Pérez-Cutillas [3], Daniela D'Agostino [1], Pietro Denisi [1], Giandomenico Foti [2] and Santo Marcello Zimbone [1]

1 Agriculture Department, Mediterranea University of Reggio Calabria, 89122 Reggio Calabria, Italy; daniela.dagostino@unirc.it (D.D.); pietro.denisi@unirc.it (P.D.); smzimbone@unirc.it (S.M.Z.)
2 Department of Civil Engineering, Energy, Environment and Mathherials, Mediterranea University of Reggio Calabria, 89122 Reggio Calabria, Italy; giuseppe.barbaro@unirc.it (G.B.); giandomenico.foti@unirc.it (G.F.)
3 Department of Geography, University of Murcia, 3001 Murcia, Spain; pedrope@um.es
* Correspondence: giuseppe.bombino@unirc.it

Abstract: The easy implementation of ecologically-sound remediation measures for the prompt stabilisation of burned areas may be crucial in Mediterranean forest environments. Manual in situ contour redirection of burned felled logs could aid in soil erosion control and facilitate forest self-regeneration. In this study, a plot-scale runoff/sediment yield survey was conducted in Calabria, Italy, within a Mediterranean pine forest that was affected by an extreme wildfire spanning over 15,000 hectares in the summer of 2021. The hydrological response to 24 rainfall events was analysed after one year of monitoring using nine Wischmeier and Smith 20% sloping plots, which were distributed into three plot-blocks representing different conditions (forested, burned with randomly directed fallen logs, and burned with fallen logs manually redirected along contour lines). The post-fire condition (with felled logs in random positions) exhibited a consistent overall increase (approximately four times) in runoff and sediment yield compared to the pre-fire situation. This degradation effect was mitigated by approximately 30% through the manual redirection of burned logs, which promoted early (three to five weeks) vegetation regeneration (including tree emergence) and enhanced coverage as vegetation spread from the log positions. The results obtained so far provide encouraging insights and warrant further research on steeper slopes and complementary aspects (regulatory, biological, mechanical, economic, etc.).

Keywords: forest wildfire; mountain areas; burned logs; soil erosion; erosion control; erosion barrier; runoff; sediment yield; forest regeneration

Citation: Bombino, G.; Barbaro, G.; Pérez-Cutillas, P.; D'Agostino, D.; Denisi, P.; Foti, G.; Zimbone, S.M. Use of Logs Downed by Wildfires as Erosion Barriers to Encourage Forest Auto-Regeneration: A Case Study in Calabria, Italy. *Water* **2023**, *15*, 2378. https://doi.org/10.3390/w15132378

Academic Editors: Vito Ferro and Alessio Nicosia

Received: 5 June 2023
Revised: 23 June 2023
Accepted: 25 June 2023
Published: 27 June 2023

1. Introduction

Mediterranean forest fires have a long history [1] and represent a relevant ecological factor [2]. Since the early 1960s, the expansion of the wildland–urban interface, rural depopulation, and the abandonment of land management practices have dramatically increased the frequency and extent of wildfires [3].

While wildfires are widely classified according to their intensity [4–6] and severity [7–9], the scientific literature shows a wide terminological variability in the definition of forest fires depending on the fire characterization criteria (e.g., fire size or burnt area distinguishing "large/mega/giga/tera-fires", fire behaviour, fire resistance to control, descending socio-economic, environmental, or human impacts). However, the differentiation between "ordinary" and "extreme" wildfire events is easy and intuitive to perceive. Extreme events of high intensity and severity can certainly produce devastating effects on the forest ecosystem. Approximately 65,000 fires take place every year in the European region, burning an average of approximately half a million forested hectares. In the EU Mediterranean area, approximately 85% of wildfires affect approximately 300,000 to almost one million ha y^{-1}, with an average burnt area of approximately 550,000 ha y^{-1} (data updated to 2021 [10–12]). Of these fires, 95% are of human origin, with arson accounting for 55.8% of cases. In the Mediterranean basin, extreme events are fortunately rare (less than 3%), but they are responsible for over 50% of the total burned area, as observed in Portugal, Spain, and Greece [13]. High-intensity and severity extreme wildfires have been relatively frequent [11] in recent decades and recur every two to three years. These extreme events often manifest as simultaneous fires that can affect large areas (more than 10,000 ha, as occurred in Italy in 2021), exhibiting extraordinary characteristics in terms of size and complexity, which usually escape human control operations.

In the European Mediterranean region, large areas have been reforested with monospecific tree species (e.g., conifers) that are more vulnerable to fire. The ongoing changes in socio-economic and climate scenarios [14] could favour the diffusion and magnitude of wildfires [15–18].

Among others, the worst consequences of high-intensity and severity wildfires are: (i) the loss of biodiversity [19–22], and (ii) the degradation of the physical–chemical properties of soil due to high heating effects [2]. Fire causes a decrease in soil porosity and an increase in bulk density due to prolonged and elevated temperatures. High-intensity fires also alter organic matter and produce ash, which forms a hydrophobic coating on the soil surface. This coating reduces soil moisture and infiltration capacity, increasing runoff propension and soil erodibility [2]. The chemical properties of the soil show changes in pH in the top soil horizons, an increase in electrolyte concentration, a decrease in cation exchange capacity and sodium adsorption ratio, and pronounced leaching of nutrients caused by precipitation. These factors result in a reduced infiltration rate, increased runoff, and soil erosion [2].

After a wildfire, the soil protection provided by forest cover is weakened [23–25]. The bare soil is exposed to the action of raindrop impact, leading to enhanced runoff and erosion processes [26–30]. In the Mediterranean area, summer wildfires can be followed by frequent and high-intensity rains [2]. Studies have measured runoff volume and sediment yield ranging from 1–4 orders of magnitude higher in burnt areas of several Mediterranean countries (e.g., [2,31–37]), both at the plot scale and to a lesser extent at the watershed level.

A high-intensity/severity fire can easily move upward, spreading through tree crowns (known as a crown fire), causing trees to fall either during the event or later (when weakened are felled by the weather). Trees felled by extreme fires may slide downstream, particularly on steep slopes. As they slide down, the logs can reach the hydrographic network, where they may (i) obstruct channel water sections, and (ii) compromise the hydraulic functionality of river works and infrastructure that interact with watercourses.

The need to restore the protective action of the forest is especially felt in the semi-arid Mediterranean area (e.g., in morphologically complex territories such as Calabria). In these regions, hydro-geomorphological, climatic, and anthropic factors accentuate erosive phenomena and hydrogeological instability [38].

Many techniques are available for the stabilisation of areas damaged by wildfires, commonly classified as "bioengineering techniques". The objective of these techniques is to trigger the protective action of the vegetation cover after the adverse event. However, these techniques have a certain level of invasiveness. Among them, log erosion barriers (LEBs) show a lower ecological impact coupled with relative affordability. The use of logs (including logs fallen by wildfires), along contour lines as a post-fire measure to control erosion processes and trigger natural forest auto-regeneration capacity, has been widely investigated by several authors [2,13,39–45]. There are few (if any) experiences in Italy, especially in the South (which is characterised by a semi-arid Mediterranean climate).

The standard installation technique involves felling burned trees and laying them on the ground along the slope contour [44,46]. Each log is anchored in place and gaps between the log and the soil surface are filled with soil to create a storage basin on the upslope side of the log. This configuration obstructs overland flow and traps sediments. The LEBs are usually laid out in staggered tiers designed to eliminate long uninterrupted flows paths [44]. A number of experiences with LEBs have been conducted in Mediterranean environments, providing highly variable (though generally positive) results across a wide range of slopes tested (25–60%) [41,47–50]. The effectiveness of LEBs depends on climatic conditions, soil characteristics, forest type, and the timing of intervention. However, Raftoyannis and Spanos [49] reported a limited post-fire ecosystem recovery, specifically regarding the regeneration of *Quercus coccifera*, which is a notoriously non-pioneer species. The slope gradient seems to have a significant influence on the effectiveness of this measure, which tends to decrease above a 30% gradient [13,42–44]. Moreover, LEBs appear ineffective for rain events with larger return periods [44]. This reclamation measure emulates the philosophy of other more common and well-tested bioengineering techniques (e.g., brush barriers, live palsied, fascines, live branches, wattles/bundles, the use of logs as construction elements, etc.) [51]. However, these measures, which can be relatively labour-intensive and not always implemented manually, do not consider the utilization of landed burnt logs as an ecological tool.

In Italy, the current legislation on forest fires (Law n. 353/2000) prohibits, among other actions, "... reforestation and environmental engineering activities supported with public financial resources ... " in areas affected by fire, except for areas with "documented hydrogeological instabilities ... " (Art. 10 par. 1). The law's main objective is to discourage any form of speculation on fire-damaged areas. It may be administratively difficult to quickly document hydrological instabilities (as the lawmaker suggests), especially potential instabilities linked to the disappearance of the forest's protective action. Environmental engineering activities would require the installation of a worksite, which could further disturb the already fragile ecosystem due to mechanised operations (e.g., soil compaction, removal of new/young plants, etc.). Even the simple removal of collapsed logs (e.g., skidding) is very expensive [52] (approximately up to 15–20 € m^{-3}, depending on local condition such a slope steepness and a 28–40 cm log diameter), and it can have significant impacts [53]. On the other hand, fallen logs can play an important ecological role as they degrade and add organic matter to the soil [54].

A feasible mitigation action that complies with the different constraints is the in situ use of collapsed burned logs, which are manually redirected along the contour lines. This elementary operation (i) appears to comply with the law, (ii) requires minimal workforce and is cost-effective, (iii) can be implemented quickly (a few weeks after the fire), (iv) is ecologically sustainable, and (v) can contribute to slope stabilization and ecosystem auto-regeneration. The implementation of this measure could also facilitate the prompt registration of forest fires, as required by the law but only in partially enforced. In this context, the aim of this work is to evaluate, the effectiveness of the in situ use of collapsed burned logs redirected along the contour lines to control erosive phenomena (in terms of runoff volume and sediment yield) and trigger a stabilisation effect that supports ecosystem auto-regeneration through a case study in Calabria, Italy.

2. Materials and Methods

2.1. Description of the Extreme Fire Events in Summer 2021 in Aspromonte Massif, Southern Calabria, Italy

In the summer of 2021, Italy, particularly the southern part of the country, was affected by extreme wildfires in terms of intensity and severity [13]. In the southernmost part of the Calabria region (Aspromonte massif), 14,840 ha were burned, with 5600 ha located within the Aspromonte National Park (data measured through ground surveys and satellite imagery, Copernicus Emergency Management Service). These wildfires resulted in the loss of two lives, several hundred livestock, and ruinous damage to many buildings and farms. The extraordinary wildfires, some of which started simultaneously, affected even the most remote areas of the Aspromonte Massif, including the centuries-old beech forest (known as "Valle infernale", a UNESCO World Heritage and site) and the pine forest (known as "Acatti" and "Afreni" woods, which are old-growth forests of more than 500 years old). Among the most affected areas by that wildfire was Roccaforte del Greco, a mountain village within the well-known "Greek island" within the Aspromonte National Park. Within a few days (approximately ten), a high-intensity/severity crown-fire destroyed hundreds of hectares of forest (mostly pine reforestations from the 1950s), resulting in the felling of thousands of trees (Scheme 1). This provided the occasion to implement the experimental monitoring site and campaign described herein.

Scheme 1. Representative photo of the Aspromonte areas affected by the wildfires in the summer of 2021.

2.2. *Study Area Description*

The study area is located in the southernmost part of the Calabria (Southern Italy, Figure 1), specifically within the mountain municipality of Roccaforte del Greco (Aspromonte, Metropolitan area of Reggio Calabria, Figure 1). It is located at the headwaters of the fiumara Amendolea watershed (150.4 km^2), with an average hillslope gradient of 97.2%. The hydro-geomorphological, ecological, and climatic conditions in this region are typical of Southern Italy [55]. The headwater areas consist of igneous and metamorphic rocks, with sequences of grey and black biotite schist [56,57]. High-magnitude erosion processes are common after heavy storms, and large amounts of coarse material are transported to the valley reach of the fiumara Amendolea, which has a large riverbed (up to 1 km) [55]. These hydro-geomorphological activities result in significant land instability and vulnerability, which are subject to specific regulatory hydro-geological restrictions that cover a large part of the Aspromonte Massif. The environment is characterised by dense Mediterranean scrub and forests with a wide variety of trees, both deciduous and evergreen. Coniferous reforestation projects have been implemented since the middle of the last century to control erosion phenomena. In the experimental area, the vegetation cover mainly consists of *Pinus* subsp. Afforestation, with *Pinus radiata*, an exotic species from the west coast of North America, being the dominant species. This species was imported to address the shortage of native conifers [58]. No specimens survived to the wildfires that occurred in 2021. The climate and forest characteristics of the study area are reported in Table 1.

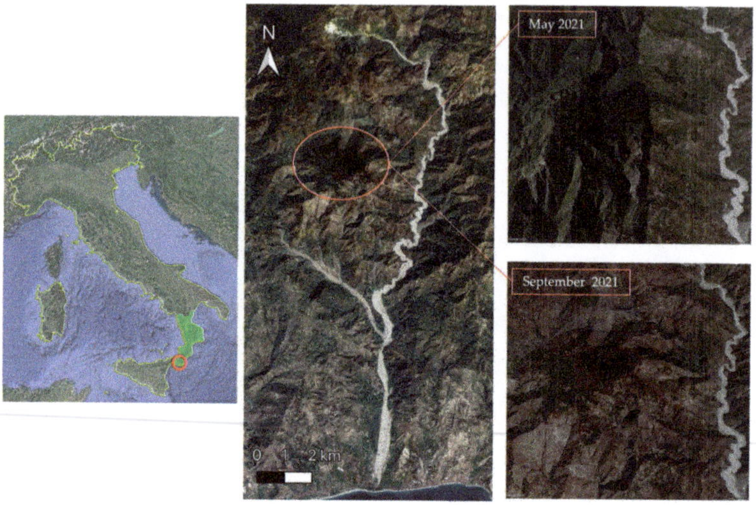

Figure 1. The study area located within the territory of Aspromonte Massif, Metropolitan Area of Reggio Calabria, Southern Calabria, Italy. Satellite imagery provided by Google Earth Pro shows the study area (red circle) before and after the wildfire that occurred in August 2021 (territory of Roccaforte del Greco).

Table 1. Main characteristics of the study area within the territory of Roccaforte del Greco (Aspromonte, Calabria, Italy).

Orography	Coordinates	38°03′ N; 15°54′ E
	Altitude	970 m a.s.l.
	Aspect	North
Climate	Condition (according to the Köppen classification [59])	Csa [1]
	Mean monthly air temperature *	5 °C (January)–23 °C (August)
	Mean annual rainfall depth *	1036 mm
	Mean monthly rainfall depth range *	13.7 mm (July)–153 mm (December)
Forestsetting	Species	*Pinus radiata*
	Density	800–1000 per ha
	Height	14–20 m
	Crown diameter	4–7.5 m

Notes: [1] Hot Mediterranean climate and semiarid conditions. * Database ARPACAL, meteorological station of Roccaforte del Greco (time series covering the period 1940–2022, n = 57, variation coefficient = 29%).

2.3. Experimental Scheme

Three plot-blocks (up to 120 m apart) were established in the north-facing study area in September 2021, with a down slope of 20% ± 2% (Figure 2a). Each plot-block represented a specific condition as follows (Figure 2b):

- "Forested" (F): representative of the pre-fire condition;
- "Burned" (B): casually directed downed logs, representative of the post-fire asset;
- "Burned and rearranged" (BR): contour-redirected downed logs, resembling log erosion barriers.

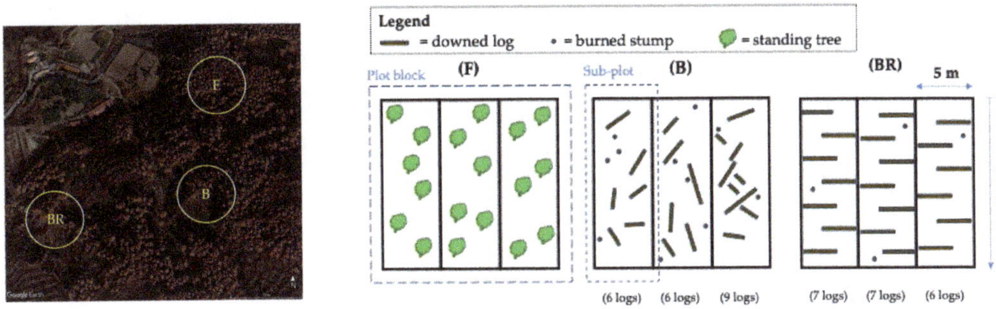

(**a**) (**b**)

Figure 2. (**a**) Location of the three plot-blocks in Roccaforte del Greco, Southern Calabria, Italy; (**b**) experimental sketch showing the three conditions studied: F, forested; B, burned with casually directed downed logs; BR, burned rearranged with manually redirected downed logs along contour lines.

Each plot-block was further divided into three 22 × 5 m sub-plots of the Wischmeier and Smith type (Figure 2b). These sub-plots were hydraulically isolated using aluminium foil embedded into the soil to a depth of 20 cm.

In the B sub-plots, there were casually directed downed logs, with lengths ranging between 1.5 and 4 m. Two cases had 6 logs, while one case had 9 logs (Figure 2b). In the BR plot-block, the downed logs were manually redirected (by a local rotation) along the contour lines within each sub-plot (Figure 3). The number of downed logs redirected within the BR sub-plots was 7 in 2 cases and 6 in one case. The logs, approximately 4 m long with an average diameter of 33 cm (ranging from 21 to 49 cm), were reallocated at an average downslope distance of 4 m, resulting in partial fragmentation of the slope path (Scheme 2a). The rotation and locking of the logs only required manual work and light equipment (hammer) to fix the wooden stakes (Scheme 2b,c). The ratio between the sub-plot area (110 m^2) and the total length of the contour logs was within the range of 3.9–4.6 m.

Figure 3. Sketch of the manual redirection along the contour lines (brown lines) and hammering of downed burned logs (dashed yellow lines indicate the rotation of logs).

Scheme 2. (**a**) Overall view of the BR sub-plots under construction (in orange, plastic gutters), (**b**) view of the burned log hammering, and (**c**) rearranged burned sub-plots (BR).

The necessary manpower for the in situ manual redirection of the felled logs can be extrapolated at 4 days ha^{-1} for two operators.

2.4. Data Collection and Processing

The experimental campaign took place from September 2021 (approximately 1 month after the wildfire ended and before the first autumn rains) to September 2022, covering a period of 13 months.

2.4.1. Soil

Within the 5 cm of the topsoil, 27 soil samples (3 × 9 sub-plot) were collected, located at the top, middle, and bottom of the nine sub-plots along their median axis (Table 2).

Table 2. Main physical–chemical characteristics of the soil in the three plot-blocks.

Physical/Chemical Features				**Plot-Block**		
				F	B	BR
= downed log; • = burned stump; = standing tree; = sampling point				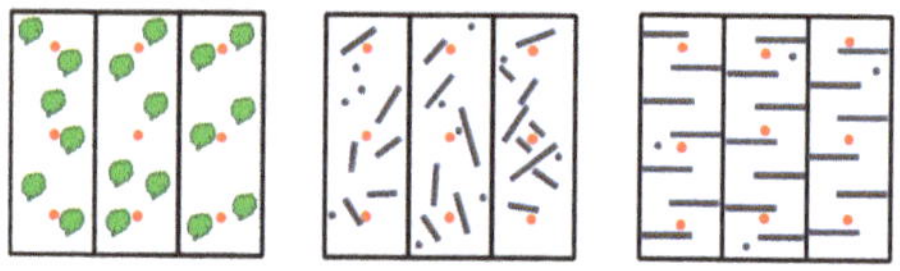		
Silt			(%)		9.4 ± 1.1	
Clay			(%)		11.7 ± 0.1	
Sand			(%)		78.9 ± 0.94	
pH				5.76 ± 0.2	6.19 ± 0.3	6.21 ± 0.3
$EC_{1:1}$ (dS/m)				0.199 ± 0.043	0.202 ± 0.031	0.274 ± 0.047
Organic matter			(%)	3.38	2.98	2.96
C			(%)	1.96 ± 0.2	1.72 ± 0.1	1.71 ± 0.1
N			(%)	0.79 ± 0.1	0.83 ± 0.0	0.83 ± 0.0
C:N				22.8	25.3	26.1
Mean infiltration rate			(mm h^{-1})	23.8	16.1	15.8
Soil water repellency				39.6	93.6	94.3
Ash cover			(%)	0.0	47.0 ± 4.8	46.6 ± 5.1
AS (n = 3)	Size fraction (mm)	0.25–0.50	(%)	78 ± 1.0	72 ± 1.3	72 ± 1.2
		0.50–1.0	(%)	83 ± 1.4	76 ± 1.4	76 ± 1.3
		1.0–2.0	(%)	31 ± 1.4	28 ± 1.2	28 ± 1.3

Soil texture was measured using the hydrometer method with sodium hexametaphosphate as a dispersant [60] and classified using the USDA triangle method.

pH values were measured using a soil/water suspension ratio of 1/2.5 (w/v) with a glass electrode.

Electrical conductivity (EC) was determined in distilled water using a 1/5 residue/water suspension. The suspension was mechanically agitated at 15 rpm for 1 h to dissolve soluble salts, and EC was then measured using a Hanna conductivity meter.

Organic carbon (C) content was determined using the dichromate oxidation method according to Walkley and Black [61] and titration with iron sulphate ($FeSO_4$, 0.2 N).

Total nitrogen (N) was measured using the method described by Kjeldahl [62].

Water infiltration rate (IR) was determined using an Eijkelkamp® portable rainfall simulator [63].

Soil water repellency (SWR) was estimated using the Water Drop Penetration Test (WDPT) method [64], conducted in the vicinity (approximately 0.25 m) of the IR measurement point. In this test, 15 drops of distilled water were released on the soil surface using a pipette. The time taken for the droplets to penetrate the soil was measured to estimate the WDPT.

Before measuring IR and SWR, the litter was removed, and the soil surface was lightly levelled.

Soil aggregate stability (AS) was determined using the rain simulator method (working with an intensity of 6 mm min^{-1}). One sample per sub-plot was tested for the three size fractions (0.25–0.50, 0.5–1.0, 1.0–2.0 mm), following the method described by Roldán et al. [65] based on the method of Benito et al. [66].

According to the USDA Soil Texture Classification System, the analyzed soil falls into the sandy-loamy category (Table 2). It is acid, low in organic matter, and exhibits a high infiltration rate and a moderate aggregate stability.

2.4.2. Rainfall

Rainfall data were continuously recorded at the meteorological station located in Roccaforte del Greco (approximately 3 km away from the plot-blocks). In addition, cumulative rainfall depth was gauged within each plot-block. Due to logistical limitations (mountainous context, distance), it was generally not possible to collect runoff volume and sediment yield immediately after each rainfall event. Consequently, rainfall data were cumulated (except in one case) in 24 rainfall event groups.

2.4.3. Runoff

Runoff volume was measured on 24 occasions using a tank connected to a plastic gutter (with a 2% gradient towards the outlet) at the bottom of each sub-plot. The collected volume was then divided by the sub-plot area.

2.4.4. Sediment Yield

Sediment dry weight was measured by collecting sediment from the plastic gutter at the bottom of each sub-plot and sampling the sediment concentration in the stored runoff volume. Sediments in the plastic gutters were collected at the same frequency as the runoff volume. The wet sediment samples were weighed and dried in an oven at 105 °C for 24 h. Organic matter was removed from the samples using the LOI Method (at approximately 375 °C for 16 h [67]), and the dried samples were weighed. Sediment concentrations were collected by mixing the stored volume in the tank and taking three successive samples, totalling approximately 0.5 L. The dried sediment from the samples was weighed and related to the sample volume in order to evaluate the sediment concentration. The sediment concentration was multiplied by the runoff volume to estimate the dissolved sediment yield component. Finally, the sediment yield was obtained by adding to the amount of sediment deposited in the plastic gutter (averaging 10% ± 6 of the total).

2.4.5. Vegetation Cover

The vegetation cover in each B and BR sub-plot was monitored twice a month using the grid method [68] with a 0.75 × 0.75 m square. The abundance–dominance index of each species collected within the sub-plots was determined according to the Braun–Blanquet method [69]. In order to assess the effect on vegetation development, the degree of vegetation cover in plot-blocks B and BR was compared, and the relationships between vegetation cover, runoff volume, and sediment yield were investigated.

2.4.6. Statistical Analysis

The statistical analyses of runoff volume and sediment yield data were carried out using Jamovi v.2.3.21 (2022). A descriptive assessment was first carried out to determine whether the data set was normally distributed using the Shapiro–Wilk test. Based on this result, a correlation matrix was applied using Spearman's correlation coefficient to determine the degree of relationship between the "runoff" and "sediment" measures with the "rainfall depth" and "rainfall intensity" factors.

In order to assess the differences between the variables recorded in the "runoff" and "sediment" data collection plots, a repeated measures ANOVA was applied using Friedman's method (non-parametric). Pairwise comparisons were obtained using the Durbin–Conover test, which was employed to determine the differences between treatments in the analysed plots with a confidence interval of 95%.

Furthermore, the effects of rainfall depth and intensity on runoff volume and sediment yield were evaluated using a factorial analysis. A One-Way ANOVA following the Kruskal–Wallis test (non-parametric) was performed, with three categories used as grouping variables based on the following parameters: rainfall depth—first category: values between 0–35 mm, second category: values between 35–70 mm, third category: values greater than 70 mm; rainfall intensity—first category: values between 0–10 mm h^{-1}, second category: values between 10–30 mm h^{-1}, third category: values greater than 30 mm h^{-1}. Pairwise comparisons of the categories were carried out using the Steel–Dwass–Critchlow–Fligner method, with a 95% confidence interval used to establish the thresholds of change in the effect of the variables analysed. Finally, the existence of statistical differences between the average values of the degree of vegetation cover in the two different burned plot-blocks (B and BR) was verified.

3. Results and Discussion

3.1. General View of Plot Hydrological Responses

The recorded rainfall data for the 13-months monitoring period are reported in Table A1 in Appendix A. There were a total of 88 rainy days, and the 24 rainfall event groups ranged in depth from 14 mm (April 2022) to 249.4 mm (October 2021). The maximum 5-min intensity varied from 2.2 mm h^{-1} (April 2022) to 35.4 mm h^{-1} (August 2022). The consequent total annual rainfall sum (10 September 2021 to August 2022) was 1278.6 mm, which was significantly above the mean historical annual depth. This higher value was within the range of variability observed in the historical series (variation coefficient of 29%). The cumulative rainfall measured within the plot-blocks was systematically lower, likely due to lower altitude (with respect to the main meteorological station) and the location of the rain gauges at the base of the slope. Except for plot-block F, which experienced expected canopy interference, the average rainfall deficit compared to the Roccaforte del Greco station was 3.09% and 2.92% for B and BR, respectively. Since there were no statistical differences between the rainfall values recorded at the meteorological station (Roccaforte del Greco) and those collected in the plot-blocks (except for F), the subsequent analysis of the plot's hydrological response was based on the Roccaforte del Greco data set, which also provided maximum intensity data. The summary values of runoff volumes and sediment yields measured for each event group are reported in Table 3.

Table 3. Descriptive values (n = 24) of sub-plot runoff volume and sediment yields. The Shapiro–Wilk test shows the assumption of normality at p values > 0.05.

	Runoff Volume (mm)			Sediment Yield (kg)		
	Plot F	Plot B	Plot BR	Plot F	Plot B	Plot BR
Mean	0.73	3.84	2.87	0.196	1.028	0.791
Median	0.40	2.29	1.73	0.171	1.040	0.615
Standard deviation	0.93	3.77	2.70	0.141	0.838	0.746
Minimum	0.02	0.30	0.20	0.007	0.060	0.018
Maximum	3.84	15.82	10.87	0.575	2.434	2.312
Shapiro–Wilk p	<0.001	<0.001	0.002	0.199	0.007	0.007

Two particularly severe events occurred during the observation period (Table A1):

- Event no. 7, which occurred in November 2021, had a cumulative rainfall depth of 249.4 mm and a maximum 5-min rainfall intensity of 30.6 mm h^{-1}. This event occurred over the plots that were only partially covered by vegetation, producing the highest values of runoff volume (15.77 mm) and sediment yield (2076.25 g) from plot-block B.
- Event no. 23, which occurred in August 2022 (105.4 mm, 35.4 mm h^{-1}), generated a runoff volume of 7.03 mm and a sediment yield of 403.65 g (plot-block B).

Nevertheless, the absolute highest sediment value (2.434 kg corresponding to 0.022 kg m^{-2}, plot B) was recorded after event no. 1, despite its relatively moderate characteristics (36.4 mm, 18.0 mm h^{-1}). This event occurred immediately after the wildfire when the ground was completely bare. For the same events, slightly lower values of runoff volume and sediment yield (33% ± 12% and 76% ± 80% on average, respectively) were observed in plot-block BR.

The correlation analysis conducted between rainfall and runoff/sediment data showed statistically significant relationships in every sub-plot (Table 4). The Spearman's correlation coefficient indicated that the highest values for runoff were observed between rainfall depth and intensity. Regarding the sediment data, the correlation was stronger with intensity data compared to rainfall depth.

Table 4. Correlation matrix of runoff and sediment with precipitation and intensity using Spearman's correlation coefficient.

		Runoff			Sediments		
		Plot F	Plot B	Plot BR	Plot F	Plot B	Plot BR
Rainfall depth	Spearman Rho	0.93 ***	0.79 ***	0.78 ***	0.57 **	0.47 *	0.43 *
	p value	< 0.001	< 0.001	<0.001	0	0.02	0.04
Rainfall intensity	Spearman Rho	0.66 ***	0.78 ***	0.77 ***	0.67 ***	0.58 **	0.6 **
	p value	<0.001	<0.001	<0.001	<0.001	0	0

Note: * $p < 0.05$, ** $p < 0.01$, *** $p < 0.001$.

A direct relationship between runoff and rainfall volumes was observed (Figure 4a). In contrast, sediment yield was more closely associated with rainfall intensity (Figure 4d), although no cohesive trend was observed. The recorded data showed an increase in runoff volume and sediment yield depending on the loss of forest vegetation coverage. This effect was somewhat mitigated by the redirection of burned logs in plot-block BR. The decreasing distribution of sediment yield, regardless of the rainfall or intensity patterns reported in the rainfall events, suggests the possibility of studying how the presence of additional factors may explain the evolution of erosion processes.

(a)

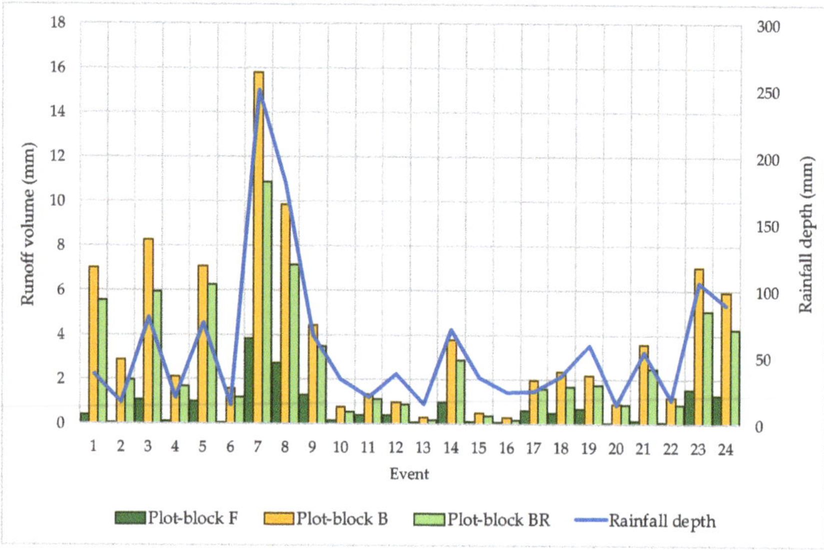

(b)

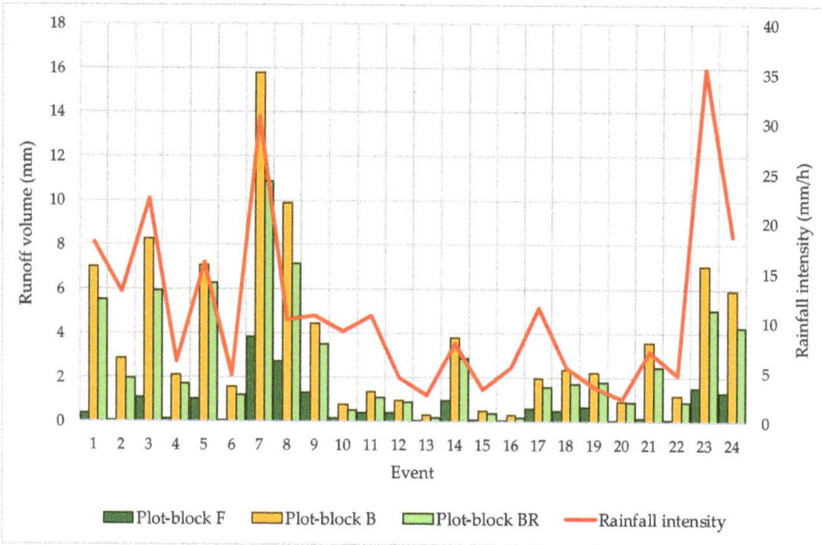

Figure 4. *Cont.*

(**c**)

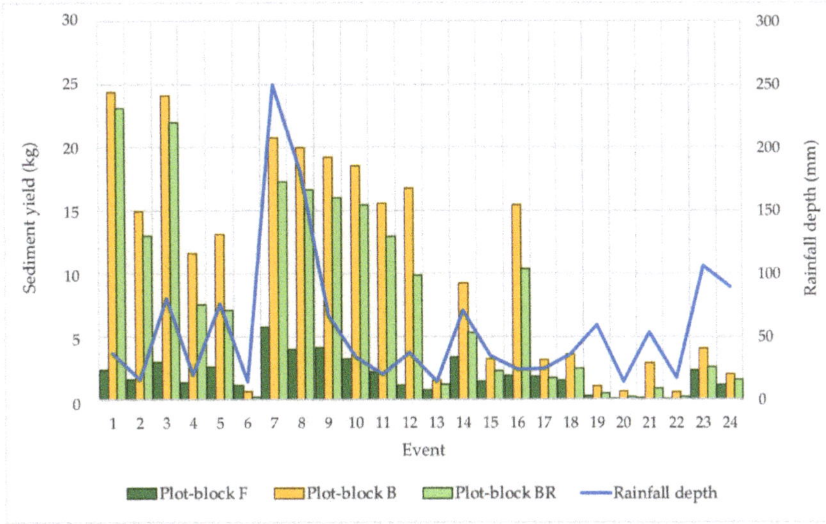

(**d**)

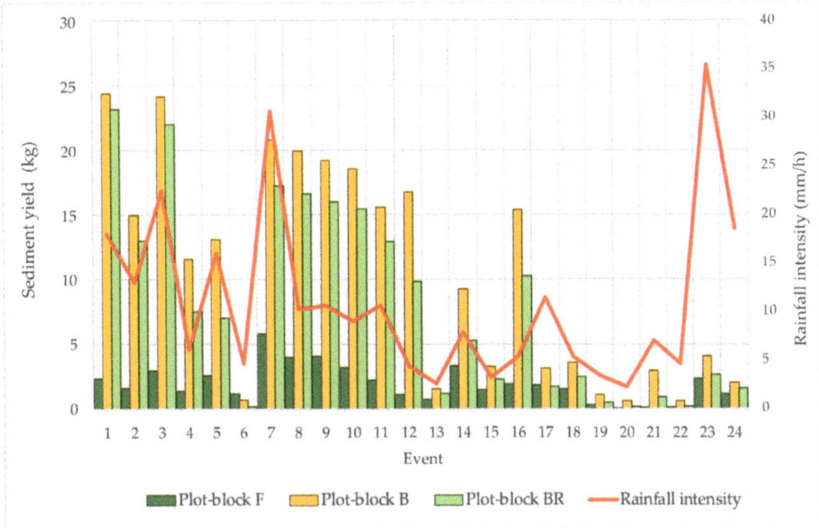

Figure 4. Evolution of mean values (n = 3), runoff volumes (**a**,**b**), and sediment yields (**c**,**d**) in sub-plots related to rainfall depth and maximum 5-min intensity.

3.2. Relations between Vegetation Cover Evolution, Runoff, and Sediment Yield after Wildfire

The analysis of the values shown in Figure 4 (Table A1) highlights that the runoff volume and sediment yield were always lower in the BR plots compared to the B plots, with an average (n = 24) reduction of approximately 30% (standard deviation of 80%, which includes events with differences ranging from 5% in event 1 to values greater than 200% in events 6, 20, 22, 23) and 29% (±14%), respectively. However, the minor differences occured in the first three events. The first rains immediately after the wildfire encountered soil without vegetation and covered by a blanket of ash (Table 2), leading to a drastic reduction in soil infiltration [70]. In the first three events, the hydrological response of the BR and B plots appeared high, even though the rainfall characteristics (rainfall depth and intensity) were not particularly severe. A significant reduction in runoff and sediment yield in the

BR plots was recorded from the fourth rainfall event. On the other hand, the absence of tree cover resulted in accentuated insolation; this encouraged the entry and development of wild vegetation (mainly bracken and bramble), which tends to cover the plots within a short time. Monitoring (twice a month) of the ground cover exerted by spontaneous vegetation showed consistently higher values (35% on average; n = 48) in the BR plots [71]. The combined effect of redirected logs and vegetation became evident from the 13th event onwards, with a drastic decrease in runoff volume and sediment yield even during the heaviest rainfall events (for example, see events 17, 19, 23, and 24 in Figure A1).

The mean values (n = 3) of vegetation cover were consistently higher in the BR plots, where the vegetation settled and established itself immediately upstream of the logs (Scheme 3b). From there, it gradually spread, occupying larger areas until it uniformly covered the entire plot (Scheme 3c). This phenomenon (although less pronounced) was also observed in the B plots.

The positive effects of the vegetation were quantified by correlating the differences in vegetation cover ($VC_{BR} - VC_B$) with those of runoff volume ($R_B - R_{BR}$) and sediment yield ($S_B - S_{BR}$) for each rainfall event group. The analysis showed that the increased stage of vegetation cover in the BR plots contributed to the reduction in both runoff volume (R^2 = 0.63) and sediment yield (R^2 = 0.66) (Figure 5).

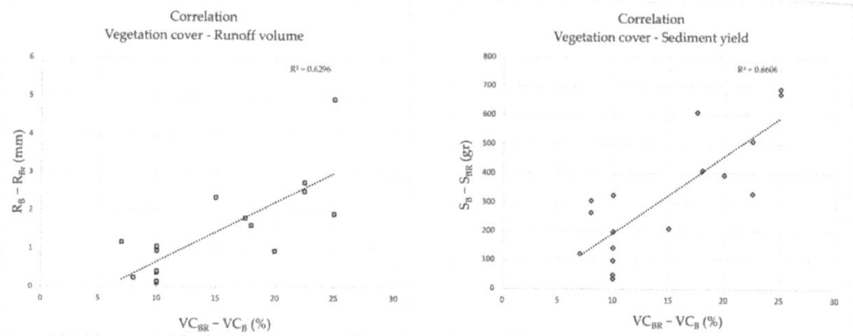

(a) (b)

Figure 5. Correlation between the differences in the values of vegetation cover ($VC_{BR} - VC_B$) and runoff volume ($R_B - R_{BR}$) (**a**) and sediment yield ($S_B - S_{BR}$) (**b**) for each event group.

This observation highlights the positive control action of the redirected contour logs, which stabilize the slope and promote the rapid colonization of herbaceous vegetation (*Rubus ulmifolius* L., *Pteridium aquilinum* L., *Cytisus villosus* L., *Lolium* sp. *pl.*, *Hypochaeris achyrophorus* L., *Hypochaeris laevigata* L., *Geranium robertianum* L., and *Myosotis arvensis* L., as indicated by their abundance–dominance value). This vegetation quickly protects the soil after the fire. This propitious action creates favourable conditions for the auto-regeneration of the wood, as demonstrated by the presence of young pine seedlings (Scheme 3d).

This result can be explained by the partial "breaking effect" of the slope length due to the contour-directed logs. The presence of contour logs, which fragment the slope path, limits the runoff velocity (which also encourages infiltration and runoff volume control) and reduces the detach/transport capacity of sediments. These effects arise as the result of complex interacting processes and representative relationships [72,73]. This effect is influenced by the incomplete logging of the transversal section, which results in the presence of zig-zag overland flow between logs, log overflow, underpass, and local sediment trapping. The effect is partially appreciable in the case of casually directed burned logs (B plot-block) and may be somewhat linked to the extent of their geometric contour projection (maximized in the case of contour-directed logs).

Scheme 3. View of (**a**) the BR sub-plot during initial setup, (**b**) vegetation established immediately upstream of the contour logs after a few months, (**c**) vegetation cover spreading, and (**d**) young pine seedlings.

3.3. Comparative Analysis of Runoff and Sediment Yields

The results obtained from the repeated measures ANOVA revealed evidence of differences among all plots in terms of hydrological aspects. Pairwise comparisons showed significant differences between plots F vs. B, F vs. BR, and B vs. BR, which indicates a high variability in the runoff results. The analysis of the effect of rainfall depth and intensity provided by the one-way ANOVA showed significant differences in all plots, highlighting a greater effect of rainfall (Table 5).

Table 5. Runoff analysis on rainfall depth and intensity using a one-way ANOVA following the Kruskal–Wallis test (non-parametric).

Kruskal-Wallis	Rainfall Depth				Rainfall Intensity			
	χ^2	gl	*p*	ε^2	χ^2	gl	*p*	ε^2
Plot F	16.6	2	**<0.001**	0.724	10.4	2	**0.005**	0.453
Plot B	16.6	2	**<0.001**	0.723	12.3	2	**0.002**	0.537
Plot BR	16.2	2	**<0.001**	0.706	12.2	2	**0.002**	0.530

Notes: Bold *p* values show statistically significant relationships greater than a 95% confidence interval.

Furthermore, differences in runoff volume were observed among plots across all ranges of rainfall, highlighting the strong correlation between runoff and rainfall volume (Table 6). On the other hand, rainfall intensity only provided differences in all plots within the 0–10 mm h^{-1} range. Beyond that range, the runoff data recorded in plots B and BR dd not show significant differences, except for plot F at rainfall intensities above 30 mm h^{-1}. This finding can be explained by the interception effect of the canopy cover as well as by the wetting processes of the vegetation, which can store significant volumes of rainfall. The decrease in runoff occurs at low rainfall intensities, whereas lower vegetation coverage at higher rainfall intensities decreases the water retention effect [74].

Table 6. Pairwise comparison of runoff volume for different categories of rainfall depth and intensity using the Dwass–Steel–Critchlow–Fligner method. Rainfall depth: 1st category, values between 0–35 mm; 2nd category, values between 35–70 mm; 3rd category, values greater than 70 mm. Rainfall intensity: 1st category, values between 0–10 mm h^{-1}; 2nd category, values between 10–30 mm h^{-1}; 3rd category, values greater than 30 mm h^{-1}.

Pairwise Comparisons (Dwass Steel Critchlow-Fligner)		Rainfall Depth						Rainfall Intensity					
		Plot F		Plot B		Plot BR		Plot F		Plot B		Plot BR	
		W	*p*	W	*p*	W	*p*	W	*p*	W	*p*	W	*p*
Category 1	Category 2	3.63	**0.028**	3.56	**0.032**	3.50	**0.036**	3.59	**0.030**	4.30	**0.007**	4.30	**0.007**
Category 1	Category 3	4.94	**0.001**	4.93	**0.001**	4.94	**0.001**	3.13	**0.069**	3.13	0.070	3.14	0.068
Category 2	Category 3	3.54	**0.033**	3.64	**0.027**	3.43	**0.040**	2.67	0.142	2.00	0.334	1.67	0.466

Notes: Bold *p* values show statistically significant relationships greater than a 95% confidence interval.

Compared to the F plot-block, the B plot-block shows an average increase in sediment equal to an order of magnitude (about five times higher). In comparison to the B plot-block, the BR plot-block demonstrated a control effect of approximately 20% (Table 3). This distinction is evident from the average plot-block sediment yield of 0.196 kg (corresponding to 0.002 kg m^{-2} and to 0.40 ton ha^{-1} y^{-1} in the F plot-block), 0.791 kg (0.007 kg m^{-2}; 2.07 ton ha^{-1} y^{-1} in the B plot-block) and 1.028 kg (0.009 kg m^{-2}; 1.59 ton ha^{-1} y^{-1} in the BR plot-block). Similar sediment yields to those produced by the F plot-block were found in Spain, Israel, and Italy by other authors [33,37,75–78] under similar climatic conditions.

In order to analyse the aspects related to soil erosion, a repeated measures ANOVA was used to determine the variances among the plots under study. The results showed the existence of significant differences between all three cases, as confirmed by Durbin–Conover pairwise comparisons. To establish the effect of rainfall depth and intensity on each plot, a more detailed study was carried out by applying a one-way ANOVA based on the Kruskal–Wallis test. The results showed that the effect of rainfall was not involved in plots B and BR, whereas rainfall intensity was a determining factor (Table 7).

The factorial analysis results made it possible to determine the thresholds at which soil loss was affected, in which changes were only observed for plot F when rainfall increased considerably (pairwise comparison of Category 1 vs. Category 3). The effect of intensity on sediment production was higher, revealing significant differences between categories 1 and 2 in all three analysed plots (*p* values: 0.033, 0.047, and 0.027 for plots F, B, and BR, respectively). However, for the rest of the pairwise comparisons, there were no differences (Table 8). Therefore, rainfall intensities below 10 mm h^{-1} determined differences in sediment yield depending on the rainfall events. However, higher rainfall intensity resulted in erosion processes to varying degrees in every plot due to the rise in kinetic energy of raindrops [79].

Table 7. Sediment statistic analysis on rainfall depth and intensity using a one-way ANOVA following the Kruskal–Wallis test (Non-parametric). Bold values show statistically significant relationships greater than a 95% confidence interval.

Plot	Rainfall Depth				Rainfall Intensity			
	χ^2	gl	*p*	ε^2	χ^2	df	*p*	ε^2
Plot F	6.34	2	**0.042**	0.276	8.36	2	**0.015**	0.364
Plot B	3.12	2	0.211	0.135	6.35	2	**0.042**	0.276
Plot BR	2.44	2	0.295	0.106	7.26	2	**0.027**	0.316

Notes: Bold *p* values show statistically significant relationships greater than a 95% confidence interval.

Table 8. Rainfall depth and intensity category pairwise comparison in sediments using the Dwass–Steel–Critchlow–Fligner method. Rainfall depth: 1st category, values between 0–35 mm; 2nd category, between 35–70 mm; 3rd category, greater than 70 mm. Rainfall intensity: 1st category, values between 0–10 mm h^{-1}; 2nd category, between 10–30 mm h^{-1}; 3rd category, greater than 30 mm h^{-1}.

Pairwise Comparisons (Dwass Steel Critchlow-Fligner)		Rainfall Depth						Rainfall Intensity					
		Plot F		Plot B		Plot BR		Plot F		Plot B		Plot BR	
		W	*p*	W	*p*	W	*p*	W	*p*	W	*p*	W	*p*
Category 1	Category 2	0.142	0.994	1.706	0.450	1.14	0.701	3.54	**0.033**	3.35	**0.047**	3.64	**0.027**
Category 1	Category 3	3.522	**0.034**	2.241	0.252	2.11	0.294	2.64	0.148	1.92	0.363	1.92	0.363
Category 2	Category 3	2.424	0.200	0.808	0.836	1.01	0.755	1.33	0.614	0.00	1.000	0.00	1.000

Notes: Bold *p* values show statistically significant relationships greater than a 95% confidence interval.

4. Conclusions

The results of the plot-scale experimental investigation conducted after the extreme wildfires in August 2021 in Calabria, Italy, generally support the effectiveness of the in situ contour re-direction of felled burned logs, inspired by log erosion barriers, in terms of runoff and sediment yield control. This approach shows promise as a means of promptly stabilizing topsoil and triggering forest auto-regeneration.

The reference forest asset (F) with *Pinus radiata* shows a runoff volume and sediment yield conservative control under natural rainfall (about 1300 mm in one year from 24 event groups of 14 to about 250 mm) similar to that previously found in Spain, Israel, and Italy. However, the presence of randomly felled burned logs (B plot-block) following the high-intensity/severity wildfire exacerbates the hydrological response, resulting in approximately five times higher runoff volume and sediment yield. The implemented remediation measure in September 2021, targeting the 20% sloping plots, showed an overall significant reduction in runoff volume and sediment yield (30% and 29%, respectively, compared to condition B). A direct relationship linked the runoff and rainfall volume, while the sediment yield was more dependent on the maximum 5-min rainfall intensity (ranging from 2.2 to 35.4 mm h^{-1}). With respect to the existing post-fire situation conditions (bare soil with randomly directed fall-down burned logs) and the pre-fire *Pinus radiata* asset, a prompt colonization of spontaneous herbaceous vegetation was observed three weeks after the treatment. This was followed by the establishment of natural forest renewal within five weeks. There was a positive correlation between the increased vegetation cover stage in the areas behind the re-directed logs and the reduction in both runoff volume and sediment yield. The implemented measure was exclusively manual and has the potential to be economically/environmentally sustainable. It also complies with the stringent Italian law on burned forests, making it well-suited for timely intervention in hydrogeologically fragile contexts.

An extended monitoring period and further in-depth research are required to further understand the effects of the treatment at higher rainfall intensities and operational slopes (e.g., 30% or more), as well as to explore other aspects (e.g., regulatory, biological, mechanical, economic. These efforts aim to establish a comprehensive framework that is effective in developing protocols and facilitating the dissemination of this measure.

Author Contributions: Conceptualization, G.B. (Giuseppe Bombino); methodology, G.B. (Giuseppe Bombino) and S.M.Z.; field surveys and data collection, G.B. (Giuseppe Bombino), P.D. and D.D.; validation, G.B. (Giuseppe Bombino) and S.M.Z.; data curation, P.P.-C., P.D., G.B. (Giuseppe Barbaro), G.F. and D.D.; formal analysis, G.B. (Giuseppe Bombino), P.P.-C. and S.M.Z.; writing—original draft preparation, G.B. (Giuseppe Bombino), P.P.-C. and S.M.Z.; writing—review and editing, G.B. (Giuseppe Bombino), D.D. and S.M.Z.; supervision, G.B. (Giuseppe Bombino), P.P.-C. and S.M.Z.; project funding, G.B. (Giuseppe Bombino) and S.M.Z. All authors have read and agreed to the published version of the manuscript.

Funding: This work was funded by the Next Generation EU—Italian NRRP, Mission 4, Component 2, Investment 1.5, specifically the call for the creation and strengthening of 'Innovation Ecosystems' and the building of 'Territorial R&D Leaders' (Directorial Decree n. 2021/3277). The project Tech4You—technologies for climate change adaptation and quality of life improvement, was assigned the identification n. ECS0000009. This work reflects only the authors' views and opinions and do not necessarily reflect those of the Ministry for University and Research or the European Commission, who can be considered responsible for them.

Data Availability Statement: Not applicable.

Acknowledgments: The authors are grateful to the Francesco Saccà agro-forestry farm in Roccaforte del Greco for providing access to the experimental system and logistic support. The authors also extend their thanks to the research fellow Giuseppina Lofaro for her insights on the regulatory aspects.

Conflicts of Interest: The authors declare no conflict of interest.

Appendix A

Table A1. Main characteristics of rainfall events recorded by the meteorological station located in Roccaforte del Greco, Southern Calabria, Italy.

Month	Event Group	Number of Rainy Days	Cumulated Rainfall Depth (mm)	Maximum 5-min Rainfall Intensity (mm h^{-1})
September 2021	1	2	36.4	18.0
	2	3	15.8	13.0
November	3	6	79.6	22.4
	4	3	19.4	6.0
	5	3	75.2	16.0
	6	2	14.0	4.6
	7	5	249.4	30.6
	8	5	180.2	10.2
November	9	5	66.4	10.6
	10	2	33.4	9.0
	11	2	20.0	10.6
	12	4	37.2	4.4
December	13	5	14.4	2.6
	14	5	70.4	7.8
	15	3	34.2	3.2
January 2022	16	6	23.8	5.4
February	17	3	24.4	11.4
	18	2	35.6	5.4
March	19	5	58.8	3.4
April	20	3	14.0	2.2
May	21	4	53.2	7.0
June	22	1	17.4	4.6
August 2022	23	7	105.4	35.4
September 2021–August 2022 (12 months)		86	1278.6	-
September 2022	24	2	89.0	18.6
Overall period	24	88	1367.6	(2.2–35.4)

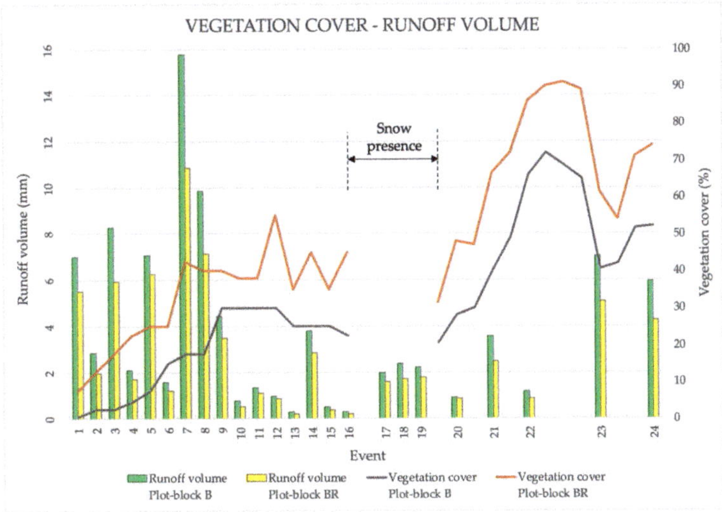

(**a**)

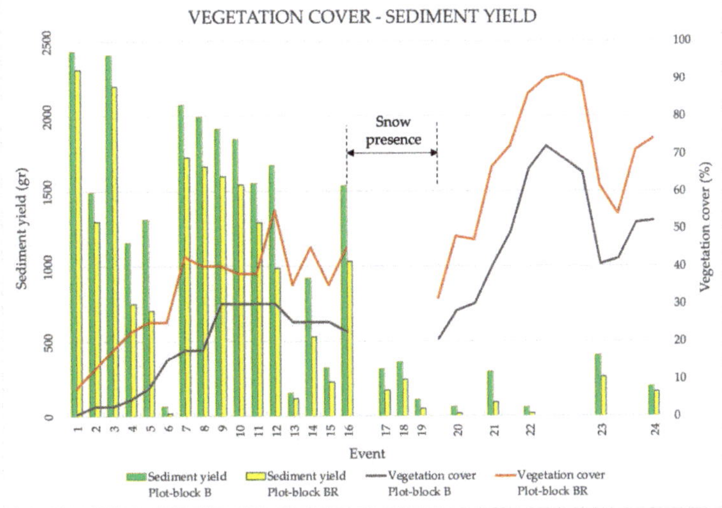

(**b**)

Figure A1. Evolution of vegetation cover in relation to runoff volume (**a**) and sediment yield (**b**).

References

1. Carrión, J.S.; Sánchez-Gómez, P.; Mota, J.F.; Yll, R.; Chaín, C. Holocene Vegetation Dynamics, Fire and Grazing in the Sierra de Gádor, Southern Spain. *Holocene* **2003**, *13*, 839–849. [CrossRef]
2. Hosseini, M.; Keizer, J.J.; Pelayo, O.G.; Prats, S.A.; Ritsema, C.; Geissen, V. Effect of Fire Frequency on Runoff, Soil Erosion, and Loss of Organic Matter at the Micro-Plot Scale in North-Central Portugal. *Geoderma* **2016**, *269*, 126–137. [CrossRef]
3. Fernandez-Anez, N.; Krasovskiy, A.; Müller, M.; Vacik, H.; Baetens, J.; Hukić, E.; Kapovic Solomun, M.; Atanassova, I.; Glushkova, M.; Bogunović, I.; et al. Current Wildland Fire Patterns and Challenges in Europe: A Synthesis of National Perspectives. *Air Soil Water Res.* **2021**, *14*, e117862212110281. [CrossRef]
4. Whelan, R.J. *The Ecology of Fire*; Cambridge University Press: Cambridge, UK, 1995.
5. Morgan, P.; Hardy, C.C.; Swetnam, T.W.; Rollins, M.G.; Long, D.G. Mapping Fire Regimes across Time and Space: Understanding Coarse and Fine-Scale Fire Patterns. *Int. J. Wildland Fire* **2001**, *10*, 329–342. [CrossRef]

6. DeBano: Fire Effects on Ecosystems—Google Scholar. Available online: https://scholar.google.com/scholar_lookup?hl=en&publication_year=1998&pages=%00empty%00&author=LF+De+Bano&author=DG+Neary&author=PF+Ffolliott&isbn=%00null%00&title=Fire%27s+Effects+on+Ecosystems (accessed on 25 May 2023).
7. Neary, D.G.; Klopatek, C.C.; DeBano, L.F.; Ffolliott, P.F. Fire Effects on Belowground Sustainability: A Review and Synthesis. *For. Ecol. Manag.* **1999**, *122*, 51–71. [CrossRef]
8. Key, C.H.; Benson, N.C. Post-Fire Assessment by Remote Sensing on National Park Service Lands. In Proceedings of the Second U.S. Geological Survey Wildland Fire Workshop, Los Alamos, NM, USA, 31 October–3 November 2000.
9. Smith, A.M.S.; Hudak, A.T.; Smith, A.M.S.; Hudak, A.T. Estimating Combustion of Large Downed Woody Debris from Residual White Ash. *Int. J. Wildland Fire* **2005**, *14*, 245–248. [CrossRef]
10. Ascoli, D.; Marchetti, M. Territorio, bioeconomia e gestione degli incendi: Una sfida da raccogliere al più presto. *For.-J. Silvic. For. Ecol.* **2018**, *15*, 71.
11. San-Miguel-Ayanz, J.; Durrant, T.; Boca, R.; Maianti, P.; Libertá, G.; Artés-Vivancos, T.; Oom, D.; Branco, A.; de Rigo, D.; Ferrari, D.; et al. *Forest Fires in Europe, Middle East and North Africa 2021*; Publications Office of the European Union: Luxembourg, 2022.
12. Shakesby, R. Post-Wildfire Soil Erosion in the Mediterranean: Review and Future Research Directions. *Earth-Sci. Rev.* **2011**, *105*, 71–100. [CrossRef]
13. Dimitrakopoulos, A.; Gogi, C.; Stamatelos, G.; Mitsopoulos, I. Statistical Analysis of the Fire Environment of Large Forest Fires (>1000 Ha) in Greece. *Pol. J. Environ. Stud.* **2011**, *20*, 327–332.
14. Cos, J.; Doblas-Reyes, F.; Jury, M.; Marcos, R.; Bretonnière, P.-A.; Samsó, M. The Mediterranean Climate Change Hotspot in the CMIP5 and CMIP6 Projections. *Earth Syst. Dynam.* **2022**, *13*, 321–340. [CrossRef]
15. Bowman, D.M.; Balch, J.K.; Artaxo, P.; Bond, W.J.; Carlson, J.M.; Cochrane, M.A.; D'Antonio, C.M.; DeFries, R.S.; Doyle, J.C.; Harrison, S.P. Fire in the Earth System. *Science* **2009**, *324*, 481–484. [CrossRef]
16. Mavsar, R.; Varela, E.; Corona, P.; Barbati, A.; Marsh, G. Economic, Legal and Social Aspects of Post-Fire Management. In *Post-Fire Management and Restoration of Southern European Forests*; Springer: Dordrecht, The Netherlands, 2012; pp. 45–78.
17. Diakakis, M.; Xanthopoulos, G.; Gregos, L. Analysis of Forest Fire Fatalities in Greece: 1977–2013. *Int. J. Wildland Fire* **2016**, *25*, 797–809. [CrossRef]
18. Elia, M.; Giannico, V.; Spano, G.; Lafortezza, R.; Sanesi, G. Likelihood and Frequency of Recurrent Fire Ignitions in Highly Urbanised Mediterranean Landscapes. *Int. J. Wildland Fire* **2020**, *29*, 120–131. [CrossRef]
19. Barbati, A.; Arianoutsou, M.; Corona, P.; De Las Heras, J.; Fernandes, P.; Moreira, F.; Papageorgiou, K.; Vallejo, R.; Xanthopoulos, G. Post-Fire Forest Management in Southern Europe: A COST Action for Gathering and Disseminating Scientific Knowledge. *iForest-Biogeosci. For.* **2010**, *3*, 5. [CrossRef]
20. Francos, M.; Úbeda, X.; Tort, J.; Panareda, J.M.; Cerdà, A. The Role of Forest Fire Severity on Vegetation Recovery after 18 Years. Implications for Forest Management of Quercus Suber L. in Iberian Peninsula. *Glob. Planet. Chang.* **2016**, *145*, 11–16. [CrossRef]
21. Caon, L.; Vallejo, V.R.; Ritsema, C.J.; Geissen, V. Effects of wildfire on soil nutrients in Mediterranean ecosystems. *Earth-Sci. Rev.* **2014**, *139*, 47–58. [CrossRef]
22. Robinne, F.-N.; Hallema, D.W.; Bladon, K.D.; Buttle, J.M. Wildfire Impacts on Hydrologic Ecosystem Services in North American High-Latitude Forests: A Scoping Review. *J. Hydrol.* **2020**, *581*, 124360. [CrossRef]
23. Scrinzi, G.; Gregori, E.; Giannetti, F.; Galvagni, D.; Zorn, G.; Colle, G.; Andrenelli, M. Un Modello Di Valutazione Della Funzionalità Protettiva Del Bosco per La Pianificazione Forestale: La Componente Stabilità Dei Versanti Rispetto Ai Fenomeni Franosi Superficiali. *For.-J. Silvic. For. Ecol.* **2006**, *3*, 98. [CrossRef]
24. Corona, P.; Barbati, A.; Ferrari, B.; Portoghesi, L. *Pianificazione Ecologica Dei Sistemi Forestali: 2a Edizione*; Compagnia delle Foreste Srl: Arezzo, Italy, 2019.
25. Portoghesi, L.; Iovino, F.; Certini, G.; Travaglini, D. Il Bosco e La Custodia Del Territorio: Il Ruolo Della Selvicoltura. *L'Italia For. E Mont.* **2019**, *74*, 263–276. [CrossRef]
26. Cerdà, A.; Borja, M.E.L.; Úbeda, X.; Martínez-Murillo, J.F.; Keesstra, S. *Pinus Halepensis* M. versus *Quercus Ilex* subsp. *Rotundifolia* L. Runoff and Soil Erosion at Pedon Scale under Natural Rainfall in Eastern Spain Three Decades after a Forest Fire. *For. Ecol. Manag.* **2017**, *400*, 447–456. [CrossRef]
27. Ebel, B.A.; Moody, J.A. Synthesis of Soil-Hydraulic Properties and Infiltration Timescales in Wildfire-Affected Soils. *Hydrol. Process.* **2017**, *31*, 324–340. [CrossRef]
28. García-Comendador, J.; Fortesa, J.; Calsamiglia, A.; Calvo-Cases, A.; Estrany, J. Post-Fire Hydrological Response and Suspended Sediment Transport of a Terraced Mediterranean Catchment. *Earth Surf. Process. Landf.* **2017**, *42*, 2254–2265. [CrossRef]
29. Cavalli, M.; Vericat, D.; Pereira, P. Mapping Water and Sediment Connectivity. *Sci. Total Environ.* **2019**, *673*, 763–767. [CrossRef] [PubMed]
30. Shakesby, R.A.; Doerr, S.H. Wildfire as a Hydrological and Geomorphological Agent. *Earth-Sci. Rev.* **2006**, *74*, 269–307. [CrossRef]
31. Béguin, P.; Millet, J.; Aubert, J.-P. Cellulose Degradation by Clostridium Thermocellum: From Manure to Molecular Biology. *FEMS Microbiol. Lett.* **1992**, *100*, 523–528. [CrossRef]
32. Lasanta, T.; Cerdà, A. Long-Term Erosional Responses after Fire in the Central Spanish Pyrenees: 2. Solute Release. *CATENA* **2005**, *60*, 81–100. [CrossRef]
33. Inbar, M.; Wittenberg, L.; Tamir, M. Soil Erosion and Forestry Management after Wildfire in a Mediterranean Woodland, Mt. Carmel, Israel. *Int. J. Wildland Fire* **1997**, *7*, 285–294. [CrossRef]

34. Inbar, M.; Tamir, M.I.; Wittenberg, L. Runoff and Erosion Processes after a Forest Fire in Mount Carmel, a Mediterranean Area. *Geomorphology* **1998**, *24*, 17–33. [CrossRef]
35. Badía, D.; Martí, C. Plant Ash and Heat Intensity Effects on Chemicaland Physical Properties of Two Contrasting Soils. *Arid Land Res. Manag.* **2003**, *17*, 23–41. [CrossRef]
36. Gimeno-García, E.; Andreu, V.; Rubio, J.L. Influence of Vegetation Recovery on Water Erosion at Short and Medium-Term after Experimental Fires in a Mediterranean Shrubland. *Catena* **2007**, *69*, 150–160. [CrossRef]
37. Mayor, A.G.; Bautista, S.; Llovet, J.; Bellot, J. Post-Fire Hydrological and Erosional Responses of a Mediterranean Landscpe: Seven Years of Catchment-Scale Dynamics. *Catena* **2007**, *71*, 68–75. [CrossRef]
38. Bedia, J.; Herrera, S.; Camia, A.; Moreno, J.M.; Gutiérrez, J.M. Forest Fire Danger Projections in the Mediterranean Using ENSEMBLES Regional Climate Change Scenarios. *Clim. Chang.* **2014**, *122*, 185–199. [CrossRef]
39. Robichaud, P.R.; Ashmun, L.E.; Sims, B.D. *Post-Fire Treatment Effectiveness for Hillslope Stabilization*; Gen. Tech. Rep. RMRS-GTR-240; U.S. Department of Agriculture, Forest Service, Rocky Mountain Research Station: Fort Collins, CO, USA, 2010; Volume 240, 62p. [CrossRef]
40. Figueiredo, T.D.; Fonseca, F.; Lima, E.; Fleischfresser, L.; Hernandez Hernandez, Z. Assessing Performance of Post-Fire Hillslope Erosion Control Measures Designed for Different Implementation Scenarios in NE Portugal: Simulations Applying USLE. In *Wildfires: Perspectives, Issues and Challenges of the 21st Century*; Nova Science Publisher: New York, NY, USA, 2017; pp. 201–227.
41. Fernández-García, V.; Fulé, P.Z.; Marcos, E.; Calvo, L. The Role of Fire Frequency and Severity on the Regeneration of Mediterranean Serotinous Pines under Different Environmental Conditions. *For. Ecol. Manag.* **2019**, *444*, 59–68. [CrossRef]
42. Robichaud, P.R.; Lewis, S.A.; Wagenbrenner, J.W.; Brown, R.E.; Pierson, F.B. Quantifying Long-Term Post-Fire Sediment Delivery and Erosion Mitigation Effectiveness. *Earth Surf. Process. Landf.* **2020**, *45*, 771–782. [CrossRef]
43. Robichaud, P.R.; Wagenbrenner, J.W.; Lewis, S.A.; Ashmun, L.E.; Brown, R.E.; Wohlgemuth, P.M. Post-Fire Mulching for Runoff and Erosion Mitigation Part II: Effectiveness in Reducing Runoff and Sediment Yields from Small Catchments. *Catena* **2013**, *105*, 93–111. [CrossRef]
44. Robichaud, P.R.; Wagenbrenner, J.W.; Brown, R.E.; Wohlgemuth, P.M.; Beyers, J.L. Evaluating the Effectiveness of Contour-Felled Log Erosion Barriers as a Post-Fire Runoff and Erosion Mitigation Treatment in the Western United States. *Int. J. Wildland Fire* **2008**, *17*, 255–273. [CrossRef]
45. Rey, F.; Bifulco, C.; Bischetti, G.B.; Bourrier, F.; De Cesare, G.; Florineth, F.; Graf, F.; Marden, M.; Mickovski, S.B.; Phillips, C. Soil and Water Bioengineering: Practice and Research Needs for Reconciling Natural Hazard Control and Ecological Restoration. *Sci. Total Environ.* **2019**, *648*, 1210–1218. [CrossRef]
46. USDA Forest Service, C. Emergency Stabilization—Burned-Area Emergency Response (BAER). In *Forest Service Manual FSM 2500, Watershed and Air Management*; USDA Forest Service: Washingotn, DC, USA, 2004.
47. Gómez-Sánchez, E.; Lucas-Borja, M.E.; Plaza-Álvarez, P.A.; González-Romero, J.; Sagra, J.; Moya, D.; De Las Heras, J. Effects of Post-Fire Hillslope Stabilisation Techniques on Chemical, Physico-Chemical and Microbiological Soil Properties in Mediterranean Forest Ecosystems. *J. Environ. Manag.* **2019**, *246*, 229–238. [CrossRef]
48. López-Vicente, M.; Kramer, H.; Keesstra, S. Effectiveness of Soil Erosion Barriers to Reduce Sediment Connectivity at Small Basin Scale in a Fire-Affected Forest. *J. Environ. Manag.* **2021**, *278*, 111510. [CrossRef]
49. Raftoyannis, Y.; Spanos, I.; Raftoyannis, Y.; Spanos, I. Evaluation of Log and Branch Barriers as Post-Fire Rehabilitation Treatments in a Mediterranean Pine Forest in Greece. *Int. J. Wildland Fire* **2005**, *14*, 183–188. [CrossRef]
50. Fernández, C.; Vega, J.A.; Jiménez, E.; Fonturbel, T.; Fernández, C.; Vega, J.A.; Jiménez, E.; Fonturbel, T. Effectiveness of Three Post-Fire Treatments at Reducing Soil Erosion in Galicia (NW Spain). *Int. J. Wildland Fire* **2011**, *20*, 104–114. [CrossRef]
51. Preti, F.; Capobianco, V.; Sangalli, P. Soil and Water Bioengineering (SWB) Is and Has Always Been a Nature-Based Solution (NBS): A Reasoned Comparison of Terms and Definitions. *Ecol. Eng.* **2022**, *181*, 106687. [CrossRef]
52. Proto, A.R.; Bernardini, V.; Cataldo, M.F.; Zimbalatti, G. Whole Tree System Evaluation of Thinning a Pine Plantation in Southern Italy. *Ann. Silvic. Res.* **2020**, *45*, 44–52.
53. de Pagter, T.; Lucas-Borja, M.E.; Navidi, M.; Carra, B.G.; Baartman, J.; Zema, D.A. Effects of Wildfire and Post-Fire Salvage Logging on Rainsplash Erosion in a Semi-Arid Pine Forest of Central Eastern Spain. *J. Environ. Manag.* **2023**, *329*, 117059. [CrossRef] [PubMed]
54. Lucas-Borja, M.E.; Heydari, M.; Miralles, I.; Zema, D.A.; Manso, R. Effects of Skidding Operations after Tree Harvesting and Soil Scarification by Felled Trees on Initial Seedling Emergence of Spanish Black Pine (Pinus Nigra Arn. Ssp. Salzmannii). *Forests* **2020**, *11*, 767. [CrossRef]
55. Bombino, G. A (New) Suggestive Hypothesis about the Origin of the Term Fiumara (Seasonally-Flowing and High Hazard Streams of Southern Italy). *Mediterr. Chron.* **2020**, *10*, 163–173.
56. Ibbeken, H.; Schleyer, R. *Source and Sediment, a Case Study of Provenance and Mass Balance at an Active Plate Margin (Calabria)*; Springer Science & Business Media: Berlin/Heidelberg, Germany, 1991.
57. Sorriso-Valvo, M. Natural Hazards and Natural Heritage—Common Origins and Interference with Cultural Heritage. *Geogr. Fis. E Din. Quat.* **2008**, *31*, 231–237.
58. Palmeri, V.; Pulvirenti, A.; Zappalà, L. La Processionaria Dei Pini Nei Boschi Della Dorsale Appenninica Della Calabria. *For.-J. Silvic. For. Ecol.* **2005**, *2*, 345. [CrossRef]

59. Kottek, M.; Grieser, J.; Beck, C.; Rudolf, B.; Rubel, F. World Map of the Köppen-Geiger Climate Classification Updated. *Meteorol. Z.* **2006**, *15*, 259–263. [CrossRef]
60. Bouyoucos, G.J. Hydrometer Method Improved for Making Particle Size Analyses of Soils1. *Agron. J.* **1962**, *54*, 464–465. [CrossRef]
61. Walkley, A.; Black, I.A. An Examination of the Degtjareff Method for Determining Soil Organic Matter, and A Proposed Modification of the Chromic Acid Titration Method. *Soil Sci.* **1934**, *37*, 29. [CrossRef]
62. Kjeldahl, J. Neue Methode zur Bestimmung des Stickstoffs in organischen Körpern. *Z. Für Anal. Chem.* **1883**, *22*, 366–382. [CrossRef]
63. Tomaščík, M.; Németová, Z.; Danacova, M. Analysis of Factors Influencing the Intensity of Soil Water Erosion. *Acta Hydrol. Slovaca* **2021**, *22*, 70–77. [CrossRef]
64. Tinebra, I.; Alagna, V.; Iovino, M.; Bagarello, V. Comparing Different Application Procedures of the Water Drop Penetration Time Test to Assess Soil Water Repellency in a Fire Affected Sicilian Area. *Catena* **2019**, *177*, 41–48. [CrossRef]
65. Roldán, A.; García-Orenes, F.; Lax, A. An Incubation Experiment to Determine Factors Involving Aggregation Changes in an Arid Soil Receiving Urban Refuse. *Soil Biol. Biochem.* **1994**, *26*, 1699–1707. [CrossRef]
66. Benito Rueda, E.; Gomez Ulla, A.; Diaz Fierros, F. Descripción de Un Simulados de Lluvia Para Estudios de Erodabilidad Del Suelo y Estabilidad de Los Agregados al Agua. *An. Edafol. Agrobiol* **1986**, *45*, 1115–1126.
67. Ball, D.F. Loss-on-Ignition as an Estimate of Organic Matter and Organic Carbon in Non-Calcareous Soils. *J. Soil Sci.* **1964**, *15*, 84–92. [CrossRef]
68. Vogel, K.P.; Masters, R.A. Frequency Grid-a Simple Tool for Measuring Grassland Establishment. *J. Range Manag.* **2001**, *54*, 653–655. [CrossRef]
69. Braun-Blanquet, J. *Zur Kenntnis Der Vegetationsverhältnisse Des Grossen Atlas*; Buchdruckerei Gebr. Fretz AG: Zürich, Switzerland, 1928.
70. Keesstra, S.D.; Maroulis, J.; Argaman, E.; Voogt, A.; Wittenberg, L. Effects of Controlled Fire on Hydrology and Erosion under Simulated Rainfall. *Cuad. Investig. Geográfica* **2014**, *40*, 269–294. [CrossRef]
71. Lucas-Borja, M.E.; Van Stan II, J.T.; Heydari, M.; Omidipour, R.; Rocha, F.; Plaza-Alvarez, P.A.; Zema, D.A.; Muñoz-Rojas, M. Post-Fire Restoration with Contour-Felled Log Debris Increases Early Recruitment of Spanish Black Pine (*Pinus Nigra* Arn. ssp. *Salzmannii*) in Mediterranean Forests. *Restor. Ecol.* **2021**, *29*, e13338. [CrossRef]
72. Ferro, V. Evaluating Overland Flow Sediment Transport Capacity. *Hydrol. Process.* **1998**, *12*, 1895–1910. [CrossRef]
73. Pampalone, V.; Carollo, F.G.; Nicosia, A.; Palmeri, V.; Di Stefano, C.; Bagarello, V.; Ferro, V. Measurement of Water Soil Erosion at Sparacia Experimental Area (Southern Italy): A Summary of More than Twenty Years of Scientific Activity. *Water* **2022**, *14*, 1881. [CrossRef]
74. Mendes, T.A.; Alves, R.D.; Gitirana, G.D.F.N., Jr.; Pereira, S.A.D.S.; Rebolledo, J.F.R.; da Luz, M.P. Evaluation of Rainfall Interception by Vegetation Using a Rainfall Simulator. *Sustainability* **2021**, *13*, 5082. [CrossRef]
75. Wittenberg, L.; Inbar, M. The Role of Fire Disturbance on Runoff and Erosion Processes—A Long-Term Approach, Mt. Carmel Case Study, Israel. *Geogr. Res.* **2009**, *47*, 46–56. [CrossRef]
76. Rulli, M.C.; Spada, M.; Bozzi, S.; Bocchiola, D.; Rosso, R. Erosion and Runoff Generation from Fire Disturbed Mediterranean Forest Area. Ph.D. Thesis, Colorado State University, Fort Collins, CO, USA, 2005. [CrossRef]
77. Rubio, J.L.; Forteza, J.; Andreu, V.; Cerni, R. Soil Profile Characteristics Influencing Runoff and Soil Erosion after Forest Fire: A Case Study (Valencia, Spain). *Soil Technol.* **1997**, *11*, 67–78. [CrossRef]
78. Campo, J.; Andreu, V.; Gimeno-García, E.; González, O.; Rubio, J.L. Occurrence of Soil Erosion after Repeated Experimental Fires in a Mediterranean Environment. *Geomorphology* **2006**, *82*, 376–387. [CrossRef]
79. Li, L.; Yang, J.; Wu, J. A Method of Watershed Delineation for Flat Terrain Using Sentinel-2A Imagery and DEM: A Case Study of the Taihu Basin. *ISPRS Int. J. Geo-Inf.* **2019**, *8*, 528. [CrossRef]

water

MDPI

Article

A Random Forest Machine Learning Approach for the Identification and Quantification of Erosive Events

Lorenzo Vergni * and **Francesca Todisco**

Department of Agricultural, Food and Environmental Science, University of Perugia, 06124 Perugia, Italy; francesca.todisco@unipg.it
* Correspondence: lorenzo.vergni@unipg.it

Abstract: Predicting the occurrence of erosive rain events and quantifying the corresponding soil loss is extremely useful in all applications where assessing phenomenon impacts is required. These problems, addressed in the literature at different spatial and temporal scales and according to the most diverse approaches, are here addressed by implementing random forest (RF) machine learning models. For this purpose, we used the datasets built through many years of soil loss observations at the plot-scale experimental site SERLAB (central Italy). Based on 32 features describing rainfall characteristics, the RF classifier has achieved a global accuracy of 84.8% in recognizing erosive and non-erosive events, thus demonstrating slightly higher performances than previously used (non-machine learning) methodologies. A critical performance is the percentage of erosive events correctly recognized to the observed total (72.3%). However, since the most relevant erosive events are correctly identified, we found only a slight underestimation of the total rainfall erosivity (91%). The RF regression model for estimating the event soil loss, based on three event features (runoff coefficient, erosivity, and period of occurrence), demonstrates better performances (RMSE = 2.30 Mg ha^{-1}) than traditional regression models (RMSE = 3.34 Mg ha^{-1}).

Keywords: soil water erosion; USLE models; plot scale; artificial intelligence; data-driven approach; SERLAB experimental site

Citation: Vergni, L.; Todisco, F. A Random Forest Machine Learning Approach for the Identification and Quantification of Erosive Events. *Water* **2023**, *15*, 2225. https://doi.org/10.3390/w15122225

Academic Editor: Maria Mimikou

Received: 22 May 2023
Revised: 9 June 2023
Accepted: 12 June 2023
Published: 13 June 2023

1. Introduction

In its accelerated forms, soil water erosion leads to several adverse on-site (loss of nutrients and organic matter, permeability) and off-site (pollution of water bodies, destruction of habitats, siltation of water bodies and flood risk increase) effects [1,2]. The effectiveness of environmental programs and policies or specific soil erosion control strategies strictly depends on the knowledge of the complex mechanism and factors conditioning the occurrence and the extent of erosive processes. This knowledge necessarily derives from the availability of adequate datasets (with the fundamental contribution of experimental and monitoring activities) and appropriate modelling approaches. Among the numerous model types (empirical, conceptual, physically based, or process-oriented) proposed in the literature to estimate soil loss at different spatial and temporal scales, empirical models are still the most used due to their capacity to combine simplicity and reliability [3,4]. In particular, the Universal Soil Loss Equation (USLE, [5]) or its revised version (Revised USLE, RUSLE, [6]) are the most widely applied models to predict long-term average soil loss values. They are also considered the standard applications of soil conservation workers [7,8]. However, since a large proportion of soil loss is due to a few particularly erosive events [9–11], mitigation and management strategies should rely on models capable of predicting event-scale erosion rather than the long-term average. Various studies have moved in this direction [12–15], proposing changes to the USLE approach (or subsequent revisions) to make it reliable on the single-event scale. These changes usually consist of the use of an erosivity index given by the combination of the event runoff coefficient, Q_R (dimensionless), and the single-storm erosion index, E·I30 = R (MJ mm ha^{-1} h^{-1}) by Wischmeier and Smith [5]. For example, the studies conducted to date at various Italian plot scale experimental sites [4,10,16] have proposed and tested a model (named USLE-MM) of the following type:

$$A_e = (Q_R \cdot R)^{b1} K_{UMM} LS C_{UMM} P_{UMM} \quad (1)$$

where A_e (Mg ha^{-1}) is the event soil loss per unit area, L and S (dimensionless) are, respectively, the USLE slope length and steepness factors, K_{UMM} (Mg ha h ha^{-1} MJ^{-1} mm^{-1}) is the soil erodibility factor, C_{UMM} (dimensionless) is the cover and management factor, P_{UMM} (dimensionless) is the support practice factor, and b1 is an exponent to be parametrized based on collected data. Although the model of Equation (1) demonstrates decidedly superior performance at the event scale to the original USLE (which suffers from a systematic bias, [12]), there is still a large portion of unexplained variance, which stimulates both the continuation of the monitoring activities and the search for new modelling approaches.

In addition to modelling soil loss, the literature has long investigated the possibility of identifying erosive events based only on rainfall characteristics. This information may be helpful in various research and practical applications. For example, it allows us to determine the triggering of erosion processes of different entities and consequently to understand their dynamics better; reduces the work necessary to manage and process erosive events; allows us to study the spatio-temporal frequency of erosive events [17]. The first reference in this context is that in [5], which indicated as erosive the rainfalls capable of exceeding the depth thresholds of 12.7 mm or 6.35 mm in 15 min. More recent studies [17–19] have searched for such a type of thresholds in large datasets of variables characterizing, not only considering the overall characteristics of rainfall (e.g., depth, duration), but also their internal structure (e.g., presence and duration of high-intensity showers). For example, Todisco et al. [17], based on the data observed at the Masse (central Italy) and Sparacia (south Italy) experimental sites, demonstrated that some of these variables could effectively classify non-erosive and erosive events and even separate erosive events that produce sheet or rill erosion.

Both the classification and the quantification of erosion processes, which are physically very complex and depend on many concomitant and interacting factors, can certainly be tackled with good prospects using artificial intelligence and machine learning technologies. These approaches have been increasingly applied in all scientific areas [20] as a result of their ability to process large amounts of data in a relatively simple way and identify complex patterns and relationships from the data itself to make reliable predictions on data not seen before. This strength is, at the same time, also a weakness, as without large enough datasets to train on, such models provide inaccurate estimates. Applications in the field of soil erosion processes mainly refer to the use of artificial neural network (ANN) models, usually applied to predict runoff and soil loss [21–25] or other erosion-related factors [26]. Among the various machine learning techniques, the random forest (RF) is increasingly applied as a result of its several advantageous characteristics, such as the relative ease of managing large datasets, the possibility of using nominal and numerical data, the high accuracy of the predictions, and the possibility of being used for both classification and regression problems [27]. Overfitting, which happens when a model is too focused on the training data and fails to generalize well to new data, is less likely to occur in RF than in ANN [28]. This is due to the bootstrap and randomized feature sampling used in RF algorithm. Moreover, although both RF and ANN appear as "black boxes", RF allows a more straightforward interpretation of the factors affecting the prediction. RF algorithms have been used with promising results in several research fields related to hydraulics and hydrology [29], such as pipe breaks in water supply networks [30], water flow in porous media [31], flooding [32] and drought [33] events. RF applications in soil water erosion studies are limited in the literature and, therefore, can be considered quite innovative. One of the few examples is the recent work by Tarek et al. [34], demonstrating the higher accuracy of the RF approach compared to other machine learning methods in the classification of erosive events.

This work aims to develop and evaluate RF algorithms for classifying erosive and non-erosive rain events and quantifying the soil loss at the single event time scale and plot spatial scale. The study is based on the large datasets obtained over many years of soil erosion monitoring at the Soil Erosion LABoratory (SERLAB) experimental site. By comparison with the results obtained from more traditional empirical methods, the good potential of RF algorithms in this type of application is demonstrated.

2. Materials and Methods

2.1. Experimental Site

In the study, we used the data collected at the Masse experimental site (42°59′34″ N 12°17′27″ E) of the Department of Agricultural, Food and Environmental Sciences of the University of Perugia. This site (Figure 1), also known as Soil Erosion LABoratory (SERLAB), was established in 2007 to monitor and characterize erosive processes at plot scale in central Italy's typical agricultural context dominated by hills. The average annual rainfall is about 900 mm, and the soil is silt–clay–loam (clay = 34%, silt = 59%, and sand = 7%). Organic matter content is about 1%, and the gravel content is negligible. There are currently 10 Wischmeier-type plots of various sizes (four plots 22 × 8 m^2, two plots 22 × 4 m^2, two plots 11 × 4 m^2 and two plots 11 × 2 m^2) with a 16% slope. The plots are kept in cultivated fallow through frequent tillage operations to remove vegetation and restore soil surface structure after erosive events. Each plot is equipped in its terminal part with a channel conveying solid and liquid runoff into 1-m^3 storage tanks, the number of which varies with the plot size. After each erosive event (or occasionally, after a series of events if they are close to one another), the runoff and soil loss are measured with a specially calibrated sampling technique [35], and the tanks are emptied and cleaned to be ready to receive the runoff of a new event. The SERLAB meteorological station has several weather instrumentations, including a tipping bucket rain gauge set to record rainfall depths with a 5 min time step.

(a) (b) (c)

Figure 1. The SERLAB experimental site for soil loss monitoring at plot scale: (**a**) pan view; (**b**) meteorological station; (**c**) detail of the storage tanks.

2.2. Datasets

The study relies on two datasets (DB1 and DB2), built with rainfall and soil loss data collected over several years of monitoring at the SERLAB site. The characteristics of the two datasets are detailed in the following sections.

2.2.1. Dataset (DB1) Used for the RF Classifier

For the development of an RF algorithm capable of classifying erosive and non-erosive events, we used the same dataset (DB1) considered in previous studies aimed at identifying the best rainfall variables (and the corresponding thresholds), able to separate erosive and non-erosive events [17,19,36]. The DB1 dataset includes 528 events (158 erosive and 370 non-erosive). The decision to consider the same dataset allows for a more straightforward comparison of the RF algorithm with other classification techniques applied recently. The dataset DB1 was built according to the steps detailed below.

The 5 min rainfall records from 1 January 2008 to 31 December 2017 were analysed to identify the individual storms, i.e., the rain events preceded and followed by 6 h or more of no rain (according to Wischmeier and Smith [5]).

Each individual storm was therefore categorized as erosive (*E*) if a measurable soil loss was found in the storage tanks, while it was considered non-erosive (*NE*) if it did not produce runoff or if the soil loss was so irrelevant that it could not be measured. Sometimes, when individual events are very close, soil loss refers to a sequence of individual storms. In this case, the individual storms included in the sequence were classified as non-erosive (erosive) if their rainfall depth was lower (higher) than the minimum depth observed for the individual erosive rainfalls [17].

For each individual storm, 31 variables (Table A1) describing the hyetograph's overall and pattern characteristics were quantified. In particular, the internal storm structure is represented by the characteristics (number, duration, severity, etc.) of bursts (i.e., intervals of continuous rain) and runs (i.e., intervals of continuous rain exceeding a predetermined truncation value p0). The identification of the truncation level, p0, was based on the frequency analysis of rainfall records by excluding zero values. The selected p0 value corresponds to a cumulative frequency of 95% [17], giving p0 = 0.8 mm in 5 min (9.6 mm/h). More calculation details about the rainfall variables can be found in Todisco et al. [17].

2.2.2. Dataset (DB2) Used for the RF Regression Model

The dataset DB2 used to develop an RF regression model for predicting the event soil loss is new and derives from an update of previous datasets used for the same purpose, based on more traditional methods. The DB2 consists of 667 records of soil loss at the plot scale measured at the SERLAB site between February 2008 and June 2022.

The rainfall depth, P (mm), and the single-storm erosion index, R (MJ mm ha^{-1} hr^{-1} [5]), were determined for each erosive event. The event total runoff, Q (mm), the runoff coefficient, Q_R = Q/P, and soil loss A_e (Mg ha^{-1}) were quantified for each plot. *A* normalized value of the soil loss was then obtained as follows:

$$A_{e,N} = \frac{A_e}{LS} \quad (2)$$

where *L* and *S* are the plot length and steepness factors by Renard et al. [6] and Nearing [37], respectively. The summary statistics related to these variables are given in Table 1.

Table 1. Summary statistics of runoff coefficient, Q_R, single-storm erosion index, *R*, and normalized soil loss, $A_{e,N}$, related to the 667 records of the DB2 dataset.

Statistic	Q_R (-)	R (MJ mm ha^{-1} h^{-1})	$Q_R \cdot R$ (MJ mm ha^{-1} h^{-1})	$A_{e,N}$ (Mg ha^{-1})
Mean	0.121	115.538	20.262	2.799
Median	0.059	70.841	3.461	0.289
CV	1.182	1.018	1.881	1.966
Min	0.001	3.894	0.076	0.002
Max	0.955	629.903	269.549	42.476

Using $A_{e,N}$ as the dependent variable and considering C_{UMM} and P_{UMM} equal to 1 (bare soil without conservation practices), the USLE_MM (Equation (1)) assumes the following form:

$$A_{e,N} = K_{UMM}(Q_R R)^{b1} \quad (3)$$

More details about the measurement methods used at the SERLAB can be found in Bagarello et al. [4], Vinci et al. [38], and Todisco et al. [35].

2.3. *Random Forest*

RF is a supervised machine learning technique that can be used for both classification and regression problems. It relies on a large number of decision trees that work as an ensemble. The final decision is based on the majority of votes for classification, while the average prediction is considered the solution to regression problems [27].

Usually, the original dataset, consisting of a response variable and one or more predictor variables (features), is subsetted to form training and validation datasets. Each decision tree of the forest is then obtained from a bootstrap sample of the training dataset and uses only some randomly chosen features during tree growth. The out of bag (OOB) set is the data not selected in the sampling process of a specific tree. The subsequent steps differ depending on whether the objective is to obtain a classification or a regression model. The following sections provide details of the procedures and evaluations applied in the two cases. All the RF analyses were implemented in the R environment using the library randomForest [39].

2.3.1. Random Forest Algorithm for Classification of Erosive and Non-Erosive Events

The initial DB1 dataset consisted of 528 storm events described by a dichotomous response variable (i.e., *NE* or *E*) and by the 31 rainfall variables used as features. The total number of features has been brought to 32 (Table A1) with the addition of a dichotomous nominal variable roughly describing the hydrological conditions of the period in which the storm event occurs ("dry" between Aprile and September, "wet" in the other months). The DB1 was randomly split into 50% for training and 50% for validation. A random split (unlike a chronological split) minimizes the chances that the two datasets (training and validation) have differences in the records due to long-term or seasonal trends. The number of features considered at each node split, M, was set as M = 6 by applying an iterative tuning method which minimizes the OOB error [39]. The number of trees was set to 5000 after a preliminary evaluation based on the classification accuracy convergence [40].

The relative importance of each feature was evaluated by quantifying the corresponding mean decrease in accuracy (MDA). The MDA of a specific feature is the average (over all trees) decrease in the accuracy obtained in the prediction of the OOB datasets before and after the permutation of that feature. The features with higher MDA are relatively more important than the others for the overall accuracy of the RF classifier.

The performance of the RF classifier was evaluated by applying the trained classifier to the corresponding validation dataset and computing the overall accuracy *OA* (%), the producer's accuracy *PA* (%), and the user's accuracy *UA* (%), [41].

The *OA* is obtained as:

$$OA = \frac{TC}{N} \tag{4}$$

where *TC* is the number of events truly classified, and *N* is the total number of events considered in the validation dataset. The *PA*, also known as recall, is given by:

$$PA_k = \frac{TC_k}{O_k} \tag{5}$$

where TC_k is the number of events of category *k* truly classified and O_k is the number of events observed in category *k*. The *UA*, also indicated as precision, is given by:

$$UA_k = \frac{TC_k}{C_k} \tag{6}$$

where C_k is the number of events classified in the category *k*.

Both *UA* and *PA* can be quantified for each category (in this study, for both *E* and *NE* rainfall events). A classifier is considered highly accurate when obtaining a combination of high recall and precision values. In particular, high recall (*PA*) corresponds to a low representation of omission errors and high precision (*UA*) to a low value of commission errors [42]. In previous works based on different classification techniques [17] other indicators had been used among the performance measures, including the Correct Selection Index (CSI), which is mathematically identical to the PA_E and the Wrong Selection Index (WSI), which is mathematically equal to 1-UA_E. Figure 2a shows the steps carried out to develop and assess the RF classifier.

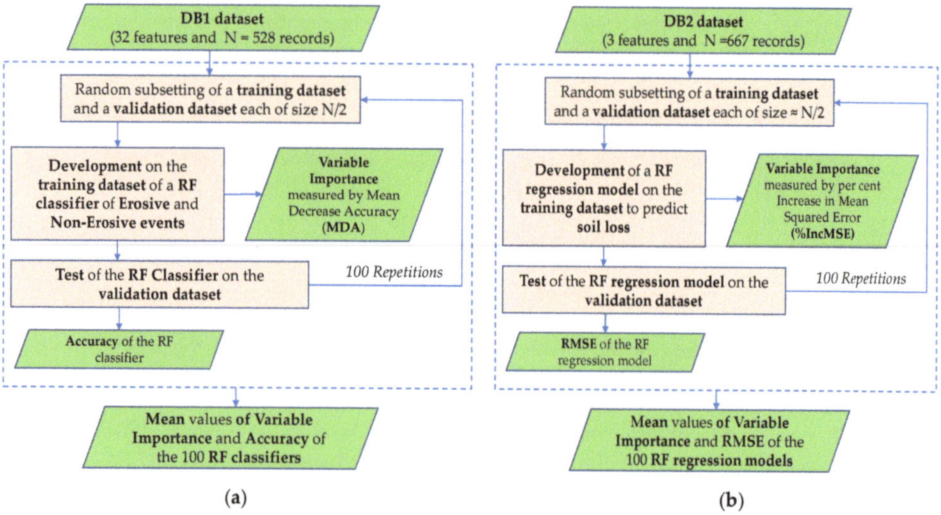

Figure 2. Workflow of the analyses carried out for the development and assessment of (**a**) an RF classifier of erosive and non-erosive rainfall events; (**b**) an RF regression model to predict the event soil loss at the plot spatial scale.

2.3.2. Random Forest Regression to Predict Event Soil Loss

The initial dataset (DB2) included 667 records. The $A_{e,N}$ variable is the response variable, and the features considered were Q_R, R, and the period of occurrence of the erosive event, defined as illustrated in Section 2.3.1. The initial dataset was randomly split into 50% for training and 50% for validation. The number of features at each split and the number of trees were set to 2 and 5000, respectively, based on the same preliminary tunings described in Section 2.3.1. The relative importance of a generic feature j was evaluated by quantifying the corresponding average per cent increase in mean squared error (*%IncMSE*). The *%IncMSE* of a specific feature in a particular tree is computed as:

$$\%IncMSE_j = \frac{(MSE_j - MSE0)}{MSE0} \tag{7}$$

where *MSE0* is the *MSE* of that tree in the OOB dataset, and MSE_j is the *MSE* in the same OOB dataset after the permutation of the feature j. The features with higher *%IncMSE* are relatively more important than the others for the accuracy of the model prediction. Figure 2b shows the steps carried out to develop and assess the RF regression model.

3. Results

3.1. RF Classifier of Erosive and Non-Erosive Events

As explained, the RF classifier was trained on a random 50% of the DB1 dataset and then applied to the remaining validation dataset. Since the results depend on the specific observations included in the random training and validation datasets, this procedure was repeated 100 times (Figure 2) to enable a more general and objective assessment. The average feature importance, based on the MDA criterion, is presented in Figure 3.

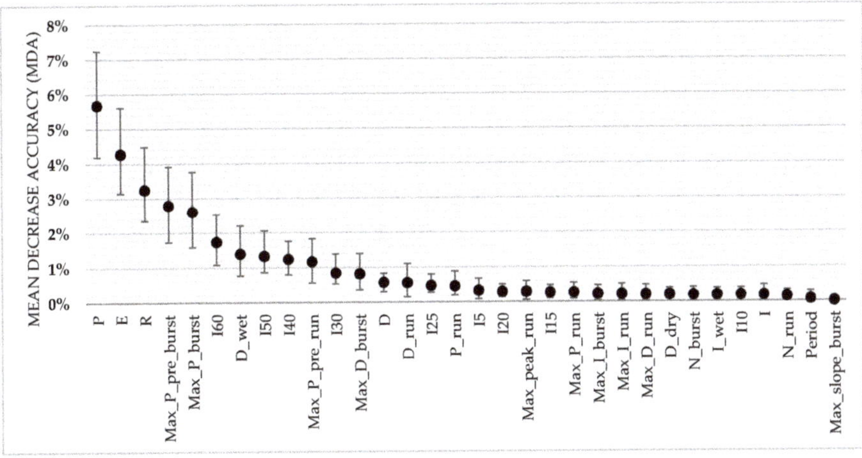

Figure 3. Importance of the features considered in the RF classifier, based on the mean decrease accuracy (MDA) criterion. Black dots and grey bars represent the mean and 90% confidence interval of the MDA values obtained in 100 random repetitions. The definition of the variables is given in Table A1.

From Figure 3, six features are found with decidedly higher relative importance than the other variables: the total precipitation *P*, the kinetic energy *E*, the erosivity *R*, the maximum rainfall depth from the start of the event to a burst Max_P_pre_burst, the maximum rainfall amount in a burst Max_P_burst, the maximum intensity over 60 min I60. The "period" feature does not seem helpful for the RF classifier.

The results shown in Figure 4 (dark blue bars) indicate the following accuracy measures: *OA* 84.8%, PA_E 72.3%, PA_{NE} 91.0% UA_E 79.2% and UA_{NE} 87.2%. For comparison purposes, Figure 4 also shows the accuracy values (light blue bars) obtained with a dataset reduced to the 6 top features based on the ranking given Figure 3. It is evident that there is only a slight reduction in the accuracy performances.

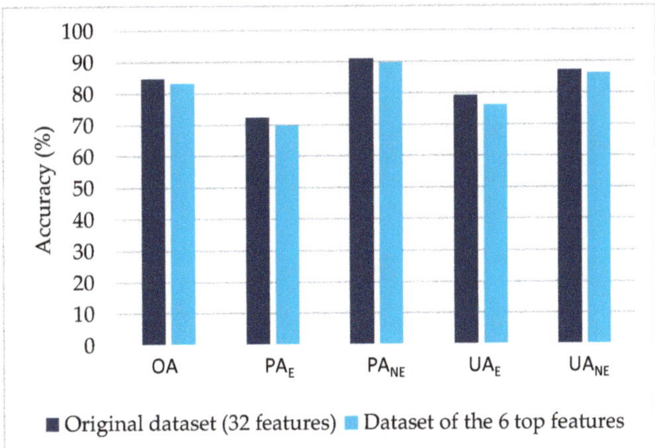

Figure 4. Mean accuracy measures of RF classifiers trained on the original dataset (32 features) and on the top 6 variables (based on the Mean Decrease Accuracy criterion, Figure 3). *OA* = Overall Accuracy; *UA* = User Accuracy; *PA* = Producer Accuracy; *E* = Erosive events; *NE*: Non-erosive events. Mean values obtained from 100 training and validation datasets.

3.2. RF Regression Model for Prediction of Event Soil Loss

The RF regression model has been trained on a random 50% of the DB2 (training dataset) and then applied to predict the soil loss $A_{e,N}$ on the remaining data (validation dataset). For comparison purposes, the same datasets were used to parameterize and validate the USLE_MM model of Equation (3). Since the model performances vary according to the random datasets, this procedure has been repeated 100 times (Figure 2) in order to have a more global and objective evaluation. As for the RF regression model, the variable importance indicates the predominant role of Q_R (mean *%IncMSE* = 30.2%, CV = 19%) followed by *R* (*%IncMSE* = 6.8%, CV = 25%) and the "Period" (*%IncMSE* = 5.5%, CV = 24%). The average root mean square error (RMSE) of the RF regression model is 2.30 Mg ha^{-1} (CV = 11%).

The fitting of the USLE_MM model of Equation (3) on the 100 random validation datasets did not lead to a relevant variability in the model parameters ($b1$ and K_{UMM}): the average values of $b1$ and K_{UMM} are 1.067 and 0.081 with CV of 1.9% and 5.0%, respectively, and they are nearly equal to those obtained on the whole 667-record dataset. Moreover, these values are aligned with those determined in the previous SERLAB 532-record dataset ($b1$ = 1.0479, K_{UMM} = 0.0896, [4]) and even with those recently estimated from a small SERLAB 47-record dataset ($b1$ = 1.10, K_{UMM} = 0.032, [16]). This indicates that the SERLAB dataset has now reached a size that makes it relatively stable to new data and subsetting. The performance of the USLE_MM model is decidedly lower than the RF regression model, with an average RMSE of 3.34 Mg ha^{-1} (CV = 9%).

A graphical comparison of the performance of the two models is presented in Figure 5, which shows the scatter plots of observed (x-axis) and predicted (y-axis) $A_{e,N}$ against the 1:1 line for 1 of the 100 random training/validation datasets.

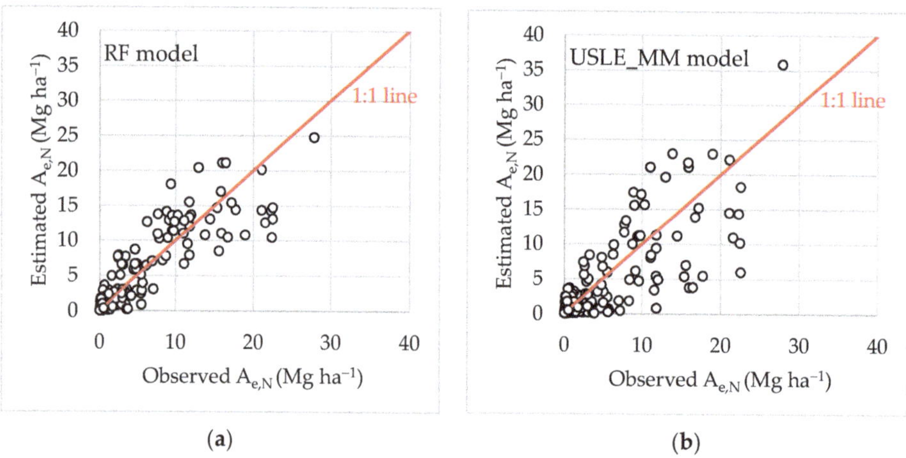

Figure 5. Comparison between observed and estimated normalized soil loss $A_{e,N}$ obtained using (**a**) the RF regression model; (**b**) the USLE mm regression model. Both models are trained on the same random 50% dataset and then applied to the remaining data.

4. Discussion

The analysis demonstrates that an RF algorithm can be remarkably effective for classifying *NE* and *E* storm events based on rainfall characteristics alone. Regarding the variable importance, the results (Figure 3) are consistent with those of previous studies, particularly with those of Todisco et al. [17], which indicated P, R, E, P_max_burst and I60 among the best variables. The importance of the rainfall depth P is also confirmed, as initially proposed by Wischmeier and Smith [5]. The accuracy measures of the classifier (Figure 4) are satisfactory, especially considering that the evaluation refers to a part of the dataset not used for the model training. The most critical accuracy value is the PA_E (72.3%), i.e., the percentage of erosive events correctly recognized to the observed total. Conversely, when the model assigns the "erosive" category to an event, it has a higher success rate (UA_E = 79.2%). These percentages decrease slightly when the model works with the six most significant features (PA_E = 70.0% and UA_E = 76.1%).

The RF classifier appears to be a valid alternative compared to other techniques previously used for the same purpose and on the same dataset. In Todisco et al.'s work [17], the best results (obtained by applying compound rainfall thresholds) indicate higher PA_E (about 82–85%) but lower UA_E (about 72–74%). Furthermore, in that case, the methodology for identifying the thresholds to separate *NE* and *E* events had not been made considering the separation between training and validation datasets.

The underestimation of the number of erosive events by the RF classifier does not significantly affect the estimation of the long-term erosivity of rainfall. In fact, the events correctly classified as erosive represent about 91% of the erosivity of all erosive events. This happens because, as known [10,11], the total erosivity is largely due to a few particularly intense or long-lasting storm events that are correctly identified by the RF classifier. The model could therefore be applied in real-time or on past time series to obtain an almost correct identification of the occurrence of the most significant erosion events. Indeed, although it is possible to analyse the variable importance, it is not straightforward to understand how the RF classifier uses the variables (and the corresponding thresholds) to arrive at the final classification.

The RF regression model, developed on only three features (Q_R, R, and "Period"), demonstrates decidedly superior performances (average RMSE = 2.30 Mg ha^{-1}) compared to the traditional USLE_MM power relationship (average RMSE = 3.34 Mg ha^{-1}) tested on the same datasets. This difference can depend on several factors. The first is undoubtedly attributable to the fact that, unlike the USLE_MM, the RF model uses the supplementary explanatory nominal variable "Period". This feature, useless in the classification algorithms (Figure 3), here proves to be relevant, with importance only slightly lower than the erosivity factor R. As further proof of its importance, an RF regression model, trained only on the Q_R and R features, led to a significant increase in the average RMSE (2.67 Mg ha^{-1}), which is, however, still lower than that of the USLE_MM. The interpretation of the reason of why a simple variable such as the "period" (with only two modes "wet" and "dry") can improve the soil loss prediction is not accessible, especially in such a type of "black box" model. An influence of the season on the reliability of soil loss prediction models has already been observed by Todisco et al. [43], which attributed this to the different conditions of the soil surface. The climatic conditions (particularly precipitation and air temperature) roughly described by the variable "period" can affect the dynamics of various soil properties such as roughness [44], infiltration and bulk density [45,46], and hydraulic conductivity [47]. Therefore, even if the RF model already explicitly considers essential features such as Q_R and R, the "period" becomes important to allow the model to evaluate the seasonal variability of their effect on soil loss. Finally, the "period" can directly contribute to describing the soil erodibility dynamics not explicitly considered in the model. Indeed, soil erodibility could vary with the "period" due to different environmental and climatic conditions (e.g., different soil moisture and soil temperature), capable of influencing the soil aggregate stability. This type of effect has been widely demonstrated in the literature, such as in [48–50]. Therefore, it would seem that the model's behaviour is consistent with the physical interpretation of the process.

Even without the "period" variable, the RF model is still superior to the USLE_MM one. This is undoubtedly due to the greater flexibility of RF in the use of input variables. In traditional regression, the independent variables are constrained in a precise functional form. In the case of the USLE_MM, using the product variable $Q_R R$ further reduces the possibility of differentiating the effect of Q_R and R.

5. Conclusions

The random forest approach has proven to be more effective than traditional (i.e., non-machine learning) methods in classifying erosive and non-erosive events and in quantifying soil loss at the event temporal scale and plot spatial scale. In this second application, it is conceivable that further performance improvements could be obtained by developing models with feature sets not constrained in the mathematical form of Equation (1). An interesting perspective of this study will be to compare different machine-learning techniques. In general, the possibility of applying such models still strongly depends on the availability of relatively large datasets. This indicates the need to continue investing in experimental and monitoring activities to increase the databases available in the most diverse environmental conditions and make machine learning approaches even more efficient.

Author Contributions: Conceptualization, L.V. and F.T.; Data curation, L.V. and F.T.; Methodology, L.V.; Software, L.V.; Supervision, F.T.; Writing—original draft, L.V.; Writing—review and editing, L.V. and F.T. All authors have read and agreed to the published version of the manuscript.

Funding: This research received no external funding.

Data Availability Statement: The data presented in this study are available on request from the corresponding author.

Conflicts of Interest: The authors declare no conflict of interest.

Appendix A

The list of the thirty-two features considered in the RF classifier is given in Table A1.

Table A1. List of the features used in the training of the random forest classifier for the separation of erosive and non-erosive storms.

Variable	Symbol	Unit
Total rainfall	P	mm
Total duration	D	h
Wet duration	D_wet	h
Dry duration	D_dry	h
Mean intensity	I	mm h^{-1}
Mean wet intensity	I_wet	mm h^{-1}
Rainfall erosivity	*R*	MJ mm ha^{-1} h^{-1}
Rainfall kinetic energy	*E*	MJ ha^{-1}
Maximum intensity over 30 min	I30	mm h^{-1}
Maximum intensity over 5 min	I5	mm h^{-1}
Maximum intensity over 10 min	I10	mm h^{-1}
Maximum intensity over 15 min	I15	mm h^{-1}
Maximum intensity over 20 min	I20	mm h^{-1}
Maximum intensity over 25 min	I25	mm h^{-1}
Maximum intensity over 40 min	I40	mm h^{-1}
Maximum intensity over 50 min	I50	mm h^{-1}
Maximum intensity over 60 min	I60	mm h^{-1}
Rainfall amount above truncation level p0	P_run	mm
Rainfall duration above truncation level p0	D_run	h
Number of runs	N_run	-
Maximum duration of an individual run	Max_D_run	h
Maximum rainfall amount of an individual run	Max_P_run	mm
Maximum peak of the run (p-p0)	Max_peak_run	mm

Table A1. *Cont.*

Variable	Symbol	Unit
Maximum rainfall depth from the start of the storm to a run	Max_P_pre_run	mm
Maximum slope of the rising limb of a burst	Max_slope_burst	%
Maximum mean run intensity	Max_I_run	mm h^{-1}
Number of bursts in a storm	N_burst	-
Maximum rainfall amount in a burst	Max_P_burst	mm
Maximum burst duration	Max_D_burst	h
Maximum rainfall depth from the start of the event to a burst	Max_P_pre_burst	mm
Maximum mean burst intensity	Max_I_burst	mm h^{-1}
Period of occurrence	Period	-

References

1. Pimentel, D. Soil erosion: A food and environmental threat. *Environ. Dev. Sustain.* **2006**, *8*, 119–137. [CrossRef]
2. Borrelli, P.; Van Oost, K.; Meusburger, K.; Alewell, C.; Lugato, E.; Panagos, P. A step towards a holistic assessment of soil degradation in Europe: Coupling on-site erosion with sediment transfer and carbon fluxes. *Environ. Res.* **2018**, *161*, 291–298. [CrossRef] [PubMed]
3. Risse, L.M.; Nearing, M.A.; Nicks, A.D.; Laflen, J.M. Error assessment in the universal soil loss equation. *Soil Sci. Soc. Am. J.* **1993**, *57*, 825–833. [CrossRef]
4. Bagarello, V.; Ferro, V.; Pampalone, V.; Porto, P.; Todisco, F.; Vergni, L. Statistical check of USLE-M and USLE-MM to predict bare plot soil loss in two Italian environments. *Land Degrad. Dev.* **2018**, *29*, 2614–2628. [CrossRef]
5. Wischmeier, W.H.; Smith, D.D. *Predicting Rainfall-Erosion Losses: A Guide to Conservation Farming*; U.S. Department of Agriculture, Science and Education Administration: Washington, DC, USA, 1978; p. 537.
6. Renard, K.G.; Foster, G.R.; Weesies, G.A.; McCool, D.K.; Yoder, D.C. Predicting soil erosion by water: A guide to conservation planning with the revised universal soil loss equation (RUSLE). In *U.S. Department of Agriculture Agricultural Handbook. No. 703*; US Department of Agriculture: Washington, DC, USA, 1997.
7. Zhou, P.; Luukkanen, O.; Tokola, T.; Nieminen, J. Effect of vegetation cover on soil erosion in a mountainous watershed. *Catena* **2008**, *75*, 319–325. [CrossRef]
8. Morgan, R.P.C. *Soil Erosion and Conservation*, 3rd ed.; Blackwell: Malden, MA, USA, 2005; p. 320.
9. Larson, W.E.; Lindstrom, M.J.; Schumacher, T.E. The role of severe storms in soil erosion: A problem needing consideration. *J. Soil Water Conserv.* **1997**, *52*, 90–95.
10. Bagarello, V.; Ferro, V.; Pampalone, V.; Porto, P.; Todisco, F.; Vergni, L. Predicting soil loss in central and south Italy with a single USLE-MM model. *J. Soils Sediments* **2018**, *18*, 3365–3377. [CrossRef]
11. Di Stefano, C.; Pampalone, V.; Todisco, F.; Vergni, L.; Ferro, V. Testing the Universal Soil Loss Equation-MB equation in plots in Central and South Italy. *Hydrol. Process.* **2019**, *33*, 2422–2433. [CrossRef]
12. Kinnell, P.I.A.; Risse, L.M. USLE-M: Soil Empirical modeling rainfall erosion through runoff and sediment concentration. *Sci. Soc. Am. J.* **1998**, *62*, 1667–1672. [CrossRef]
13. Kinnell, P.I.A. Event soil loss, runoff and the Universal Soil Loss Equation family of models: A review. *J. Hydrol.* **2010**, *385*, 384–397. [CrossRef]
14. Bagarello, V.; Di Piazza, G.V.; Ferro, V.; Giordano, G. Predicting unit plot soil loss in Sicily, south Italy. *Hydrol. Process.* **2008**, *22*, 586–595. [CrossRef]
15. Bagarello, V.; Ferro, V.; Pampalone, V. A new version of the USLEMM for predicting bare plot soil loss at the Sparacia (South Italy) experimental site. *Hydrol. Process.* **2015**, *29*, 4210–4219. [CrossRef]
16. Todisco, F.; Vergni, L.; Ortenzi, S.; Di Matteo, L. Soil Loss Estimation Coupling a Modified USLE Model with a Runoff Correction Factor Based on Rainfall and Satellite Soil Moisture Data. *Water* **2022**, *14*, 2081. [CrossRef]
17. Todisco, F.; Vergni, L.; Vinci, A.; Pampalone, V. Practical thresholds to distinguish erosive and rill rainfall events. *J. Hydrol.* **2019**, *579*, 124173. [CrossRef]
18. Xie, Y.; Liu, B.; Nearing, M.A. Practical thresholds for separating erosive and non-erosive storms. *Trans. Am. Soc. Agric. Eng.* **2002**, *45*, 1843–1847.
19. Todisco, F. The internal structure of erosive and non-erosive storm events for interpretation of erosive processes and rainfall simulation. *J. Hydrol.* **2014**, *519*, 3651–3663. [CrossRef]
20. Mukhamediev, R.I.; Popova, Y.; Kuchin, Y.; Zaitseva, E.; Kalimoldayev, A.; Symagulov, A.; Levashenko, V.; Abdoldina, F.; Gopejenko, V.; Yakunin, K.; et al. Review of Artificial Intelligence and Machine Learning Technologies: Classification, Restrictions, Opportunities and Challenges. *Mathematics* **2022**, *10*, 2552. [CrossRef]
21. Licznar, P.; Nearing, M.A. Artificial Neural Networks of Soil Erosion and Runoff Prediction at the Plot Scale. *Catena* **2003**, *51*, 89–114. [CrossRef]

22. Kim, M.; Gilley, J.E. Artificial Neural Network Estimation of Soil Erosion and Nutrient Concentrations in Runoff from Land Application Areas. *Comput. Electron. Agric.* **2008**, *64*, 268–275. [CrossRef]
23. Albaradeyia, I.; Hani, A.; Shahrour, I. WEPP and ANN Models for Simulating Soil Loss and Runoff in a Semi-Arid Mediterranean Region. *Environ. Monit. Assess.* **2011**, *180*, 537–556. [CrossRef]
24. de Farias, C.A.S.; Santos, C.A.G. The Use of Kohonen Neural Networks for Runoff–Erosion Modeling. *J. Soils Sediments* **2014**, *14*, 1242–1250. [CrossRef]
25. Arif, N.; Danoedoro, P. Hartono Analysis of Artificial Neural Network in Erosion Modeling: A Case Study of Serang Watershed. *IOP Conf. Ser. Earth Environ. Sci.* **2017**, *98*, 012027. [CrossRef]
26. Yusof, M.F.; Azamathulla, H.M.; Abdullah, R. Prediction of Soil Erodibility Factor for Peninsular Malaysia Soil Series Using ANN. *Neural Comput. Appl.* **2014**, *24*, 383–389. [CrossRef]
27. Breiman, L. Random forests. *Mach. Learn.* **2001**, *45*, 5–32. [CrossRef]
28. Wang, S.; Aggarwal, C.; Liu, H. Random-Forest-Inspired Neural Networks. *ACM Trans. Intell. Syst. Technol.* **2018**, *9*, 69. [CrossRef]
29. Tyralis, H.; Papacharalampous, G.; Langousis, A. A Brief Review of Random Forests for Water Scientists and Practitioners and Their Recent History in Water Resources. *Water* **2019**, *11*, 910. [CrossRef]
30. Konstantinou, C.; Stoianov, I. A comparative study of statistical and machine learning methods to infer causes of pipe breaks in water supply networks. *Urban Water J.* **2020**, *17*, 534–548. [CrossRef]
31. Konstantinou, C.; Biscontin, G. Experimental Investigation of the Effects of Porosity, Hydraulic Conductivity, Strength, and Flow Rate on Fluid Flow in Weakly Cemented Bio-Treated Sands. *Hydrology* **2022**, *9*, 190. [CrossRef]
32. Schoppa, L.; Disse, M.; Bachmair, S. Evaluating the performance of random forest for large-scale flood discharge simulation. *J. Hydrol.* **2020**, *590*, 125531. [CrossRef]
33. Park, H.; Kim, K.; Lee, D.K. Prediction of Severe Drought Area Based on Random Forest: Using Satellite Image and Topography Data. *Water* **2019**, *11*, 705. [CrossRef]
34. Tarek, Z.; Elshewey, A.M.; Shohieb, S.M.; Elhady, A.M.; El-Attar, N.E.; Elseuofi, S.; Shams, M.Y. Soil Erosion Status Prediction Using a Novel Random Forest Model Optimized by Random Search Method. *Sustainability* **2023**, *15*, 7114. [CrossRef]
35. Todisco, F.; Vergni, L.; Mannocchi, F.; Bomba, C. Calibration of the soil loss measurement method at the Masse experimental station. *Catena* **2012**, *91*, 4–9. [CrossRef]
36. Vergni, L.; Vinci, A.; Todisco, F. Influence of the rainfall Time Step on the Thresholds for Separating Erosive and Non-Erosive Events. In *AIIA 2022: Biosystems Engineering towards the Green Deal*; Lecture Notes in Civil Engineering; Ferro, V., Giordano, G., Orlando, S., Vallone, M., Cascone, G., Porto, S.M.C., Eds.; Springer Nature: Berlin, Germany, 2023; Volume 337, *in press*.
37. Nearing, M.A. A single continuous function for slope steepness influence on soil loss. *Soil Sci. Soc. Am. J.* **1997**, *61*, 917–919. [CrossRef]
38. Vinci, A.; Todisco, F.; Mannocchi, F. Calibration of manual measurements of rills using terrestrial laser scanning. *Catena* **2016**, *140*, 164–168. [CrossRef]
39. Liaw, A.; Wiener, M. Classification and Regression by random Forest. *R News* **2002**, *2–3*, 18–22.
40. Breiman, L. *Manual on Setting up, Using, and Understanding Random Forests v3.1*; Statistics Department University of California: Berkeley, CA, USA, 2002; pp. 1–58.
41. Congalton, R.G. A review of assessing the accuracy of classifications of remotely sensed data. *Remote Sens. Environ.* **1991**, *37*, 35–46. [CrossRef]
42. Weaver, J.; Moore, B.; Reith, A.; McKee, J.; Lunga, D. A comparison of machine learning techniques to extract human settlements from high resolution imagery. In Proceedings of the International Geoscience and Remote Sensing Symposium (IGARSS), Valencia, Spain, 22–27 July 2018.
43. Todisco, F.; Brocca, L.; Termite, L.F.; Wagner, W. Use of satellite and modeled soil moisture data for predicting event soil loss at plot scale. *Hydrol. Earth Syst. Sci.* **2015**, *19*, 3845–3856. [CrossRef]
44. Vinci, A.; Todisco, F.; Vergni, L.; Torri, D. A comparative evaluation of random roughness indices by rainfall simulator and photogrammetry. *Catena* **2020**, *188*, 104468. [CrossRef]
45. Todisco, F.; Vergni, L.; Vinci, A.; Torri, D. Infiltration and bulk density dynamics with simulated rainfall sequences. *Catena* **2022**, *218*, 106542. [CrossRef]
46. Todisco, F.; Vergni, L.; Ceppitelli, R. Modelling the dynamics of seal formation and pore clogging in the soil and its effect on infiltration using membrane fouling models. *J. Hydrol.* **2023**, *618*, 129208. [CrossRef]
47. Todisco, F.; Vergni, L.; Iovino, M.; Bagarello, V. Changes in soil hydrodynamic parameters during intermittent rainfall following tillage. *Catena* **2023**, *226*, 107066. [CrossRef]
48. Torri, D.; Ciampalini, R.; Gil, P.A. The Role of Soil Aggregates in Soil Erosion Processes. In *Modelling Soil Erosion by Water*; Boardman, J., Favis-Mortlock, D., Eds.; NATO ASI Series; Springer: Berlin/Heidelberg, Germany, 1998; Volume 55, pp. 247–257.

49. Lavee, H.; Sarah, P.; Imeson, A.C. Aggregate Stability Dynamics as Affected by Soil Temperature and Moisture Regimes. *Geogr. Ann. A* **1996**, *78*, 73–82.
50. Imeson, A.C.; Lavee, H.; Calvo, A.; Cerda, A. The erosional response of calcareous soils along a climatological gradient in Southeast Spain. *Geomorphology* **1998**, *24*, 3–16. [CrossRef]

water

MDPI

Article

Agricultural Practices for Hillslope Erosion Mitigation: A Case Study in Morocco

Jean Marie Vianney Nsabiyumva [1,2], Ciro Apollonio [3,4,*], Giulio Castelli [1,5,6], Andrea Petroselli [7], Mohamed Sabir [8] and Federico Preti [1]

1 Department of Agriculture, Food, Environment and Forestry (DAGRI), University of Florence, Piazzale delle Cascine, 18, 50144 Firenze, Italy
2 Polytechnic University of Gitega, Route Nationale No 8, km 4, Gitega BP 490, Burundi
3 Department of Agriculture and Forestry Sciences (DAFNE), Tuscia University, 01100 Viterbo, Italy
4 Associazione Italiana per L'Ingegneria Naturalistica (AIPIN), via San Bonaventura 13, 50145 Firenze, Italy
5 UNESCO Chair in Hydropolitics, University of Geneva, 66 Boulevard Carl-Vogt, 1205 Geneva, Switzerland
6 Environmental Governance and Territorial Development Hub (GEDT), University of Geneva, 66 Boulevard Carl-Vogt, 1205 Geneva, Switzerland
7 Department of Economics, Engineering, Society and Business Organization (DEIM), Tuscia University, 01100 Viterbo, Italy
8 National School of Forest Engineering, BP 511, Avenue Moulay Youssef, Tabriquet-Salé 11 000, Morocco
* Correspondence: ciro.apollonio@unitus.it

Abstract: In the last decades, the Rif area in Morocco has been frequently affected by soil erosion due to intense rainfall events. In order to help farmers improve their lives and avoid damages caused by this phenomenon, a management project (the MCA Project) aiming to grow fruit trees has been realized. The objective of this study was to evaluate, in three provinces of Morocco, the effect on the hydrological response of selected erosion control management techniques combined with olive tree plantations. The investigated variables were the final infiltration (If), the imbibition of rainwater (Pi), the runoff coefficient (Kr), and the soil detachment (D). In particular, for each investigated soil utilization, three replications of a rain simulation test (80 mm/h) and soil sampling were conducted. Results for surface conditions demonstrate that under vegetation in matorral and fallow, the surface is covered at more than 75% with a high content of organic matter (OM) at 4.5% and 2.6%, respectively. Despite the compaction observed in those land uses, the surface area opened exceeded 90% in the study area. Regarding the soils physical properties, they were rich in silt at more than 40%; the rate of porosity is high where bulk density is low. At the Taounate site, low porosity was at 62% in fallow and at 55% in plowing, with high densities of 1.01 g/cm^2 and 1.2 g/cm^2, respectively. Tests also demonstrate that vegetation has an important role in moisture conservation in the depths of 0 to 10 cm at all sites with macroagregate stability (MA%) compared to plowing sites. In terms of soil hydrology, vegetation reduces the runoff because, under matorral (it was at 0%), it avoids soil erosion.

Keywords: erosion control management; hydrological parameters; runoff risk; Rif; agro-sylvo-pastoral systems; olive trees; agroforestry; soil erosion management; Mediterranean

Citation: Vianney Nsabiyumva, J.M.; Apollonio, C.; Castelli, G.; Petroselli, A.; Sabir, M.; Preti, F. Agricultural Practices for Hillslope Erosion Mitigation: A Case Study in Morocco. *Water* **2023**, *15*, 2120. https://doi.org/10.3390/w15112120

Academic Editors: Vito Ferro and Alessio Nicosia

Received: 2 May 2023
Revised: 25 May 2023
Accepted: 30 May 2023
Published: 2 June 2023

1. Introduction

In the Mediterranean basin, soil erosion due to intense rainfall events, i.e., water erosion, is one of the major processes of land degradation, hence constituting a great source of desertification [1]. Moreover, there is an increasing trend of aridity that accelerates such a process [2]. According to FAO studies [3], water erosion affects 50% of the soil in Turkey, 40% in Morocco, and 35% in Greece.

The factors favoring water erosion in this area are generally the fragility of the soil, the irregularity of intense rainfall events, the soil slope, the high air temperatures, which accelerate the mineralization of organic matter, and the vegetation cover reduction caused by climate and man [4]. The most visible form of erosive action is represented by the gullies, which cause the formation of "badlands", which have no agricultural interest, and mudslides, which clog reservoirs and cause a reduction in their storage capacity. Given the problems caused by water erosion, it is necessary to take adequate measures to slow down this phenomenon and ensure sustainable management of water and soil resources [5].

In this regard, many countries of the southern Mediterranean have started, in the last decades, a proactive policy of land development and conservation. Several anti-erosion techniques have been introduced and popularized on sloping land [6–10]. Some countries adopted traditional practices or improved them; others took inspiration from technologies developed in Europe or the United States [11].

In Morocco, there are many national intervention plans dealing with water erosion. Many Soil Defense and Restoration (DRS) intervention actions use fruit trees (such as olive trees, almond trees, and vines) to conserve soil and water, improve the incomes of farmers, and create jobs in the areas of intervention [12].

In the western Rif, Heusch [13] showed that water erosion causes each year more than 3000 t/km^2 of soil to be poured into the sea, and 60% of the eroded soils in Morocco come from this area, which covers only 6% of the national territory. The causes of this degradation are deforestation due to land cultivation and other natural causes such as climate and the nature of the substrate.

Under these constraints, a considerable effort has been provided by the population of the western Rif, thanks to the developed agro-forestry-pastoral farming systems. Various fruit tree species have been planted on the terraces, on the boundaries of plots and properties, and in the gullies [14]. The growing fruit is of capital interest to the population, not only for economic aspects but also for the fruit trees' capacity against water erosion [15].

However, knowledge of the variation of the soil hydrological parameters (e.g., infiltration and runoff) associated with the soil's physical properties (e.g., apparent density, roughness, and porosity) is very important to define the susceptibility of soils to water erosion [16,17]. Equally important is to develop experimental studies for the analysis of erosion phenomena at the basin scale [18]. For instance, the infiltration capacity of surface horizons essentially depends on the state of the soil surface, which is affected by its roughness, biological activity, plant cover [19,20], humidity, cracking, stoniness, aggregation, and water repellency [21].

In the western Rif region (Larache) and in the pre-Rif regions (Taounate and Taza) studies evaluating the impact of fruit tree DRS on the hydrological soil characteristics are rare. Most of such studies have shown only that the production of runoff from the soil is strongly linked to its physical characteristics, surface condition, and vegetation cover [22,23]. In 2015, Simonneaux et al. [24] provided an interesting study on soil erosion in Morocco, where the "soil loss" was estimated by indirect measures.

Due to the limited availability of peer-reviewed experimental studies on erosion in the Moroccan Rif area, the present work represents a novel contribution, especially considering that the majority of experiments on soil conservation based on experimental plots are located in China and the USA, with very limited availability in the South Mediterranean [25]. Most of the recent contributions focus on modeling applications [26] or trackers [27]. At the same time, the Rif is one of the most critical areas for environmental development and is a hotspot for erosion [28].

The study was carried out as a part of the Millennium Challenge Account (MCA)—Morocco Project for the fruit tree component in the Moroccan Rif. The component aims to stimulate the growth of the agricultural sector in this area through the transition from annual extensive crops, and in particular cereals, to multi-annual fruit tree crops that are market-oriented (e.g., olives, almonds, and figs).

Specifically, we first aim at determining the effects of the development on hydrodynamics, surface condition, and other soil characteristics; and second, we study the relationships between hydrological parameters and selected soil characteristics that can serve as indicators of runoff and soil erosion.

2. Materials and Methods

2.1. The Study Area

The study area was included in the MCA—Morocco Project and is part of the Rif in northern Morocco. It extends over the western, central, and eastern Rif.

The three provinces of Larache, Taounate, and Taza were selected as investigation areas. In each of these provinces, some of the perimeters developed in the Fruit Arboriculture Project (FAP-MCA) have been selected as our experimental sites. Table 1 summarizes the investigated study perimeters, and their location is represented in Figure 1.

Table 1. FAP-MCA perimeters retained in each province of our study area.

Project Area	Zone 1	Zone 2	Zone 3
Provinces	**Larache**	**Taounate**	**Taza**
Study Perimeters	Ain Maabad Dar Lkhil	El Gara	Ahl Zaouia
	Ain Hadid	Faytora	Lkassibat
	Sidi Ait Atmane		Ait Maalla

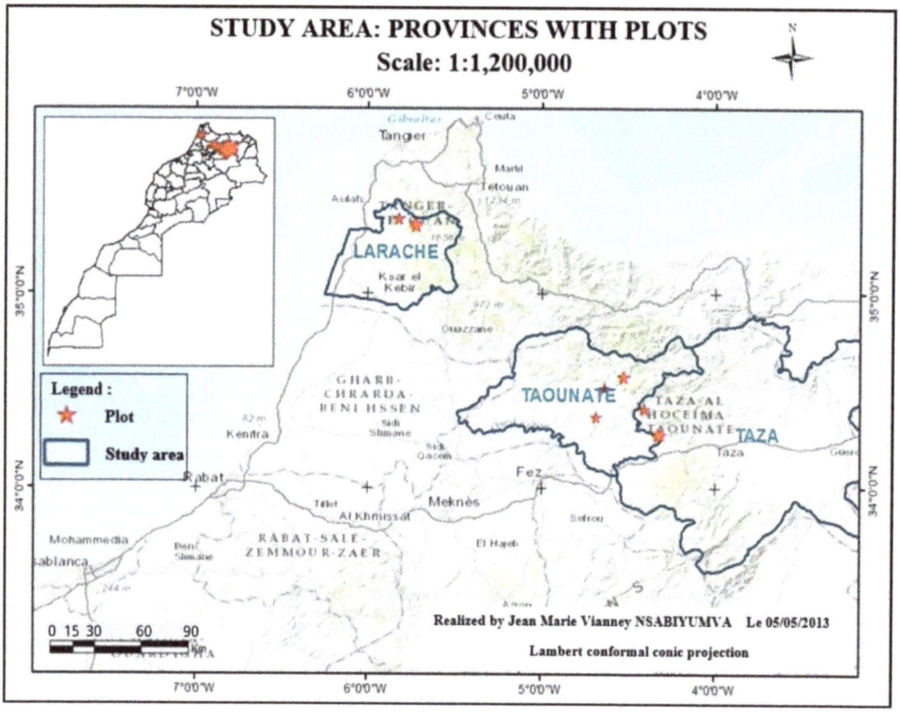

Figure 1. Location of study areas in the provinces of Larache, Taounate, and Taza.

Additionally, for comparative studies, other study perimeters have been added in the first two provinces despite not being developed under the FAP. In Larache, we chose a matorral and a perimeter plowed at Ain Maabid Dar Lkhil; and in Taounate, we chose a third Kifah perimeter developed as part of the "Economic and rural development of the western Rif Project" known in Morocco as the DERRO Project.

2.1.1. Description of the Study Area's Physical Environment

The geological and pedological situation of the study area is reported in Table 2 [29].

Table 2. Comparison among three study sites on geological and pedological conditions.

Sites	Geology	Soils
Larache	marly material of the Cretaceous; schistose scales and quartzite debris	little evolved soils of erosion; calcimagnesic soils with vertic caracter and vertic soils
Taounate	clayey marls	vertic soils eroded
Taza	Soft bedrock or hard geological bedrock and marly material	mineral soils, little evolved soils, calcimagnesic soils, isohumic soils, and soils with iron sesquioxides

About the bioclimate, the province of Larache has a subhumid bioclimate with a warm winter, an average yearly rainfall of 689 mm, an average maximum temperature of the hottest month equal to 33 °C, an average minimum temperature of the coldest month of 7 °C, and a rainfall quotient of 91.9 mm. The province of Taounate has a subhumid bioclimate with a temperate winter; the yearly rainfall is 505 mm; the mean maximum temperature of the hottest month is 36 °C; the average minimum temperature of the coldest month is 5 °C; and the rainfall is 56.4 mm. The province of Taza has a semi-arid to temperate winter; the average cumulative yearly rainfall is 557 mm; the mean maximum temperature of the hottest month is 45 °C; the average minimum temperature of the coldest month is 4 °C; and the rainfall quotient is 45.7 mm. Moreover, the plots studied in those three provinces had slopes ranging from 13% to 40%. In particular, at Larache, the slope of the plots varies between 16% and 37%; in Taounate, the slope is between 16% and 24%. Finally, in Taza, the plots studied were located on a slope of between 13% and 40%.

2.1.2. Land Uses

Regarding land use, forest trees are almost absent except for a few reforested Eucalyptus and pine trees in Taounate, but in the fallow land there were a few trees (juniper) scattered around. *Chamærops humilis* (doum) was present in these perimeters because they were traditionally areas of large crops. Oleander was along the banks of the wadis.

Since the investigated project areas are non-irrigated agricultural areas that were adapted to the olive tree species, the farming system is dominated by cereals and extensive farming to take advantage of pasture stubble and straw during the lean season to feed livestock. This cereal usually grows from medium to steep slopes; hence, it accentuates soil erosion and degradation. According to the beneficiaries of the studied perimeters, farmers in the area tend to abandon extensive breeding (cattle, sheep, and goats) for fruit growing, especially olive trees. This fruit arboriculture project encourages the association of olive trees and cereals in these regions, which will make it possible to conserve water and soil in addition to maintaining livestock farming. Moreover, the discussions with farmers in the field revealed that there are also vegetables, market gardening, and fallow land.

2.1.3. Choice of Experimental Perimeters

The objects of experiments were chosen according to the empirical method according to the knowledge of the field, and the objectives were fixed within the framework of this research. These perimeters are part of those planted by the FAP; it was deemed appropriate to take two types per perimeter, namely plowing and fallowing, to conduct comparative studies to better understand the impact of the cultivation techniques used at each cultivated site on soil erosion. Thus, fallow and plowing were considered when simulating rainfall in these three provinces for each zone. Therefore, in Larache, Taounate, and Taza, there were, respectively, 6, 4, and 6 sites for the simulation of rainfall. However, all the perimeters added for the comparison were not concerned by rainfall, except the Larache matorral, which showed no runoff. In total, the rainfall simulation concerned 17 sites in 3 repetitions, therefore 51 simulation plots.

2.2. *Methods*

2.2.1. Rainfall Simulation Tests

Before conducting the rainfall simulation, the choice of the location of the experimental plot was made according to the following criteria: the representativeness and homogeneity of the soil surface, and the degree and direction of the slope. In each selected plot, there was the installation of the simulation plot, which has an area of 1.66 m × 0.6 m (1 m^2). Two 1.66-m metal angles were inserted into the ground for about 5 cm to avoid lateral water losses (leaks). Downstream of these two angles, a triangular metal platform was installed to direct runoff water and sediment into a container. Afterward, the ground surface condition was described by the quadrat point method [30] using the diagonals and a stake every 2 cm. Thus, the surfaces covered at ground level (SC, %) include all that is litter (L, %), vegetation (Veg, %), and pebbles not integrated into the soil mass (CNI, %). The open surfaces (SO, %) mainly concern the cracks (Fiss, %), the galleries, and the clods (Mt, %), while the closed surfaces (SF, %) correspond to the areas closed off by a film (Pel, %) or to the visible pebbles embedded in the ground (CI, %). In addition, the total soil organic matter (OM, %) was obtained after the determination of the carbon content (C, %) since the ratio of organic matter to carbon is almost constant and equal to OM/C = 1.72 [31]. In addition, other parameters such as the soil penetration resistance (PEN, kg/cm^2) and the shear strength to the force used by the water in detaching the aggregates (SS, kg/cm^2) related to the surface condition were measured, respectively, by a pocket penetrometer and a scissometer in nine measurements (3 repetitions × 3 for each simulation plot). The simulation plot (Figure 2) was divided into three parts along the angles delimiting the plot. Finally, the roughness index (Ir, %) was measured by the chain method [30], which consists of using a metal chain laid in a straight line. The roughness index is determined by Ir = $(L_d - l)/l$) where L_d (m) is the length of the extended chain and l (m) is the width of the plot over the width of the plot. Three repetitions were carried out to also measure the roughness index.

Figure 2. Images related to the construction of the experimental plots.

After the evaluation of the surface condition, the preparation of the rainfall simulation device has been carried out. This consisted of a watering ramp 50 cm wide, less than that of the plot, and connected by a flexible pipe to a tank that was filled with 60 L of water and that was located a few meters higher than the plot. Subsequently, the adjustment of the simulated rainfall intensity (on average 80 mm/h) was carried out after several tests using a valve at the outlet of the tank. A graduated cylinder and a stopwatch were used. The measurement of the simulated rainfall intensity was carried out at the beginning and end of the simulation.

After obtaining the desired rainfall intensity, the rainfall simulator was used. The rainfall simulator can be considered a simple manual irrigator [30], which allows drops with relatively little energy to be projected onto a surface of 1 m^2. Watering was performed at an average height of 50 cm above the soil surface. It was a question of distributing the water and avoiding watering outside the plot. The watering was carried out simultaneously with the timing: at the beginning, the elapsed time (reaction time: Tr, min) was recorded for the triggering of the runoff to quantify the height of the imbibition rain (Pi, mm).

Then, the recording of the amount of runoff water every 5 min was performed until the stabilization of the runoff; otherwise, until the time reached one hour. The collected runoff water has been put in a container and then stored in a transparent jerrycan with a capacity of 10 L to later measure the quantities of sediments in the laboratory.

On a previously prepared simulation sheet, the various measurements were noted: location of the site and coordinates, degree of slope, date, surface condition, runoff trigger time, runoff volumes, and so on. A simulation test required, on average, 3 h of work for a team of 3 people. Through these simulation tests, we managed to determine the final infiltration (If, mm/h), which corresponds to the average of the last two permanent infiltration values, and the imbibition rain (*Pi*, mm), which is the height of water infiltrated before the start of runoff. *Pi* is calculated from the soil reaction time (*Tr*, min) and the simulation rainfall intensity (*i*, mm/h). It was given by the following formula:

$$Pi = (i \times Tr)/60.$$

2.2.2. Soil Sampling

Two soil samples (1 kg each) were taken next to the rainfall simulation plot at a depth ranging from 0 to 5 cm to determine the soil structural stability (rate of water-stable macroaggregates) and to carry out other soil analyses (particle size and organic matter) in the laboratory.

The undermining facilitated their removal. At three points around the rainfall simulation plot, a soil sample is taken in depth increments of 0–5, 5–10, and 10–15 cm to assess the soil bulk density using a 10 cm long cylinder and 4 cm in diameter. All the samples were kept away from heat and transported to the laboratory for the following analysis.

2.2.3. Physical Analyses of Samples

The physical analyses of the samples concerned the following parameters: bulk density, total porosity, soil moisture, structural stability measured by macroaggregate stability, and detachability. The analyses were conducted in the laboratories of the National Forestry School of Engineers and the Agronomic and Veterinary Institute (IAV) Hassan II. In particular:

- Bulk density (Da, g/cm^3):

After having placed the soil samples taken at different depths in boxes for drying, the samples have been weighed fresh and dried after 24 h in an oven at 105 °C. According to Blake [32], the formula for calculating the bulk density is:

$$Da = ma/V$$

With *ma* = anhydrous mass of soil (g); *V* = volume of sampling cylinder (cm^3).

- Total porosity (*P*, %) is deduced by the following formula [32]:

$$P = 100 - [(Da/Dr) \times 100]$$

With Dr = actual soil density, which is a constant = 2.65 g/cm^3

- Soil moisture (H, %)

The collected samples were weighed to determine their fresh weight (Pf, g). In the laboratory, these samples were dried in an oven at 105 °C for 24 h and weighed again to find their dry weight (Ps, g). The difference between the fresh weight (Pf) and the dry weight (Ps), which gives the weight of water (Pwater, g) and the soil moisture (H, %), was obtained by this formula [33]:

$$\mathrm{H} = \mathrm{Pwater}/\mathrm{Ps} \times 100$$

- Rate of water-stable macroaggregates (MA, %)

For this study, the structural stability is estimated by a stability test that is based on disaggregation combining wetting and sieving. This method [34] is used to estimate the rate of water-stable macroaggregates with a diameter greater than 1 mm but between 1 and 2 mm. Aggregates of 1 to 2 mm in diameter are shaken vigorously in the test tube, and stable aggregates of more than 0.25 mm in diameter are measured. Each study perimeter has been considered to have a homogeneous stratum, and it has been subjected to three repetitions.

- Detachability of soils (D, g/m^2/h)

After agitation to mix the runoff collected in the simulation plot, quantification was performed after water removal by pipette. Then, the drying of the deposited sediments was carried out in the oven for 24 h. Finally, the weighing made it possible to give the quantity of sediments from which the detachability of the soils (D) is obtained by the following formula [35]:

$$D = \textit{Sediment weight}\ (\mathrm{g}) \times [60/\textit{simulation time}\ (\mathrm{min})]$$

2.3. Statistical Data Processing

To be able to compare the effects of land uses on the hydrological behavior of soils and to find the correlations that may exist between the different parameters and the investigated variables, statistical processing was carried out on the observed data. The treatment consisted of analyses of variance (ANOVA) and simple and multiple comparisons of the means according to the method of the least significant difference (P.P.D.S.).

Since the soil samples collected in Larache have different numbers depending on the land use, the Scheffé test was used [36]. For the other regions (Taounate and Taza), there are only two situations to compare (plowed land and fallow). For this, the recommended test is a T-Student for independent samples. The probability threshold considered for all the tests is 5%. Through a simple linear regression, the Pearson correlation coefficient R was obtained, and regression equations were determined to better understand the soil parameter that most influences the hydrological parameters (If, Pi, Kr, and D).

3. Results and Discussions

3.1. Results

3.1.1. Effects of Land Uses on the Surface Conditions

After the application of the statistical tests and the comparison of the average values obtained for each property of the soil in the different land uses between the three provinces (Table 3), it was noticed that the higher contents of organic matter (OM) are observed under matorral and fallow (4.50% and 2.60%, respectively). For plowed sites, we obtained values ranging between 2.03% and 3.56%. In addition, it was found that the studied surface parameters are statistically significantly different between plowing and fallow.

Table 3. Effects of land uses on the surface conditions of the FAP-MCA perimeters in Larache, Taounate, and Taza provinces in Morocco.

Provinces	Larache			Taounate		Taza	
Land Uses	Matorral	Fallow	Plowing	Fallow	Plowing	Fallow	Plowing
	Organic matter						
OM (%)	4.43	3.56	2.8	3.46	3.1	2.56	2.03
	Surface conditions						
SN (%)	1.08 a	9.00 a	83.69 b	22.83 a	75.67 b	18.25 a	90.83 b
SC (%)	99.25 b	91.00 b	16.31 a	77.17 a	24.50 b	81.75 b	9.17 a
SO (%)	99.90 b	90.78 a,b	78.67 a	96.67 a	90.17 b	92.33 b	82.50 a
SF (%)	0.10 a	9.22 a,b	19.67 b	3.34 a	9.83 b	7.58 a	17.50 a
PEN (kg/cm^2)	0.71 a	0.68 a	0.59 a	0.37 a	0.25 a	1.30 a	1.50 a
SS (kg/cm^2)	2.49 b	2.37 a,b	1.69 a	3.22 a	2.06 b	2.80 a	2.63 a
Ir (%)	3.16 a	16.94 a,b	22.17 b	21.21 a	44.33 b	16.12 a	31.26 a

Note: OM—rate of organic matter; SN—percentage of the bare surface; SC—rate of the cover surface; SO—rate of the open surface; SF—rate of the closed area; PEN—force of penetration resistance; SS—force of shear resistance; Ir—rate of roughness. The means of the same row followed by the same letter are not statistically different ($p < 0.05$).

However, the results obtained show that the soils under fallow were covered (SC) and open (SO) to more than 75%, and the soils of the plowed plots had bare and open surfaces greater than 75%. In terms of fallow land, we have covered areas of 91% in Larache, 82% in Taza, and a lower rate in Taounate (77%). The open surfaces exceed 90% for the three regions.

Moreover, the soils under fallow are more compacted and more resistant in Larache and Taounate. At Taza, the trends are reversed only for penetration (PEN) and shear strength (SS). Although the penetrometer values are not statistically different, the value is more pronounced in Taza in both fallow (PEN = 1.3 kg/cm^2) and plowing (PEN = 1.5 kg/cm^2) compared to the other two regions. This situation in Taza could be due to the presence of crusts formed on the surface during previous rains on the plowed and sun-dried land under which our experiments were carried out.

3.1.2. Effects of Land Uses on Soil Physical Properties

Among the physical properties of the soils in the three provinces is the soil's texture. Then, the soils analyzed are generally rich in silt, at more than 40%, despite the presence of a high rate of clay and a few sandy particles.

Moreover, according to the results obtained, the soils of Larache have a sandy clay texture with 40.11% clay, 43.96% silt, and 15.22% sand. Those of Taounate have a silty clay texture composed of 39.24% clay, 40.46% silt, and 20.29% sand. In Taza, the soils analyzed had 37.61% clay, 42.83% silt, and 19.58% sand, and they had a texture of fine clay silt (Table 4).

Table 4. Texture of soils analyzed of the FAP-MCA perimeters in Larache, Taounate, and Taza provinces at Morocco.

Provinces	SF (%)	SG (%)	Arg (%)	LG (%)	LF (%)	Texture
Larache	7.60	7.62	40.81	29.15	14.81	Sandy clay
Taounate	16.92	3.37	39.24	25.86	14.60	Silty-clayey
Taza	6.92	12.66	37.61	19.46	23.37	Fine clay silt

Note: SF (%): Fine sand rate; SG (%): Coarse sand; Arg (%): Clay rate; LG (%): Coarse silt rate; LF (%): Fine silt rate.

Regarding the physical properties of the soils in the three provinces (Table 5), it can be noticed that the soils are generally loosely packed because the bulk density (Da) varies from 0.79 to 1.20 g/cm^3 at the soil surface (0–10 cm). At sites covered by litter or vegetation (matorral and fallow), the density values do not show significant statistical differences and have low surface values (0.80 to 1.00 g/cm^3). The highest values (1.01 and 1.20 g/cm^3) are observed in Taounate for fallow and plowing, respectively. These density values increase from surface to depth.

Table 5. Effects of land uses on the physical properties of soils in the FAP-MCA perimeters in Larache, Taounate, and Taza provinces in Morocco.

Provinces	Larache				Taounate		Taza	
Land Uses	**Matorral**	**VPO**	**Fallow**	**Plowing**	**Fallow**	**Plowing**	**Fallow**	**Plowing**
				Soil Physical Properties				
Da10 (g/cm^3)	0.79 a	0.82 a	0.90 a	0.88 a	1.01 a	1.20 b	0.90 a	1.00 a
Da20 (g/cm^3)	0.87 a	0.93 a	0.99 a	0.88 a	1.11 a	1.31 a	1.00 a	1.14 b
Da30 (g/cm^3)	1.09 a	0.94 a	1.00 a	0.92 a	1.31 a	1.27 a	1.03 a	1.13 a
P10 (%)	70.22 a	69.14 a	65.99 a	66.65 a	61.97 b	54.64 a	66.10 a	62.25 a
P20 (%)	67.00 a	65.01 a	62.70 a	66.86 a	58.75 a	50.54 a	62.26 b	56.72 a
P30 (%)	58.71 a	64.42 a	62.10 a	65.19 a	58.30 a	52.04 a	61.03 a	57.18 a
H10 (%)	34.59 b	29.87 a,b	29.79 a,b	24.49 a	42.10 b	18.04 a	19.28 a	16.93 a
H20 (%)	32.31 b	31.70 a,b	28.51 a	28.23 a	40.54 b	17.57 a	22.03 a	20.02 a
H30 (%)	29.04 a	46.01 a	25.29 a	28.13 a	44.67 b	20.84 a	22.03 a	22.10 a
MA (%)	34.02 a,b	58.37 b	19.82 a	17.46 a	54.51 b	40.94 a	59.00 a	43.46 a

Note: Da "n": Bulk density at "n" (cm) soil depth; P "n": Rate of porosity at "n" (cm) soil depth; H "n": Soil moisture at "n" (cm) soil depth; the means of the same row followed by the same letter are not statistically different ($p < 0.05$).

For the porosity (P) at the level of these three provinces, an expected opposite trend to that of the evolution of bulk density was observed, with the lowest values in Taounate (respectively 62% and 55% on the surface for fallow and plowing) where densities were high. The porosity increased from surface to depth. However, the porosity values in the plots covered by vegetation (matorral and fallow) are not statistically different except in Taounate.

The soil moisture content (H) at the level of the provinces of study increased with depth. These values presented significant differences in covered and non-plowed plots, even if these values cannot be directly linked to erosion control.

However, on the surface and at depth (0–10 cm), this rate remains high at the level of fallow land compared to plowing. Thus, the fallow land in Taounate had 42.10% soil moisture, followed by that of Larache with a soil moisture rate of 29.79%, and finally that of Taza with a soil moisture rate of 19.28%. The lowest soil moisture levels exist at the plowing level and vary from 16% to 25% for these three regions.

As for the stable macroaggregate rate (MA), there are no significant statistical differences in terms of plowing in the three provinces. However, these values are statistically significantly different in plots covered by vegetation or litter. They are high in all fallows in the three provinces compared to plowing.

3.1.3. Effects of Land Uses on Soils Hydrological Properties

From the point of view of hydrological properties (Table 6), the values of the final infiltration (If) do not show a statistically significant difference at the level of fallows despite the existence of runoff. The final infiltration remains higher in the plowings than in the fallows at the level of the three provinces. Indeed, the plowing being freshly stirred is very loose and very open, as demonstrated by P (%) values. This state favored infiltration in all the plowing perimeters in the three regions after one hour of simulation. The furrows along contour lines have created a roughness favorable to infiltration, especially in Taounate and Taza, where the runoff coefficient is zero and without detachability.

In Larache, since the plowing was carried out on the steepest slope, this situation favors runoff (15%) and soil detachability (29 g/m^2/h).

Table 6. Effects of land uses on the hydrological properties of soil of the FAP-MCA perimeters in Larache, Taounate, and Taza provinces in Morocco.

Provinces		Larache		Taounate		Taza	
Land Uses	Fallow	Plowing	Matorral	Fallow	Plowing	Fallow	Plowing
			Soil Hydrological Properties				
If (mm/h)	39.68 a	64.70 ab	74.4 b	55.52 a	73.43 a	69.36 a	71.53 a
Pi (mm/h)	4.74 a	24.03 a	74.4 b	36.57 a	73.43 a	46.12 a	71.53 a
Kr (%)	46.61 b	14.63 a	0.00 a	13.18 a	0.00 a	2.23 b	0.00 a
D (g/m^2/h)	2.28 a	29.21 a	0.00 a	1.07 a	0.00 a	0.20 a	0.00 a

Note: If—final infiltration; Pi—imbibition of rainwater; Kr—rate of runoff coefficient; D—detachability. The means of the same row followed by the same letter are not statistically different ($p < 0.05$).

For the fallow sites, we found a higher runoff coefficient but fewer eroded particles in the direction of the slopes than in cultivated soils.

In Larache province, the vegetation under matorral seems to promote infiltration as much as possible (If = 74.4 mm/h) and reduce runoff and detachability as much as possible (D = 0 g/m^2/h).

3.1.4. Influences of Soil Parameters on Their Hydrological Properties in Larache, Taounate, and Taza Provinces of Morocco

The highlighting of the influence of soil parameters on their hydrological properties was conducted by studying the correlation between the hydrological properties and the explanatory variables (Table 7).

Table 7. Relationship between hydrological properties (If, Pi, Kr, and D) and physical parameters and soil surface condition in Larache province.

Explanatory Variables	Regression Equations	R^2
	Final infiltration: If (mm/h)	
SO	If = −50.352 + 1.289 × SO	0.785
PEN	If = 85.263 − 49.559 × PEN	0.696
SS	If = 127.589 − 33.978 × SS	0.653
P20	If = −254.402 + 4.771 × P20	0.428
	imbibition rainwater: Pi (mm/h)	
SO	Pi = −107.779 + 1.616 × SO	0.619
PEN	Pi = 62.387 − 62.319 × PEN	0.550
P10	Pi = −499.655 + 7.865 × P10	0.446
	Runoff coefficient: Kr (%)	
SO	Kr = 134.580 − 1.326 × SO	0.696
PEN	Kr = −7.292 + 55.027 × PEN	0.564
SS	Kr = − 47.579 + 34.514 × SS	0.717
	Detachability: D (g/m^2/h)	
	<60%	

Note: If—final infiltration; Pi—imbibition rainwater; Kr—runoff coefficient; D—detachability; SO—open surface; PEN—resistance to penetrometry; SS—resistance to shear strength; P "n"—porosity at "n" cm of soil depth; R^2—correlation coefficient.

Thus, the variables considered more explanatory of hydrological properties (final infiltration (If), imbibition of rainwater (Pi), runoff coefficient (Kr), and detachability (D)) are those whose coefficient of determination (R^2) is greater than 60%. The variables that influenced them were different in the investigated provinces. The final infiltration was positively correlated with soil cohesion parameters (open surface and porosity) and negatively correlated with settlement ones. The open surface explained its variation at 79%, and resistance to penetration (PEN) explained it at 70%. In the province of Taounate, the final infiltration is positively correlated with the roughness index (Ir), which explains it at 93%, and negatively correlated with the shear strength (SS), which explains it at 86%. For Taza, the final infiltration is mainly explained by the roughness index (Ir) and the surface porosity at a soil depth of 0 to 10 cm (P10).

Concerning the imbibition of rainwater (Pi), the open surface with a positive correlation is the only variable that explains it to more than 60% in Larache. In the province of Taounate, its variation is explained by three parameters at more than 74%. Thus, two of these parameters were negatively correlated with Pi, such as the shear strength (SS) and the resistance to penetrometry (PEN). They explained its variation, respectively, at 94% and 87%. However, the correlation with the roughness index (Ir) was positive and explained Pi at 75%. For Taza province, the open surface (SO), the roughness index (Ir), and the porosity at a soil depth of 0 to 10 cm (P10) exerted their influence with respective rates of 87%, 84%, and 73% with a positive correlation.

For the runoff coefficient (Kr), in Larache province, it is positively correlated to shear strength (SS) and negatively correlated to the open area (SO) that influenced it, respectively, at 72% and 70%. In Taounate, the roughness index (Ir), with its negative correlation to Kr, and the shear strength (SS), with a positive correlation, explained their influence on Kr, respectively, at 85% and 79%. For Taza province, the highest correlations for Kr were obtained with the open surface (SO) and the roughness index (Ir), at 88% and 79%, respectively.

In these three provinces, the three hydrological parameters (If, Pi, and Kr) were explained by the parameters of soil cohesion (SO, Ir, and P10), which allow water to infiltrate. In addition, they are also explained by the settlement parameters (SS and PEN), which promote runoff by reducing the infiltration time.

For detachability, all variables explained it to less than 60% in Larache (see Figure 3 and Table 7). However, in Taounate (see Figure 4 and Table 8), it is more positively correlated with shear strength (SS), surface of 0–10 cm (H10) soil moisture, and resistance to penetrometry (PEN). The latter explain it respectively at 99%, 85%, and 84%. In addition, detachability is negatively correlated with the roughness index (Ir), which explains its 82% variation. Finally, in Taza (see Figure 5 and Table 9), the correlations for detachability were negative for all three explanatory parameters. However, it is explained by the open area (SO) at 89%, the roughness index (Ir) at 79%, and the surface porosity at a soil depth of 0 to 10 cm (P10) at 66%. Apart from surface soil moisture (H10), which is the only new explanatory parameter of detachability, the others related to cohesion (SO, Ir, and P10) and compaction (SS and PN) had already explained If and Pi in these three provinces. Thus, they are those related to soil compaction that positively influence detachability.

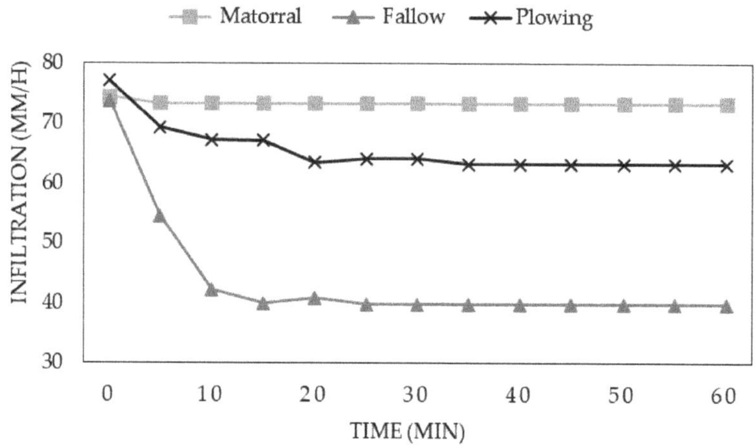

Figure 3. Effect of land use (fallow, plowing, and matorral) on the infiltration capacity [I (mm/h)] of soils in the perimeters studied in Larache province.

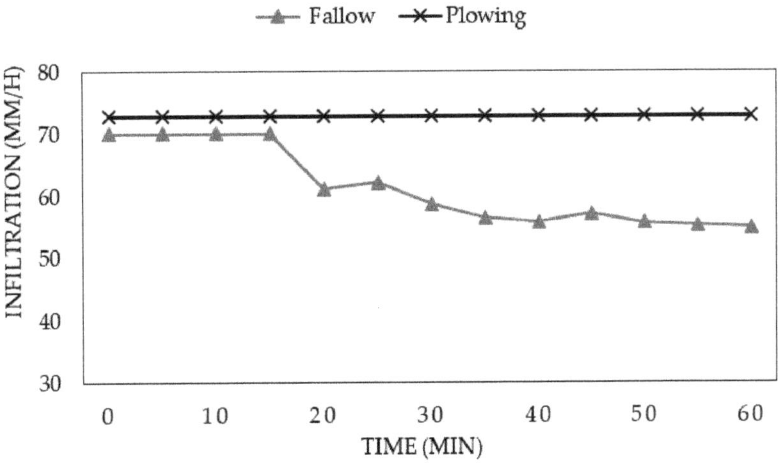

Figure 4. Effect of land use (fallow and plowing) on the infiltration capacity (I (mm/h)) of soils in the perimeters examined in Taounate province.

Table 8. Relationship between hydrological properties (If, Pi, Kr, and D) and physical parameters and soil surface condition in Taounate province.

Explanatory Variables	Regression Equations	R^2
	Final infiltration: If (mm/h)	
PEN	If = 89.09 − 80.699 × PEN	0.464
SS	If = 75.374 − 9.629 × SS	0.858
Ir	If = 0.572 + 2.729 × Ir	0.930
	Imbibition rainwater: Pi (mm/h)	
PEN	Pi = 112.150 − 187.376 × PEN	0.873
SS	Pi = 74.298 − 17.052 × SS	0.938
Ir	Pi = −42.142 + 4.148 × Ir	0.749
	Runoff coefficient: Kr (%)	
PEN	Kr = −12.165 + 61.487 × PEN	0.400
SS	Kr = −1.993 + 7.583 × SS	0.790
Ir	Kr = 56.813 − 2.145 × Ir	0.853
H10	Kr = −8.877 + 0.402 × H10	0.517
	Detachability: D ($g/m^2/h$)	
PEN	D = −1.133 + 5.463 × PEN	0.835
SS	D = −0.056 + 0.521 × SS	0.986
Ir	D = 3.561 − 0.129 × Ir	0.819
H10	D = −0.685 + 0.032 × H10	0.849

Note: If—final infiltration; Pi—imbibition rainwater; Kr—runoff coefficient; D—detachability; PEN—resistance to penetrometry; SS—resistance to shear strength; Ir—roughness index; H10—soil moisture at 10 cm depth; R^2—correlation coefficient.

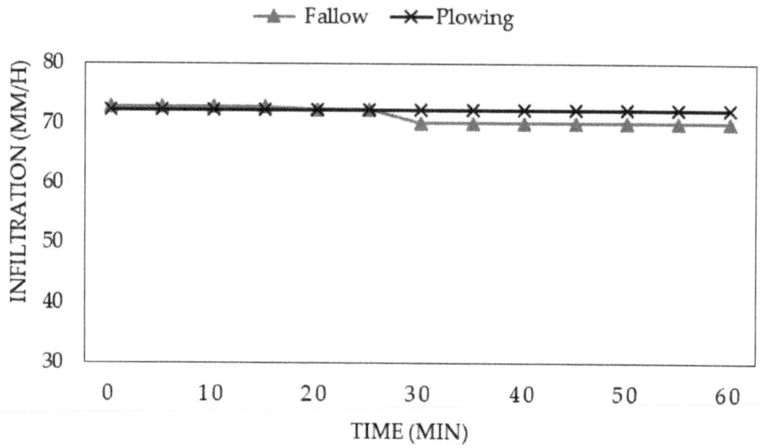

Figure 5. Effect of land use (fallow and plowing) on the infiltration capacity [I (mm/h)] of soils in the perimeters examined in Taza province.

Table 9. Relationship between hydrological properties (If, Pi, Kr, and D) and physical parameters and soil surface condition in Taza province.

Explanatory Variables	Regression Equations	R^2
	Final infiltration: If (mm/h)	
SO	If = 30.777 + 0.429 × SO	0.603
PEN	If = 77.428 − 4.427 × PEN	0.534
Ir	If = 53.539 + 0.853 × Ir	0.824
P10	If = 40.453 + 0.451 × P10	0.819
	Imbibition rainwater: Pi (mm/h)	
SO	Pi = -214.019 + 2.950 × SO	0.874
Ir	Pi = −38.642 + 4.916 × Ir	0.839
P10	Pi = −102.471 + 2.426 × P10	0.726
	Runoff coefficient: Kr (%)	
SO	Kr = 23.542 − 0.242 × SO	0.879
Ir	Kr = 8.865 − 0.391 × Ir	0.788
	Detachability: D ($g/m^2/h$)	
SO	D = 2.107 − 0.022 × SO	0.879
Ir	D = 0.793 − 0.035 × Ir	0.788
P10	D = 1.230 − 0.017 × P10	0.662

Note: If—final infiltration; Pi—imbibition rainwater; Kr—runoff coefficient; D—detachability; SO—open surface; PEN—resistance to penetrometry; Ir—roughness index; P10—porosity at 10 cm soil depth; R^2—correlation coefficient.

3.2. Discussion

3.2.1. Effects of Land Uses on Surface Conditions

In those three provinces, the results obtained on organic matter, which are superior to 2% in plowed sites and superior to 2.60% under matorral or fallow, can be justified considering the prevention of organic matter mineralization during soil aeration following tillage, confirming the previous literature [36].

In this situation, the open surfaces exceed 90% in all three regions. This situation could be explained by the plant cover of the herbaceous carpet and the litter observed, as well as the cracks and galleries created under the effect of the very intense pedofaunal activity at the level of these fallows. In addition, as vertisols are the dominant soils there, cracks due to roots explain these high values.

Indeed, many researchers [37–39] showed that in vertisols with plants, there are much smaller cracks in the rows of these plants. This pattern has been attributed to the proliferation of roots holding the soil together. In addition, vertisols are known to have open cracks at the surface or at depth during some years. These cracks can be at least 1 cm wide and up to 50 cm deep. Those cracks are absent if these soils are irrigated [40]. These high proportions of open and covered surfaces obtained at the fallow level are consistent with these studies, which have shown that fallow land allows organic restitution and stimulates the activity of soil fauna and microflora [41–43] in Africa south of the Sahara, in the Central African Republic [44], and in the Ivory Coast [45].

The highest rate of bare surfaces is observed in the Taza region at 90.83%, and that of open surfaces is obtained in the Taounate region at 90.17%. This could be explained by the fact that the plowed soils were freshly disturbed and that the clods, cracks, and embedded stones would have contributed to the increase in the opening in these regions. These results agree with those of the authors, who specified that the clods crack but that they experience significant variability in the state of cracking and their size after plowing in the compacted treatment [46]. In addition, these cracks and clods are also related to swelling clays found in vertisols [40].

The high observed penetrometer values would be signs of compaction due to trampling by livestock, which has been confirmed by other studies [47–49]. Concerning the results in Taza, previous works show that, in plowed sites, precipitation-led erosion with detaching and displacement of soil particles can lead to the clogging of the interstices of the soil surface, forming a slacking crust a few millimeters deep in thickness [50,51]. The roughness (Ir, %) was much more pronounced at the level of the plowing than the fallows in all three provinces, which is explained by the fact that the clods at the level of the plowing in this study area created microreliefs consisting of random irregularities or oriented by the tillage [52].

3.2.2. Effects of Land Uses on Soil Physical Properties

This situation on the bulk density could be explained by the compaction under the effect of the trampling of animals in matorral and fallow land as well as the effect of the plow pan at field level. In some research work, it has been shown that the passage of agricultural machinery aggravates compaction because the mechanical strength of the aggregates decreases [53,54]. Agricultural tools, in particular the plow, can induce the formation of the slaking crust down to a depth of 20 to 30 cm [55].

For porosity (P), the situation observed is in line with the research of Richard et al. [56], which attests that compaction reduces the pores in the soil. Therefore, compaction increases the bulk density of the soil [57,58] and decreases porosity. Porosity values remained higher in plots under plant cover than in plowed ones. Some studies qualify this porosity as of biological origin because the presence of plant residues on the surface and the absence of mechanical disturbance favor the installation of earthworms, which create cavities or galleries [59–61].

For the soil moisture content (H), vegetation, litter, and surface pebbles hinder evaporation, which would explain the high soil moisture rate in Taounate in fallow land. At the level of plowing, evaporation is relatively higher, which reduces the soil moisture rate. Several authors have found similar results, which testify that under vegetation cover or mulch, there is better conservation of soil moisture compared to plowed and uncovered soils because they prevent water evaporation [62,63]. The values for stable macroaggregate rate (MA) obtained in all fallows remain high in the three provinces compared to plowing. Some authors have shown that working the soil would reduce its structural stability [64].

3.2.3. Effects of Land Uses on Soils Hydrological Properties

Regarding the soil hydrological properties, the final infiltration has been influenced by the surface state. So, at Taounate and Taza, the runoff coefficient was zero. The irregularities created by the clods and the roughness due to the used tools allow the storage of the water volume in their orientation [65]. In addition, Lipiec et al. [60] testify that the macroporosity is more important at the level of the plowing and that the infiltration passes through this structural porosity. From there, these results in Larache are confirmed by Lipiec et al. [60] on the importance of macroporosity and microreliefs in plowing because infiltration reached 85%. These results are in the same direction as those of other authors who specify that, on cultivated soils, the pores are clogged by the fractionated aggregates because of the mechanical disintegration due to the raindrops. There is then the formation of the capping crust, which reduces infiltration [66] because the reduction in roughness leads to a reduction in the stored water volume [67]. The diminished microreliefs will then be exceeded, and there will be runoff oriented in the direction of the roughness [65], and indeed, in Larache, this orientation is according to the slope. This same author also explained the tearing mechanism of the particles during runoff in cultivated soil. It can be diffuse or in a concentrated way, as in the thalwegs, and the soil particles are torn by the force exerted on its bed to form rills or gullies as well as the incisions [65], from where this high detachability of 29 $g/m^2/h$ comes.

These results are similar to those obtained by other researchers who have shown that uncultivated soils are more effective in controlling erosion than runoff [68]. Additionally, uncultivated soils have low microporosity, encountered mainly at the level of plowing [61]. The latter is of biological origin [60]. This high runoff can also be explained by the fact that at depth there is already high soil moisture; hence, it should have been due to the soil, which was close to saturation [69]. In addition, the density that was observed at the level of the fallow soils would have been due to the compaction. Therefore, the decrease in porosity facilitated runoff. The low loss of sediments in these fallows cannot be explained by the fact that the raindrops' kinematic energy is intercepted by the plant cover and the litter [70], and the water reaches the ground with lower energy and is less disruptive of soil structure [71].

In Larache province, results on infiltration and runoff with detachability have been influenced by cracks due to the roots of plants, as well as the biological porosity that is created when the soil is not mechanically disturbed and dry, which favor infiltration more than runoff [60,72,73].

3.2.4. Influences of Soil Parameters on Hydrological Properties in Larache, Taounate, and Taza Provinces of Morocco

In these three provinces of the study area, the three hydrological parameters (If, Pi, and Kr) were explained by the parameters of soil cohesion (SO, Ir, and P10), which allow water to infiltrate. In addition, they are also explained by the settlement parameters (SS and PEN), which promote runoff by reducing the infiltration time. These results agree with those of Lipiec et al. [60], who show that macroporosity and microreliefs facilitate infiltration. As for the settlement parameters, Richard et al. [56] showed that settlement disconnects the pores by decreasing the path taken by the infiltration, and hence the imbibition of rainwater decreases. When the percentage of pores decreases in the soil, Assouline et al. [74] say that the water conductivity goes to saturation, which increases the runoff coefficient (Kr).

About detachability, apart from surface soil moisture (H10), which is the only new explanatory parameter of detachability, the others related to cohesion (SO, Ir, and P10) and compaction (SS and PN) had already explained If and Pi in these three provinces. Thus, they are those related to soil compaction that positively influence detachability. These reduce porosity and facilitate runoff [56], which tears soil particles from plowed soil in the direction of the slope. Additionally, on that plowed soil, the slaking crust forms, clogs the pores, and facilitates runoff, which removes the soil, leaving rills or gullies as well as scratches [65].

4. Conclusions

This research was carried out as part of the Defense and Restoration of Soils (DRS) by fruit trees using the olive tree in the Moroccan Rif.

The results of this study allowed us to highlight the importance of plant cover (herbaceous stratum) in increasing the water balance of the soil. Indeed, final infiltration (If) and imbibition of rainwater (Pi) were high in the matorral of Larache (74.4 mm/h), lower in the plowed olive orchard, and even lower in the level of the fallow planted with olive trees. In addition, the role of vegetation cover in protecting soil against erosion has been elucidated. Although the fallow soils present a higher runoff coefficient (Kr = 46.61%), they were less detachable (2.28 $g/m^2/h$) than the plowed soils (30 $g/m^2/h$).

Regarding the hydrological characteristics (If, Pi, Kr, and D), they were favored positively by a high roughness of the ground in the direction of the contour lines in Taounate and Taza. Moreover, in the three regions, they were positively influenced by the open surface and the porosity. All these soil characteristics can be used for the choice of development and for modeling the risks of runoff and erosion since they are easily measurable for future research in the study area. As a further improvement in terms of soil erosion reduction and land management, considering the role of plant cover, it could be suggested to alternate plowing and fallow in this fruit tree to promote restructuring and enrichment in organic matter and fight against erosion.

Author Contributions: All authors: Conceptualization, methodology, validation, formal analysis, investigation, data curation, writing—original draft preparation; writing—review and editing. All authors have read and agreed to the published version of the manuscript.

Funding: This research received funding from the Millennium Challenge Account project (MCA project).

Data Availability Statement: The data that support the findings of this study are available upon request from the corresponding authors.

Conflicts of Interest: The authors declare no conflict of interest.

References

1. García-Ruiz, J.M.; Nadal-Romero, E.; Lana-Renault, N.; Beguería, S. Erosion in Mediterranean landscapes: Changes and future challenges. *Geomorphology* **2013**, *198*, 20–36. [CrossRef]
2. Raclot, D.; Le Bissonnais, Y.; Annabi, M.; Sabir, M.; Smetanova, A. Main Issues for Preserving Mediterranean Soil Resources from Water Erosion Under Global Change. *Land Degrad. Develop.* **2018**, *29*, 789–799. [CrossRef]
3. FAO. Conservation des ressources naturelles en zones arides. *Cah. FAO Conserv. Des. Sols.* **1990**, *50*, 62.
4. Antipolis, S. *Les Menaces sur les Sols dans les Pays Méditerranéens. Etude Bibliographique*; Cahiers du Plan Bleu; Plan Bleu: Marseille, France, 2003; Volume 2, p. 80.
5. Chaibou, A. *Etude des Impacts des Utilisations et des Autres Usages des Terrains Forestiers sur le Comportement Hydrologique et L'érosion des Sols dans la Zone D'amsitten en Vue de Son Aménagement*; Mémoire de 3ème cycle; ENFI: Salé, Morocco, 2010; p. 104.
6. Liu, G.; Tian, F.X.; Warrington, D.N.; Zheng, S.Q.; Zhang, Q. Efficacy of grass for mitigating runoff and erosion from an artificial loessial earthen road. *Trans. ASABE* **2010**, *53*, 119–125. [CrossRef]
7. Dan, C.X.; Liu, G.; Zhao, Y.G.; Shu, C.B.; Shen, E.S.; Liu, C.; Tan, Q.H.; Zhang, Q.; Guo, Z.; Zhang, Y. The effects of typical grass cover combined with biocrusts on slope hydrology and soil erosion during rainstorms on the Loess Plateau of China: An experimental study. *Hydrol. Process.* **2022**, *37*, e14794. [CrossRef]
8. Moldovan, M.; Tăut, I.; Rebrean, F.A.; Szilard, B.; Arion, I.D.; Dîrja, M. Determining the Anti-Erosion Efficiency of Forest Stands Installed on Degraded Land. *Sustainability* **2022**, *14*, 15727. [CrossRef]
9. Girona-García, A.; Vieira, D.C.; Silva, J.; Fernández, C.; Robichaud, P.R.; Keizer, J.J. Effectiveness of post-fire soil erosion mitigation treatments: A systematic review and meta-analysis. *Earth-Sci. Rev.* **2021**, *217*, 103611. [CrossRef]
10. Parhizkar, M.; Shabanpour, M.; Lucas-Borja, M.E.; Zema, D.A.; Li, S.; Tanaka, N.; Cerda, A. Effects of length and application rate of rice straw mulch on surface runoff and soil loss under laboratory simulated rainfall. *Int. J. Sediment Res.* **2021**, *36*, 468–478. [CrossRef]
11. Al Ali, Y. Structure et Evolution de la Terre et des Autres Planètes. Ph.D. Thesis, Univ de Montpellier II Sciences et Techniques du Languedoc, Ecole Doctorale SIBAGHE, Université Montpellier 2, Montpeiller, France, 2007; p. 178.
12. Loukili, M.; Lahlou, O.; Abouyaala, M.A.; Allaoui, M.; Ozer, A.; Salmon, M. Intérêt de la carte d'aptitude des terres dans la lutte anti-érosive par la DRS fruitière, Cas de l'amandier dans le cercle d'Aknoul, Maroc, Rif oriental. *Hommes Terre l'Eau* **2006**, *133*, 50–54.
13. Heusch, B. Erosion du Pré-rif,une étude comparative de l'érosion hydraulique dans les collines marneuses du pré-rif occidental, annales de la recherche forestière au Maroc. *SRF Rabat* **1970**, *12*, 9–176.
14. Sabir, M.; Roose, E.; Merzouk, A.; Nouri, A. Techniques traditionnelles de gestion de l'eau et de lutte antiérosive dans deux terroirs du Rif occidental (Maroc). *Bull. Réseau Eros.* **1999**, *19*, 456–471.

15. Nafi, M. *Part de L'arboriculture Fruitière dans le Revenu des Exploitations Agricoles du Bassin Versant Moulay Bouchta*; Mémoire de fin d'étude; ENFI: Salé, Morocco, 2005; p. 117.
16. Petroselli, A.; Apollonio, C.; De Luca, D.L.; Salvaneschi, P.; Pecci, M.; Marras, T.; Schirone, B. Comparative Evaluation of the Rainfall Erosivity in the Rieti Province, Central Italy, Using Empirical Formulas and a Stochastic Rainfall Generator. *Hydrology* **2021**, *8*, 171. [CrossRef]
17. Todisco, F.; Vergni, L.; Ortenzi, S.; Di Matteo, L. Soil Loss Estimation Coupling a Modified USLE Model with a Runoff Correction Factor Based on Rainfall and Satellite Soil Moisture Data. *Water* **2022**, *14*, 2081. [CrossRef]
18. Chen, W.; Huang, Y.C.; Lebar, K.; Bezak, N. A systematic review of the incorrect use of an empirical equation for the estimation of the rainfall erosivity around the globe. *Earth-Sci. Rev.* **2023**, *238*, 104339. [CrossRef]
19. Tauro, F.; Cornelini, P.; Grimaldi, S.; Petroselli, A. Field studies on the soil loss reduction effectiveness of three biodegradable geotextiles. *J. Agric. Eng.* **2018**, *49*, 117–123. [CrossRef]
20. Apollonio, C.; Petroselli, A.; Tauro, F.; Cecconi, M.; Biscarini, C.; Zarotti, C.; Grimaldi, S. Hillslope Erosion Mitigation: An Experimental Proof of a Nature-Based Solution. *Sustainability* **2021**, *13*, 6058. [CrossRef]
21. Doré, L.M. *Effets de la Transformation des Ecosystèmes sur les Comportements Hydrodynamiques des Sols*; ENFI: Salé, Morocco, 2006; p. 166.
22. Al Karkouri, J.; Laouina, A.; Roose, E.; Sabir, M. Capacité d'infiltration et risques d'érosion des sols dans la vallée des Beni Boufrah-Rif central (Maroc). *Eros. Mont. Semi-Arid. Méditerranéennes* **2000**, *20*, 342–355.
23. Sabir, M.; Barthès, B.; Roose, E. Recherche d'indicateurs des risques de ruissellement et d'érosion sur les principaux sols des montagnes méditerranéennes du Rif occidental (Maroc). *Sci. Changements Planétaires/Sécheresse* **2004**, *15*, 105–110.
24. Simonneaux, V.; Cheggour, A.; Deschamps, C.; Mouillot, F.; Cerdan, O.; Le Bissonnais, Y. Land use and climate change effects on soil erosion in a semi-arid mountainous watershed (High Atlas, Morocco). *J. Arid. Environ.* **2015**, *122*, 64–75. [CrossRef]
25. Xiong, M.; Sun, R.; Chen, L. Effects of soil conservation techniques on water erosion control: A global analysis. *Sci. Total Environ.* **2018**, *645*, 753–760. [CrossRef]
26. El Assaoui, N.; Bouiss, C.E.; Sadok, A. Assessment of Water Erosion by Integrating RUSLE Model, GIS and Remote Sensing–Case of Tamdrost Watershed (Morocco). *Ecol. Eng. Environ. Technol.* **2023**, *24*, 43–53. [CrossRef]
27. Sadiki, A.; Faleh, A.; Navas, A.; Bouhlassa, S. Assessing soil erosion and control factors by the radiometric technique in the Boussouab catchment, Eastern Rif, Morocco. *Catena* **2007**, *71*, 13–20. [CrossRef]
28. Moore, H.; Fox, H.; Harrouni, M.; Alami, A. Environmental challenges in the Rif mountains, northern Morocco. *Environ. Conserv.* **1998**, *25*, 354–365. [CrossRef]
29. Henri, P. Etude de la conception des techniques paysannes de Gestion et Conservation de l'Eau et des Sols dans un territoire du pré-rif Marocain. In *Rapport de Stage en 2ème Année à l'IUT Génie Biologique de Saint Etienne*; France, 2012; p. 49.
30. Roose, E. Méthodes de mesure des états de surface du sol, de la rugosité et des autres caractéristiques qui peuvent aider au diagnostic de terrain des risques de ruissellement et d'érosion, en particulier sur les versants cultivés des montagnes. *Bull.-Réseau Eros.* **1996**, *16*, 87–96.
31. Walkley, A.J.; Black, I.A. Estimation of soil organic carbon by the chromic acid titration method. *Soil Sci.* **1934**, *37*, 29–38. [CrossRef]
32. Blake, G.R. Bulk Density. In *Methods of Soil Analysis: Part 1 Physical and Mineralogical Properties, Including Statistics of Measurement and Sampling*; Black, C.A., Ed.; American Society of Agronomy, Inc.: Washington, DC, USA, 1965; pp. 374–390.
33. Globe. Protocole de mesure d'humidité du sol par gravimétrie. 2005; 19.
34. Pojasok, T.; Kay, B.D. Assessment of a combination of wet sieving and turbidimetry to characterize the structural stability of moist aggregates. *Can. J. Soil Sci.* **1990**, *70*, 33–42. [CrossRef]
35. Roose, E.; Smolikowski, B. Comparaison de trois techniques de mesure de l'infiltration sur fortes pentes: Monocylindre et 2 simulateurs de pluies. Application à un versant de la vallée de Godim au Cap Vert. *Bull. Réseau Eros.* **1997**, *17*, 282–296.
36. Scheffé, H. *The Analysis of Variance*; John Wiley & Sons: Hoboken, NJ, USA, 1999.
37. Johnston, J.R.; Hill, H.O. A study of the shrinking and swelling properties of the rendzina soils. *Soil Sci. Soc. Am. Proc.* **1944**, *9*, 24–29. [CrossRef]
38. Johnson, W.C. Controlled soil cracking as a possible means of moisture conservation on wheatlands of the southwestern Great Plains. *Agron. J.* **1962**, *54*, 323–325. [CrossRef]
39. Renard, K.G.; Keppel, R.V.; Hickey, J.J.; Wallace, D.E. Performance of local aquifers as influenced by stream transmission losses and riparian vegetation. *Trans. ASAE* **1964**, *7*, 471–474. [CrossRef]
40. U.S. Soil Taxonomy, Soil Conservation Service. Soil Survey Staff Soil Taxonomy: A basic system of soil classification for making and interpreting soil surveys. In *USDA Agric. Handbook No. 436*; National Bureau of Soil Survey and Land Use Planning (ICAR), India Offset Press: New Delhi, India, 1978; p. 754.
41. Pieri, C. Fertilité des terres de savane. In *Bilan de Trente ans de Recherche et de Développement Agricole au Sud du Sahara*; Ministère Français de la Coopération et du Développement: Paris, France, 1989; p. 444.
42. Hien, V.; Sedogo, M.P. Étude des effets de la jachère de courte durée sur la production et l'évolution des sols dans différents systèmes de culture du Burkina Faso. In Proceedings of the La jachère en Afrique de l'Ouest. Atelier International, Montpellier, France, 2–5 December 1991; Editions ORSTOM. pp. 221–232.
43. Roose, E. *Introduction à la Gestion Conservatoire de L'eau, de la Biomasse el de la Fertilité des Sols (GCES)*; Bulletin Pédologique F.A.O.: Montpellier, France, 1994; n. 70; p. 420.

44. Morel, R.; Quantin, P. Observation sur l'évolution à long termede la fertilité des sols cultivés à Grimari (Rép. Centrafricaine). *Série* **1972**, *27*, 667–739.
45. Valentin, C. État de dégradation de 2 terroirs Sénoufo-Nord Côte-d'Ivoire, Abidjan, Iirsda, 1989, 9p. multigr.
46. Coulomb, I.; Caneill, J.; et Manichon, H. Comportement du sol au labour: Méthode d'analyse et évaluation des conséquences de l'état initial du sol sur l'état transformé par le labour Agronomie. *EDP Sci.* **1992**, *13*, 45–56. [CrossRef]
47. Lal, R.; Reicosky, D.C.; Hanson, J.D. Evolution of the plow over 10,000 years and the rationale for no (till farming. *Soil Tillage Res.* **2007**, *93*, 1–12. [CrossRef]
48. Traoré Kalifa, B.; Mc Carthy, G.; Gigou Jacques Doumbia Mamadou, D.; Bagayoko, A.; Yost Russell, S.; Konaré, H.; Dioni, L.; Coulibaly, H.; Sidibe, A.; Kablan Richard, A. Aménagement en courbes de niveau et conservation du carbone. In: Gestion de la biomasse, érosion et séquestration du carbone. Séquestration du carbone et érosion des sols. In *Land Use, Erosion and Carbon Sequestration. 2. Soil Erosion and Carbon Sequestration*; Roose, E., De Noni, G., Prat, C., Ganry, F., Bourgeon, G., Eds.; IRD, CIRAD-AMIS-AGRONOMIE: Paris, France, 2004; pp. 568–577.
49. Tribak, A. Contraintes du milieu et fragilité d'un espace montagnard marocain: Les montagnes du Prérif oriental. *Ann. Géographie* **2002**, *625*, 227–245. [CrossRef]
50. Boiffin, J. La Dégradation Structurale des Couches Superficielles du Sol sous L'action des Pluies. Ph.D. Thesis, Sciences Agronomiques, INA Paris-Grignon, Paris, France, 1984; p. 423.
51. Bresson, L.M.; Boiffin, J. Morphological characterization of soil crust development stages on an experimental field. *Geoderma* **1990**, *47*, 301–325. [CrossRef]
52. El Mazi, M.; El-fengour, M.; Houari, A. The influence of clearing followed by cultivation on the fertility and stability of a fersiallitic soil in the central Rif Morocco. *J. Appl. Sci. Environ. Stud.* **2019**, *2*, 113–125.
53. Richard, G.; Boizard, H.; Roger-Estrade, J.; Boiffin, J.; Guerif, J. Field study of soil compaction due to traffic in northern France: Pore space and morphological analysis of the compacted zones. *Soil Tillage Res.* **1991**, *51*, 151–160. [CrossRef]
54. Hamza, M.A.; Anderson, W.K. Soil compaction in cropping systems-A review of the nature, causes and possible solutions. *Soil Tillage Res.* **2005**, *82*, 121–145. [CrossRef]
55. Tebrugge, F.; During, R.A. Reducing tillage intensity-a review of results from a long-term study in Germany. *Soil Tillage Res.* **1999**, *53*, 15–28. [CrossRef]
56. Richard, G.; Cousin, I.; Sillon, J.F.; Bruand, A.; Guerif, J. Effect of compaction on the porosity of a silty soil: Influence on unsaturated hydraulic properties. *Eur. J. Soil Sci.* **2001**, *52*, 49–58. [CrossRef]
57. Voorhees, W.B. The effect of soil compaction on crop yield. In *Agricultural Research Service*; SAES: Milan, Italy, 1986; pp. 1–7.
58. Reeder, R.C.; Hakansson, I. Subsoil compaction by vehicles with high axial load extent, persistence and crop response. *Soil Tillage Res.* **1994**, *29*, 277–304.
59. Wahl, N.A.; Bens, O.; Buczko, U.; Hangen, E.; Huttl, R.F. Effects of conventional and conservation tillage on soil hydraulic properties of silty-loamy soil. *Phys. Chem. Earth* **2004**, *29*, 821–829. [CrossRef]
60. Lipiec, J.; Kus, J.; Slowinska-Jurkiewicz, A.; Nosalewicz, A. Soil porosity and water infiltration as influenced by tillage methods. *Soil Tillage Res.* **2006**, *89*, 210–220. [CrossRef]
61. Strudley, M.W.; Green, T.R.; Ascough, J.C. Tillage effects on soil hydraulic properties in space and time: State of the science. *Soil Tillage Res.* **2008**, *99*, 4–48. [CrossRef]
62. Dounias, I. Etudes et Travaux n 19, les systèmes de culture à base de couverture végétale et semis direct en zones tropicales: Synthèse bibliographique. *Cent. Natl. D'etudes Agron. Régions Chaudes* **2001**, 164.
63. Scopel, E.; Findeling, A.; Chavez-Guerra, E.; Corbeels, M. Impact of direct sowing mulch-based cropping systems on soil carbon, soil erosion, and maize yield. *Agron. Sustain. Dev.* **2005**, *25*, 425–432. [CrossRef]
64. Mesfin, S.; Oliveira Almeida, L.A.; Yazew, E.; Bresci, E.; Castelli, G. Spatial Variability of Soil Moisture in Newly Implemented Agricultural Bench Terraces in the Ethiopian Plateau. *Water* **2019**, *11*, 2134. [CrossRef]
65. Armand, R. Étude des Etats de Surface du Sol et de Leur Dynamique pour Différentes Pratiques de Travail du Sol. Mise au Point d'un Indicateur de Ruissellement. Ph.D. Thesis, Université de Strasbourg, Strasbourg, France, 2009; p. 210.
66. Cosandey, C.; Robinson, M. *Hydrologie Continentale*; Éditions Armand COLIN: Paris, France, 2000; p. 360.
67. Darboux, F.; Gascuel-Odoux, C.; Davy, P. Effects of surface water storage by soil roughness on overland flow generation. *Earth Surf. Process. Landf.* **2002**, *27*, 223–233. [CrossRef]
68. Kwaad, F.J.P.M.; Van der Zijp, M.; Van Dijk, P.M. Soil conservation and maize cropping Systems on slopping loess soils in the Netherlands. *Soil Tillage Res.* **1998**, *46*, 13–21. [CrossRef]
69. Cappus, P. Bassin expérimental d'Alrance-étude des lois de l'écoulement-application au calcul et à la prévision des débits. *Houille Blanche* **1960**, 493–520. [CrossRef]
70. MidWest Plan Service (MWPS). *Conservation Tillage Systems and Management*; Agricultural and Biosystems Engineering Department, Iowa State University: Ames, IA, USA, 2000; p. 270.
71. Abrahams, A.D.; Parsons, A.J.; Wainwright, J. Resistance to overland flow on semiarid grassland and shrubland hillslopes, Walnut Gulch, Southern Arizona. *J. Hydrol.* **1994**, *156*, 431–446. [CrossRef]
72. Hangen, E.; Buczko, U.; Bens, O.; Brunotte, J.; Huttl, R.F. Infiltration patterns into two soils under conventional and conservation tillage: Influence of the spatial distribution of plant roots structures and soil animal activity. *Soil Tillage Res.* **2002**, *63*, 181–186. [CrossRef]

73. Anken, T.; Weisskopf, P.; Zihlmann, U.; Forrer, H.; Jansa, J.; Perhacova, K. Long-term tillage system effects under moist cool conditions in Switzerland. *Soil Tillage Res.* **2004**, *78*, 171–183. [CrossRef]
74. Assouline, S.; Taveres-Filho, J.; Tessier, D. Effect of compaction on soil physical and hydraulic properties: Experimental results and modeling. *Soil Sci. Soc. Am. J.* **1997**, *61*, 390–398. [CrossRef]

water

MDPI

Article

Internal Erosion Stabilization of Cohesionless Soil Using Lime

Shaziya Banu and Mousa Attom *

Department of Civil Engineering, American University of Sharjah, Sharjah P.O. Box 26666, United Arab Emirates; g00059462@alumni.aus.edu
* Correspondence: mattom@aus.edu

Abstract: Soil embankments are valuable for the adequate reserve and supply of water to multiple industries. However, they are susceptible to internal soil erosion, which may ultimately lead to structural collapse. To counteract this issue, soil stabilization is practiced in the construction industry. This paper proposes the internal erosion stabilization of cohesionless soil using quicklime. For this research, two cohesionless soil types were investigated and treated with quicklime: poorly graded and well-graded cohesionless soils. For poorly graded soil, the lime percentage varied from 0.0% to 6.0% based on the soil's weight, while for well-graded soil, it ranged from 0.0% to 3.0%. All the soil specimens were cured for 24 h and tested using the hole erosion test (HET) to replicate the internal erosion effortlessly. The analyzed results demonstrated the efficiency of quicklime as an internal erosion stabilizer for cohesionless soils. The optimum lime content for poorly graded cohesionless soils was 5.0%; for well-graded, the percentage was approximately 3.0%. Moreover, adding lime significantly improved the strength, critical shear stress, and erosion rate index of the soil.

Keywords: hole erosion test (HET); internal erosion; piping erosion; poorly graded cohesionless soil; soil stabilization; well-graded cohesionless soil

check for updates

Citation: Banu, S.; Attom, M. Internal Erosion Stabilization of Cohesionless Soil Using Lime. *Water* **2023**, *15*, 1992. https://doi.org/10.3390/w15111992

Academic Editors: Vito Ferro and Alessio Nicosia

Received: 20 April 2023
Revised: 10 May 2023
Accepted: 19 May 2023
Published: 24 May 2023

1. Introduction

The adequate availability and supply of water is essential for the effective operation of industries such as construction and agriculture. Earthen embankment structures generally provide proper water sources to these industries by alternating, averting, or reserving water [1]. Thus, the structural failure of earthen embankments can lead to a catastrophe, destroying multiple lives and properties [2]. Luthi [3] argued that approximately 30–50% of embankment failures and mishaps are caused by the internal erosion of the soil. Even Hanneman [4] and Foster et al. [5] found that internal erosion is the most prominent reason for embankment failures. Internal soil erosion is triggered by soil particle erosion due to water penetration through the structure [3,6]. A continuous flow of water through the soil leads to the formation of holes in the structure, resulting in structural failure. Hence, seepage is considered a threat to earthen structures during construction's design and operation stages. Therefore, engineers are advised to be cautious and aware of the properties of materials used in earthen structures [2]. Due to the gravitas of this issue, recent studies [7,8] have emphasized the effects of internal erosion of soil on earthen structures and their counteractive measures. Many studies [9–12] have investigated the factors affecting internal soil erosion, such as soil type, water, and material properties, and soil-piping resistance relationships. For dispersive soil, soil resistance against erosion rate increases with higher salt content in the water [9]. To measure the embankment soil erosion parameters, hole erosion test (HET) and slot erosion test (SET) methods were established by Wan and Fell [13]. Soil investigation determined that factors such as fine soil and clay content, clay mineralogy, dry density, compaction water content, dispersity, degree of saturation, plasticity, and cementing materials (such as iron oxides) affect their shear strength parameters, permeability, settlement, and erosion rate values [14–16]. In addition, it was concluded that higher erosion with reduced critical shear stress occurs in non-cohesive and coarse-grained soils compared to fine-grained soils.

To prevent embankment dam failure, stabilization techniques are performed on the soil, increasing the soil strength and restricting the formation of holes in the structure. Soil stabilization may be performed by physical or chemical means [17]. Many research studies [18–27] have focused on the effect of adding stabilization materials to the soil. Lemaire et al. [18] investigated the inclusion of a cement–lime mix in silty soils. The silty soil was stabilized by adding 5.0% and 1.0% of cement and lime, respectively. Moreover, a significant increase in the structure's unconfined compressive strength and microporosity filling was noticed. In addition, stabilization with a lime and micro silica admixture of silty sand soils was investigated by Karimi et al. [19] in the presence of sulfates. The results demonstrated a significant increase in the soil's California Bearing Ratio (CBR) value and reduced swelling properties. Thus, micro silica waste material successfully improved the silty sand soil resistance. Khemissa and Mahamedi [20] mixed cement and lime in over-consolidated clay and determined that the addition of 8.0% of cement and 4.0% of lime into the clay improved its bearing capacity, durability, and shear strength. Al-Aghbari, Mohamedzein, and Taha [21] tested the stabilization effects of Portland cement and cement kiln dust on desert sands. The results showed remarkable improvements in the shear strength, maximum dry density, and unconfined compressive strength of the soil. Hence, the study concluded that combining cement and cement kiln dust enhances desert sand's shear strength and compressibility properties [21]. They also investigated the impact of utilizing municipal solid waste incinerator ash to strengthen desert sand soil. Incinerator ash enhanced the shear and unconfined compressive strength characteristics with lower permeability [22]. Recently, Soundarya [23] studied the effects of fly ash and ground granulated blast-furnace slag (GGBS) on lime-stabilized mud blocks. Approximately 5% of lime was added, and the fly ash and GGBS contents varied from additions of 0% to 10% (with a 2% increment). The results demonstrated a positive correlation between the amount of fly ash and GGBS and the compressive strength of the mud blocks. Wet compressive strength was significantly affected by the increase in the GGBS amount compared to the fly ash amount. Additionally, fly ash improved water absorption rates more than GGBS since fly ash is more efficient in producing cementitious materials with lime and reducing pore interconnectivity. Stabilized compressed earth blocks were also tested with lime and a combination of cement and lime by Malkanthi, Balthazaar, and Perera [24]. It was found that the cement and lime addition provided higher compressive strength than the addition of just lime. Ten percent lime gave one of the best performances, and it was concluded that lime-stabilized blocks might be used for single-story buildings. However, for Grade 2 block strength with 15% and 10% of clay and silt, respectively, a mix of 5% of cement and 5% of lime was required. For blocks with 5% clay and silt, approximately 7% cement and 3% lime were required. Studies have also used a combination of cement and rice husk ash [28] or cement and fly ash [29] to stabilize the soil. Attom and Shatnawi determined that clayey soil is strengthened with the inclusion of wheat husk [30]. Internal erosion soil stabilization using chemicals was investigated by Vinod et al. [25] and Vakili et al. [26]. Vinod et al. [25] found that adding lignosulfonate to dispersive soils results in better soil erosion coefficients and critical shear stress, ultimately providing higher soil strength and stabilization. In comparison, Vakili et al. [26] mixed lignosulfonate and reinforced polypropylene in dispersive soil to stabilize it against internal erosion. Other studies emphasizing soil stabilization include [31–39]. Although significant research has been performed using lime to stabilize clayey soils, a lack of research exists on using lime to stabilize cohesionless soils such as sand. One of the few studies on the effect of lime stabilization on cohesionless (sandy) soil with different curing times was conducted by the authors [40]. For this study, poorly and well-graded soil types were mixed with quicklime and tested after 1, 2, and 7 days of curing. Using the hole erosion test (HET), the soil types were determined to be stabilized to an optimum level with a minimum of 2 days of curing. Another study on the effect of lime stabilization of coarse sandy soil against internal erosion was performed by Elandaloussi et al. [41]. The erosion test results showed that the lime stabilization effect begins with a curing time of only 24 h due to the agglomeration of fine particles. Moreover, as per their reported results, longer curing times, such as 3 months, did not significantly increase the efficiency of lime treatment.

Thus, this research studies the effect of different lime percentages at a single curing time, and this emphasizes the effect of the internal erosion stabilization of cohesionless soils with different lime percentages and a short (24 h) curing time. As mentioned, the authors performed a similar investigative study on the effect of curing time on lime-stabilized cohesionless (sandy) soil in [40]. Poorly and well-graded soils were examined with quick lime in these studies. Different lime percentages were mixed with the soil based on the soil weight. The soil samples were compacted and prepared in a standard proctor mold with a relative compaction rate of 95.0%. Lime addition resulted in improvement in the soil strength and erosion rate index. The rest of the paper has been organized in the following manner: Firstly, the research significance is presented. Next, the methodology used to prepare and test the soil samples is provided. Following this, the results and their detailed discussion are presented, including the addition of lime effect on water flow path, critical shear stress, erosion rate index, and internal soil erosion. Lastly, the conclusion of the study is provided.

2. Research Significance

Soil embankments are essential for providing, diverting, and retaining water for different purposes. Since the primary component of any soil embankment is sand, it is vital to stabilize the sand being used for the effective and successful operation of the structure. Based on the reviewed literature, lime is found to be an efficient and cost-effective stabilizer for clayey soil. However, to the best of the author's knowledge, there has been insufficient research investigating the stabilization effect of the addition of lime to cohesionless soil. Hence, this study tries to increase the knowledge base by studying the internal erosion stabilization of cohesionless soil using lime and promoting its utilization with confidence in the industry. This research studied two cohesionless soil types (poorly and well-graded), which were cured for 24 h. The poorly graded soil was mixed with lime in percentages from 0.0% to 6.0% (based on the soil's weight), and well-graded soil was mixed with lime in percentages from 0.0% to 3.0%. The hole erosion test (HET) was adopted to replicate the internal erosion of the soil scenario. This study assists in analyzing lime-stabilized cohesionless soil properties. It also supports the raising of awareness about internal soil erosion in earth-fill dams and counteractive measures using soil stabilization. Further studies may be performed on this topic using other stabilization materials, such as solid waste and bitumen.

3. Experimental Program and Methodology

The experimental program of this study is inspired by [13] and the authors performed a similar investigation previously [40] to study the curing time effect.

3.1. Material Properties

The material properties of the soil studied are displayed in Table 1. Specific tests were performed to determine the initial soil properties in accordance with ASTM standard procedures. Based on the results in Table 1, it is clear that Soil A was poorly graded while Soil B was well-graded cohesionless soil. The manufacturer provided the quick lime properties (which can be found in [40]).

Table 1. Soil Properties (similar to [40]).

Property	Soil A	Soil B
Specific Gravity, G_s of soil	2.60	2.67
Amount of Clay (%)	0.00	4.00
Amount of Silt (%)	0.60	8.00
Amount of Sand (%)	99.40	88.00
Coefficient of Uniformity, C_u	1.60	11.05
Coefficient of Curvature, C_c	0.90	2.26
Optimum Moisture (or Water) Content, ω_{op} (%)	11.95	13.00
Maximum Dry Density, ρ_{dmax} (kg/m^3)	1690.00	1908.00
Classification	Poorly Graded	Well Graded

3.2. Specimen Preparation

Lime percentages from 0.0% to 6.0% were added for Soil A, while, for Soil B, lime percentages from 0.0% to 3.0% were used. Thus, three soil specimens were prepared and cured for each lime percentage for 24 h. The soil specimen preparation included the following steps: Firstly, the amount of soil, lime, and water (based on optimum moisture content) was measured and mixed. Then, compaction of the prepared mix in a mold was performed with a relative compaction rate of 95%. Following this, a hole was drilled with a diameter of 6 mm through the compacted soil specimen's cross-section, as shown in Figure 1a (this step was conducted to initiate soil piping erosion). Lastly, the specimen was labeled and stored for its curing time. An average of two specimens for each lime percentage was considered for higher accuracy of the results. The final results are discussed in this paper for each lime percentage. As mentioned, this study adopted the hole erosion test, HET (by Wan and Fell [13]) to analyze the soil resistance against internal erosion. This method is preferred because of its simplicity, ease, and lower cost. Moreover, it is considered the best method to replicate piping erosion behavior [42]. The HET apparatus and schematic diagram are displayed in Figure 1b,c, respectively.

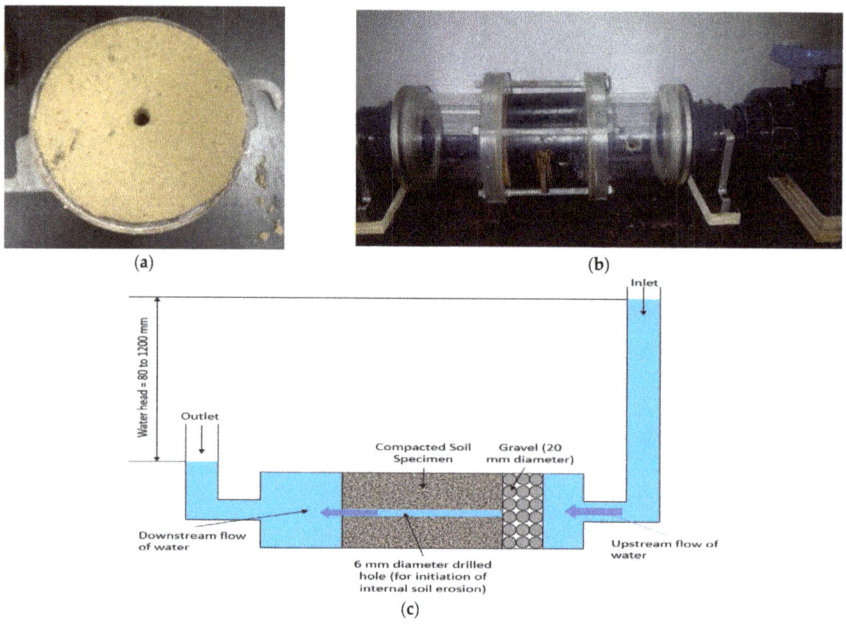

Figure 1. (**a**) Specimen top view before curing (with an initial hole diameter of 6 mm), (**b**) laboratory apparatus, and (c) schematic diagram (adapted from [13]) of hole erosion test (HET) (similar to [40]).

3.3. Specimen Testing Process

The upstream chamber of the HET apparatus was filled with gravel particles of 20 mm in diameter to regulate and filter the water flow. Subsequently, the prepared specimen was fitted between the inlet and outlet chambers with the help of O rings and bolts (as shown in Figure 1b). The water head ranged between 800 and 1200 mm, and the test was initiated by allowing the water to flow through the hole of the specimen. At different intervals of test run time, the rate of flow was measured using the outlet pipe. The test ran until the specimen's failure, with a minimum of 45 min. At the end of the test, the water flow was stopped, and the specimen was extracted. Finally, the final hole diameter was determined, and the test apparatus was cleaned before reusing it for other specimens. It is advised that future researchers interested in researching this topic should ensure that the HET apparatus is built to be leakproof to avoid any water leaks during the testing process. Additionally, during the test, the flow rate (water volume and time) must be determined accurately by measuring them at multiple time intervals throughout the test. Furthermore, the final hole diameters are advised to be measured using a vernier caliper for higher accuracy.

3.4. Process for Analysing the Data

The HET test data helped determine the soil's shear stress at the start of erosion and the erosion rate index. Erosion occurs when the water force outpaces the hydraulic shear stress [13]. The soil erosion rate index (I) parameter is a good indicator of the soil's internal erosion resistance. Stronger soil internal erosion resistance provides a higher erosion rate index. The soil erosion descriptions for different erosion rate indices are provided in Table 2.

Table 2. Internal Erosion Descriptions for Various Erosion Rate Indices [13,40].

Group Number	Erosion Rate	Description
1	Less than 2	Extremely Rapid Erosion
2	2 to 3	Very Rapid Erosion
3	3 to 4	Moderately Rapid Erosion
4	4 to 5	Moderately Slow Erosion
5	5 to 6	Very Slow Erosion
6	Greater than 6	Extremely Slow Erosion

Initially, Reynold's number (R) was utilized to determine the type of water flow as per Equations (1) and (2):

$$V_t = \frac{Q_t}{\pi\left(\frac{\varphi_t^2}{4}\right)} \quad (1)$$

$$R = \frac{V\varphi_t\rho_w}{v} \quad (2)$$

where V_t denotes the average velocity of water flowing through the hole (m/s) at time t, Q_t is the flow rate at time t (m^3/s), φ_t is the water flow path (hole) diameter at time t (m), ρ_w is the water density (kg/m^3), and v denotes the absolute water viscosity (Pa·s) = 1.00×10^{-3} [40,43]. Using the flow type, flow rate, and initial and final measured hole diameters, the friction factors (f_{Tt} or f_{Lt}) were determined at the start and end of the test with the help of Equations (3) and (4) [13,40]:

$$Turbulent\ Flow : \varphi_t = \left[\frac{64Q_t^2 f_{Tt}}{\pi^2\rho_w g s_t}\right]^{1/5} \quad (3)$$

$$Laminar\ Flow : \varphi_t = \left[\frac{16Q_t f_{Lt}}{\pi\rho_w g s_t}\right]^{1/3} \quad (4)$$

where f_{Tt} or f_{Lt} is the turbulent or laminar friction factors at any time, respectively, ρ_w is the water density, g is the acceleration due to gravity, and s_t is the soil sample hydraulic gradient. Based on the calculated initial and final friction factors, the friction factor versus the time curve was plotted, giving the friction factor value at any time (f_{Tt} or f_{Lt}). Then, the hole diameter at any time (φ_t) was calculated by reusing Equations (3) or (4) with the evaluated friction factors at any time (f_{Tt} or f_{Lt}). Further, knowing the φ_t values, the rate of diameter change with time $\left(\frac{d\varphi_t}{dt}\right)$ was determined. Next, using the evaluated values, soil dry density (ρ_d) and Equations (5) and (6), hydraulic shear stress (τ_t) and erosion rate per unit surface area ($\dot{\varepsilon_t}$) at time t were determined [13,40]:

$$\tau_t = \rho_w g s_t \frac{\varphi_t}{4} \tag{5}$$

$$\dot{\varepsilon_t} = \frac{\rho_d}{2}\frac{d\varphi_t}{dt} \tag{6}$$

$$\dot{\varepsilon_t} = C_e(\tau_t - \tau_c) \tag{7}$$

$$I = -log(C_e) \tag{8}$$

Subsequently, as illustrated in Figure 2, the erosion rate ($\dot{\varepsilon_t}$) was plotted against hydraulic shear stress (τ_t) to obtain the soil erosion coefficient (C_e) (as per Equation (7)). Lastly, utilizing the C_e value and Equations (7) and (8), the critical shear stress (τ_c) and soil erosion rate index (I) values were respectively determined. The I value was then compared with the details in Table 2 to describe the soil erosion type [13,40].

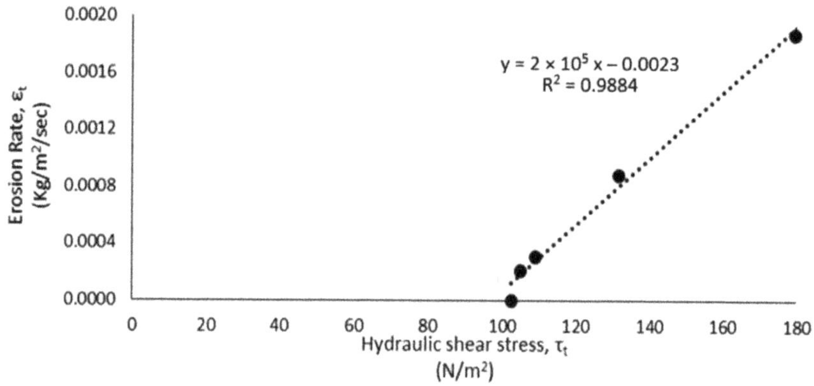

Figure 2. Erosion rate (ε_t) plot against hydraulic shear stress (τ_t) for Soil A with 5.0% lime.

4. Results and Discussion

4.1. Addition of Lime Effect on Water Flow Path

The effect of lime percentage on the water flow path (hole) diameter over their test run time (in minutes) for Soils A and B with 24 h curing time is illustrated in Figure 3. For Soil A, with an initial hole diameter of 6.0 mm, an increase in hole diameter of approximately 27.0 mm (33.0–6.0 mm) was observed at 1.0% of lime, with a test run time of approximately 7 min. However, at 2.0% of lime, the increase in the hole diameter was approximately 15.5 mm (21.5–6.0 mm), with a test run time of 10 min. Therefore, a 42.6% reduction in hole diameter occurred with the increase in lime content from 1.0% to 2.0%, along with a longer test run time. This demonstrates that the internal erosion resistance of the soil improved with the increase in the addition of lime. Moreover, it is evident from Figure 3a that with a further increase in lime percentage, the change in water flow path (hole) diameter reduces, with the lowest change and longest test run time occurring at 6.0% lime for Soil A. At 6.0% lime, only a 4.3 mm increase (10.3–6.0 mm) in the hole diameter was observed with a test run time of approximately an hour (60 min). Thus, an 84.1% reduction in the change in diameter was obtained with the increase in lime from 1.0% to 6.0%. Hence, the soil strength improved significantly with a higher lime content and reduced the internal erosion rate by resisting the change in the water flow path (hole) diameter. The results for 0.0% of lime in Soil A are not provided since the soil specimen collapsed within a few seconds of the test run time; thus, data could not be derived for this specimen. Therefore, it was noted as a failure point and was deemed insignificant. Similarly, Soil B portrayed that a higher lime percentage decreased the change in water flow path diameter (as shown in Figure 3b). An increase of 9.0 mm (15.0–6.0 mm) was noted at 0.0% lime with a test run time of 6 min. At 1.0% of lime, the hole diameter change was approximately 2.2 mm (8.2–6.0 mm), with a test run time of 51 min. Thus, the water flow path diameter change was decreased by 75.6% with a longer test run time. A minimalistic change in water flow path diameter of 2.0 mm (8.0–6.0 mm) was noticed at 3.0% lime with a test run time of more than 2 h (approximately 135 min). Hence, the inclusion of 3.0% lime in Soil B and 24 h of curing time decreased the diameter change by 77.8% compared to the Soil B specimen without any lime. Therefore, a higher amount of lime strengthened the cohesionless soil and significantly improved its internal erosion resistance ability.

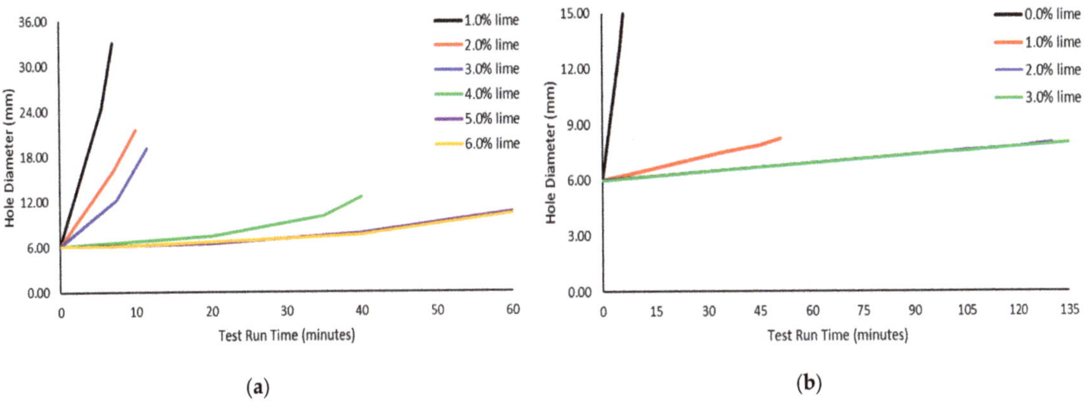

Figure 3. Hole diameter (mm) versus test run time (mins) plot with their lime contents (%) in (**a**) Soil A and (**b**) Soil B.

Additionally, Figure 4 illustrates the water flow path (hole) diameter (mm) at the end of the test against their lime percentages (%) for Soils A and B. It displays that the final hole diameter reduced with more lime. Furthermore, it demonstrates that Soil B specimens at 24 h curing time performed considerably better than Soil A specimens. Since 0.0% of lime in Soil B did not fail instantly, unlike Soil A, it resisted internal erosion briefly. Additionally, the final diameter of the hole for all lime percentages was lower in Soil B when compared to Soil A. This occurred due to the soil's physical properties. Soil A was a poorly graded cohesionless soil with lower strength due to the absence of clayey and silty soil elements, resulting in lower stability. On the other hand, Soil B (well-graded cohesionless soil) showed a significant amount of silt and clay (which react well with lime) and, therefore, had slightly higher strength and stability, and thus it showed better resistance against internal erosion. In addition, Soil A required double the amount of lime (approximately 6.0% of lime) compared to Soil B (which required only approximately 3.0% lime) to significantly reduce the change in diameter. Hence, Soil B reacted and stabilized better with lime in comparison with Soil A. Similar findings were deduced in [40].

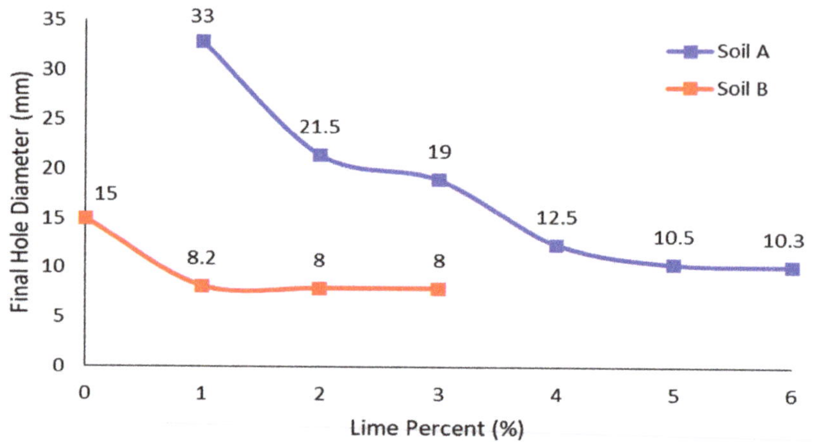

Figure 4. Final hole (water flow path) diameter (φ_f) plot against lime percentage (%) for Soils A and B with 24 h curing time.

4.2. Addition of Lime Effect on Critical Shear Stress

Figure 5 displays the critical shear stress values for Soils A and B with their respective lime contents (%) and 24 h curing time. Higher lime percentages resulted in an increase in critical shear stress values for both Soil A and Soil B, as illustrated in Figure 5. Although an increase in lime from 1.0% to 2.0% in Soil A did not improve the critical shear stress value (=85 N/m^2), a further increase in lime to 3.0% slightly improved the value to 97 N/m^2. Thus, an approximately 14.1% increase in critical shear stress occurred at 3.0% lime, compared to 1.0% lime in Soil A. According to the results, the highest critical shear stress of 115 N/m^2 for Soil A was obtained at 5.0% lime. Thus, increasing the lime from 1.0% to 5.0% resulted in 35.3% higher critical shear stress. Moreover, a further increase in lime did not increase the shear stress value, it can thus be deduced that Soil A was optimally stabilized at 5.0% lime and 24 h of curing time. In contrast, the critical shear stress of Soil B rose from 100.5 to 105 N/m^2 with an increase in lime from 0.0% to 1.0%, respectively. Therefore, for Soil B, the stress increased by 4.5% with the addition of 1.0% lime in comparison to the Soil B specimen without any lime. Moreover, a further increase in lime to 3.0% improved the critical shear stress value to 140 N/m^2, which was approximately 39.3% higher compared to the critical shear value of 0.0% lime. A lime percentage greater than 3.0% was not performed since the soil was observed to be optimally stabilized at 3.0% lime and 24 h curing time. Hence, higher lime percentages resulted in improving the critical shear stress of the soil. Previous researchers also deduced these findings [34,40,44,45]. In addition, the critical shear stress values of Soil B were determined to be higher in comparison with Soil A. As mentioned, this is attributed to the fact that Soil A was a poorly graded cohesionless soil with lower strength and stability, while Soil B was a well-graded cohesionless soil with slightly higher strength and stability due to the presence of clay and silt. Due to this, Soil B required only approximately 3.0% of lime to reach a shear stress value of 140 N/m^2, while Soil A required approximately 5.0% lime to reach a shear stress value of 115 N/m^2. Hence, it is evident that Soil B reacted well with lime, providing higher critical shear stress and strength to resist internal erosion compared to Soil A, similar to the reported results in [40].

4.3. Addition of Lime Effect on Erosion Rate Index (I_{HET})

To adequately represent the effectiveness of the internal erosion stabilization of cohesionless soil using lime, the erosion rate index (I_{HET}) value was evaluated for the two soil types. Figure 6 provides the erosion rate indices for different lime percentages in Soils A and B within 24 h of curing. This figure demonstrates that higher lime percentages result in improving the I_{HET} value until the effective lime stabilization of the soil occurs. For Soil A with 1.0% lime and 24 h curing, the I_{HET} was 3.69, representing a moderately rapid erosion type, as per Table 2. Increasing the amount of lime to 2.0% increased the I_{HET} to 4.00, a moderately slow erosion type, as per Table 2. Therefore, increasing the lime percentage increased the I_{HET} value (by 8.4%) and the soil's internal erosion resistance ability. With higher lime percentages such as 5.0% (or 6.0%) and 24 h of curing, the I_{HET} of Soil A improved to a value of 4.69, which is considered to be a moderately slow erosion type, as per Table 2. Therefore, increasing lime from 1.0% to 5.0% led to a 27.1% improvement in the I_{HET} value (as shown in Figure 6). This validates that an increase in Soil A's lime content caused the internal erosion stabilization of the soil. For Soil B at 0.0% lime and 24 h curing time, the I_{HET} was determined to be 3.69, a moderately rapid erosion type, as per Table 2. With the addition of 1.0% lime, the I_{HET} increased by 27.1%, changing the soil erosion resistance from a moderately rapid to a moderately slow erosion type. A further increase in lime to 3.0% increased the I_{HET} to a value of 5.301, representing a very slow erosion type, as per Table 2. Therefore, increasing the lime from 0.0% to 3.0% resulted in improving the I_{HET} value by 43.7%. These results also validate that increasing the lime content improves the internal erosion resistance of soil, similar to the deductions reported in [34,40,44]. Moreover, Soil B again demonstrated better results, with comparatively higher erosion rate indices with respect to Soil A (refer to Figure 6). Soil B reached a 13.0% higher I_{HET} value with a lime content of only 3.0%, in comparison to Soil A with a higher lime content of 5.0%, demonstrating the significance of the initial soil properties.

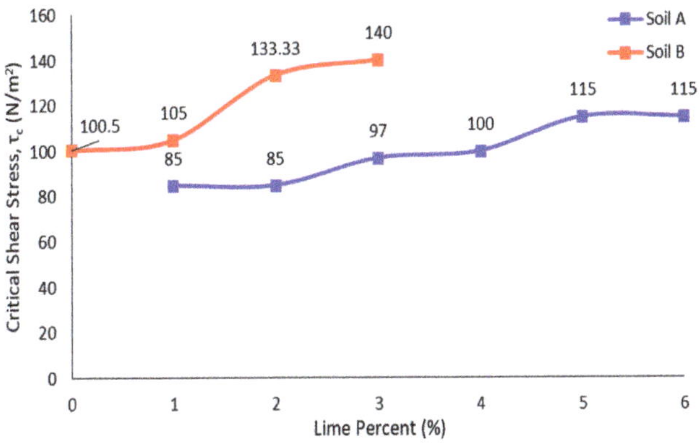

Figure 5. Critical shear stress (τ_c) plot against lime percentage (%) for Soils A and B with 24 h of curing time.

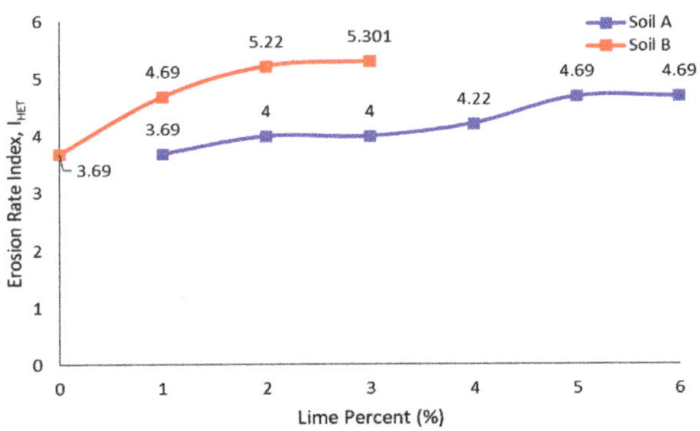

Figure 6. Erosion rate index (I_{HET}) plot against lime percentage (%) for Soils A and B with a curing time of 24 h.

4.4. Internal Erosion Types of Soil A and Soil B

The erosion parameters for Soils A and B at different lime percentages are displayed in Table 3. Increasing the lime percentage caused a reduction in the final water flow path diameter and an increase in the critical shear stress (τ_c) and erosion rate index (I_{HET}) values. Thus, higher I_{HET} means stronger soil, higher shear strength, and lower internal erosion hole diameter values. Based on the results in Table 3, Soil A required approximately 5.0% of lime to effectively stabilize against internal erosion. As Soil A achieved the highest I_{HET} and τ_c values of 4.69 and 115 N/m^2, respectively, at 5.0% lime, and as per Table 2, this soil demonstrated a moderately slow erosion type. Moreover, Soil A showed a significantly small rise in water flow path diameter at 5.0% lime with a test run time of 60 min (1 h). Table 3 shows that a further increase in lime does not improve the strength and internal erosion resistance of the soil. Thus, Soil A was optimally stabilized with the addition of 5.0% lime and curing of 24 h. For Soil B, approximately 3.0% of lime was required for its optimum stabilization with curing of 24 h. At 3.0% lime in Soil B, the I_{HET} and τ_c were equal to 5.301 and 140 N/m^2, respectively. This I_{HET} value represents a very slow erosion type, as per Table 2. In addition, Soil B displayed a minimal increase in the water flow path of 2.0 mm with a constant pressure head of 1200 mm and a test run time of 135 min (more than 2 h). Thus, it can be deduced that the soil was stabilized. Hence, Soil B was stabilized with 3.0% lime with curing of 24 h. Similar conclusions were drawn by [40]. For an illustration of some soil specimens' final water flow path holes, refer to Figure 7.

Table 3. Final hole erosion test results of Soils A and B.

Lime Percentage	Erosion Parameters	Soil A	Soil B
0.0%	Final Hole Diameter, φ_f (mm)	-	15
	Critical Shear Stress, τ_c (N/m^2)	-	100.5
	Erosion Rate Index, I_{HET}	-	3.69
	Soil Erosion Type	Failed Instantly	Moderately Rapid
1.0%	Final Hole Diameter, φ_f (mm)	33	8.2
	Critical Shear Stress, τ_c (N/m^2)	85	105
	Erosion Rate Index, I_{HET}	3.69	4.69
	Soil Erosion Type	Moderately Rapid	Moderately Slow
2.0%	Final Hole Diameter, φ_f (mm)	21.5	8
	Critical Shear Stress, τ_c (N/m^2)	85	133.33
	Erosion Rate Index, I_{HET}	4	5.22
	Soil Erosion Type	Moderately Slow	Very Slow
3.0%	Final Hole Diameter, φ_f (mm)	19	8
	Critical Shear Stress, τ_c (N/m^2)	97	140
	Erosion Rate Index, I_{HET}	4	5.301
	Soil Erosion Type	Moderately Slow	Very Slow
4.0%	Final Hole Diameter, φ_f (mm)	12.5	-
	Critical Shear Stress, τ_c (N/m^2)	100	-
	Erosion Rate Index, I_{HET}	4.22	-
	Soil Erosion Type	Moderately Slow	-
5.0%	Final Hole Diameter, φ_f (mm)	10.5	-
	Critical Shear Stress, τ_c (N/m^2)	115	-
	Erosion Rate Index, I_{HET}	4.69	-
	Soil Erosion Type	Moderately Slow	-
6.0%	Final Hole Diameter, φ_f (mm)	10.3	-
	Critical Shear Stress, τ_c (N/m^2)	115	-
	Erosion Rate Index, I_{HET}	4.69	-
	Soil Erosion Type	Moderately Slow	-

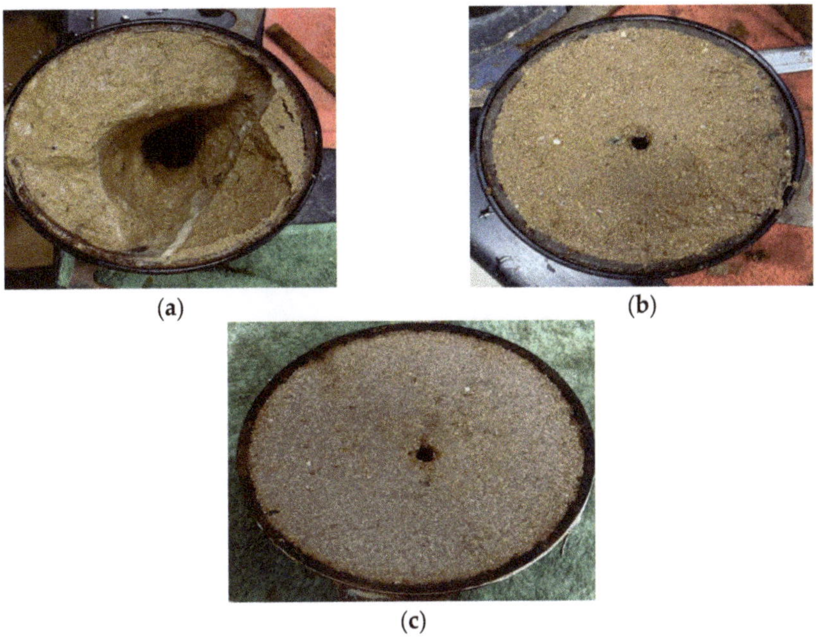

Figure 7. Hole diameters at the end of the test (φ_f) for (**a**) Soil A with 1.0% lime and Soil B with (**b**) 1.0% and (**c**) 2.0% lime (similar to [40]).

Figure 8 demonstrates the relationship between the obtained final hole diameter (φ_f) and τ_c against the I_{HET} values of Soils A and B with different lime percentages and 24 h of curing. It displays that a higher erosion rate index lowered the final internal erosion hole diameter and increased the critical shear stress (τ_c) values for both soil types. For Soil A, adding 5.0% lime resulted in a 68.2% lower final hole diameter, 35.3% higher τ_c value, and 27.1% higher I_{HET} value. For Soil B, the addition of only 3.0% lime resulted in a 46.7% lower final hole diameter, 39.3% higher τ_c value, and 43.7% higher I_{HET} value. Thus, Soil B reacted significantly better with lime than Soil A. Again, this was because Soil B contained clay and silt (refer to Table 1). Since clay and silt are binding materials and lime reacts much better with soils with binding materials, Soil B showed better results. In contrast, Soil A was poorly graded with an absence of clay and silt; consequently, it showed lower shear strength and a relatively weaker reaction with lime. Hence, Soil A required a higher lime content (5.0%) for stabilization than Soil B. Nevertheless, even with a higher lime percentage, Soil A (due to its weak properties) resulted in lower I_{HET} and τ_c values by 11.5% and 17.9%, respectively, compared to Soil B. Additionally, the internal erosion type of Soil A with 5.0% lime (moderately slow) was inferior than Soil B with only 3.0% lime (very slow). Therefore, it was determined that the initial soil type, properties, and gradation played a vital role in the internal erosion stabilization effect of the soil using lime.

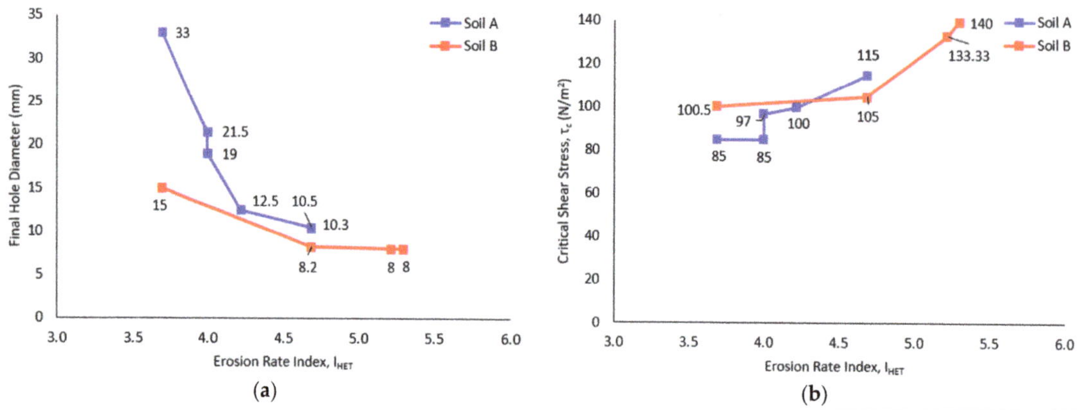

Figure 8. (**a**) Final hole diameter (φ_f) and (**b**) critical shear stress (τ_c) versus erosion rate index (I_{HET}) relationship with the increase in lime percentages for 24 h curing time.

5. Conclusions

The internal erosion stabilization of cohesionless soil using lime was investigated in this paper. The hole erosion test (HET) was adopted and conducted to execute the experimental testing of the study, as it best replicated the soil's internal erosion. Two cohesionless soil types, poorly graded and well-graded, were tested with various lime percentages and 24 h of curing time. HET test results were analyzed, and different erosion parameters, including water flow path diameter, critical shear stress (τ_c), and erosion rate index (I_{HET}), were evaluated. Based on the findings, the following conclusions were drawn:

1. Quick lime is an adequate internal erosion stabilizer for cohesionless soils;
2. The water flow path diameter can be reduced and controlled with a higher amount of lime for both types of cohesionless soil;
3. An increase in lime content leads to stronger critical shear stress (τ_c) and higher erosion rate index (I_{HET}) values for cohesionless soil;
4. The inclusion of 5.0% lime in poorly graded soil resulted in a 68.2% lower final hole diameter, 35.3% higher critical shear stress (τ_c), and 27.1% higher erosion rate index (I_{HET}) value. In contrast, for Soil B, the addition of only 3.0% lime resulted in a 46.7% lower final hole diameter, 39.3% higher τ_c, and 43.7% higher I_{HET} value;
5. For poorly graded cohesionless soil, with 5.0% optimum lime content, the internal erosion improved from a moderately rapid erosion type to a moderately slow erosion type, and for well-graded soil, with only 3.0% of lime, the internal soil erosion improved from a moderately rapid to a very slow erosion type;
6. Lower lime is required to stabilize well-graded cohesionless soil as opposed to poorly graded soil. Moreover, stabilized well-graded soil demonstrated comparatively better erosion parameters than stabilized poorly graded soil. This is related to the clay and silt (binding materials) present in well-graded soil, which gives higher strength and better supports the lime reaction. Hence, well-graded soil is preferred in construction compared to poorly graded cohesionless soil.

6. Recommendations for Future Work

This study was performed at the American University of Sharjah (AUS) as part of their research program to investigate the internal erosion stabilization of cohesionless soil using lime. Based on the experience, it is highly recommended to ensure that the testing apparatus is built to be leakproof and that multiple flow rates are noted at various time intervals to increase the accuracy of the results. Moreover, due to the limited scope of this study, not all factors could be investigated. Thus, further research may be performed to investigate internal soil erosion with different (1) soil properties such as clay, silt, and sand contents; cohesions; angles of internal friction; optimum moisture contents; maximum dry densities; and gradations; (2) hole erosion test factors such as the pressure head and test run time; and (3) internal erosion testing methods such as the slot erosion test (SET), rotating cylinder test, jet erosion test (JET), and soil dispersity tests (such as pinhole, Emerson crumb, and double hydrometer tests). Additionally, studies related to other soil stabilizers (bitumen, oil shale, fly ash) can be conducted on this topic.

Author Contributions: Conceptualization, S.B. and M.A.; methodology, S.B. and M.A.; software, S.B. and M.A.; validation, S.B. and M.A.; formal analysis, S.B. and M.A.; investigation, S.B. and M.A.; resources, S.B. and M.A.; data curation, S.B. and M.A.; writing—original draft preparation, S.B.; writing—review and editing, S.B. and M.A.; visualization, S.B. and M.A.; supervision, M.A.; project administration, M.A.; funding acquisition, M.A. All authors have read and agreed to the published version of the manuscript.

Funding: This research was financially supported by the American University of Sharjah (AUS) through the Faculty Research Grant program (FRG20-M-E61) and Open Access Program (OAP).

Data Availability Statement: Not applicable.

Acknowledgments: The authors would like to acknowledge the support and contributions of the American University of Sharjah (AUS), their Civil Engineering Department staff, and the Open Access Program (OAP) during this research. This paper represents the opinions of the authors and does not represent the opinions or position of the American University of Sharjah.

Conflicts of Interest: The authors declare no conflict of interest.

Notations

AUS	American University of Sharjah
CBR	California Bearing Ratio
C_c	Coefficient of Curvature
C_e	Soil Erosion Coefficient
C_u	Coefficient of Uniformity
f_{Lt}	Laminar Friction Factors at any time, t
f_{Tt}	Turbulent Friction Factors at any time, t
g	Gravitational Acceleration (m/s^2)
GGBS	Ground Granulated Blast-furnace Slag
G_s	Specific Gravity of Soil
HET	Hole Erosion Test
I	Soil Erosion Rate Index
I_{HET}	Erosion Rate Index determined using Hole Erosion Test
JET	Jet Erosion Test
OAP	Open Access Program
Q_t	Flow Rate at time, t (m^3/s)
R	Reynold's Number
SET	Slot Erosion Test
s_t	Soil Sample Hydraulic Gradient
V_t	Average Water Flow Velocity at time, t (m/s)
$\varepsilon_t^{'}$	Erosion Rate per Unit Surface Area at time, t (kg/m^2/s)
v	Absolute Water Viscosity (Pa·s)
ρ_d	Soil Dry Density (kg/m^3)
ρ_{dmax}	Maximum Dry Density (kg/m^3)
ρ_w	Water Density (kg/m^3)
τ_c	Critical Shear Stress (N/m^2)
τ_t	Hydraulic Shear Stress at time, t (N/m^2)
φ_f	Final hole (water flow path) diameter
φ_t	Water Flow Path (Hole) Diameter at time t (m)
$\left(\frac{d\varphi_t}{dt}\right)$	Rate of Diameter Change with Time (m/s)
ω_{op}	Optimum Moisture (or Water) Content

References

1. Chen, S.H. Introduction. In *Hydraulic Structures*; Springer: Berlin/Heidelberg, Germany; Wuhan, China, 2015; pp. 1–39.
2. Kuriqi, A.; Ardiçlioglu, M.; Muceku, Y. Investigation of seepage effect on river dike's stability under steady state and transient conditions. *Pollack Period.* **2016**, *11*, 87–104. [CrossRef]
3. Lüthi, M. A Modified Hole Erosion Test (HET-P) to Study Erosion Characteristics of Soil. Master's Thesis, The University of British Columbia, Vancouver, BC, Canada, 2011.
4. Hanneman, D. Recent examples of backward erosion piping, internal erosion along a conduit, and embankment cracking at Bureau of Reclamation dams. In Proceedings of the Internal Erosion in Embankment Dams and Their Foundations, Brno, Czech Republic, 26–29 April 2011.
5. Foster, M.; Fell, R.; Spannagle, M. A method for assessing the relative likelihood of failure of embankment dams by piping. *Can. Geotech. J.* **2000**, *37*, 1025–1038. [CrossRef]
6. Robbins, B.; Griffiths, D. Internal Erosion of Embankments: A review and Appraisal. In Proceedings of the Rocky Mountain Geo-Conference 2018, Golden, CO, USA, 2 November 2018; ASCE Library: Reston, VA, USA, 2018.
7. Klipalo, E.; Besharat, M.; Kuriqi, A. Full-Scale Interface Friction Testing of Geotextile-Based Flood Defense Structures. *Buildings* **2022**, *12*, 990. [CrossRef]
8. Kuriqi, A.; Hysa, A. Multidimensional Aspects of Floods: Nature-Based Mitigation Measures from Basin to River Reach Scale. In *Nature-Based Solutions for Flood Mitigation*; Springer: Cham, Switzerland, 2021; Volume 107, pp. 11–33.
9. Cao, Y.; Wang, J.L.; Chen, H.C.; Li, Y.; Nurtjahyo, P. EFA tests and the influence of various factors on the erodibility of cohesive soils. In Proceedings of the 1st International Conference on Scour of Foundations, College Station, TX, USA, 17–20 November 2002.
10. Hanson, G.I.; Tejral, S.H.; Temple, D. Internal Erosion and Impact of Erosion Resistance USDA Small Watershed Program. *Collab. Manag. Integr. Watersheds Trans. ASABE* **2007**, *5*, 773–784.
11. Briaud, J.-L. Case Histories in Soil and Rock Erosion: Woodrow Wilson Bridge, Brazos River Meander, Normandy Cliffs, and New Orleans Levees. *J. Geotech. Geoenviron. Eng.* **2008**, *134*, 1425–1447. [CrossRef]
12. Sherard, J.L.; Richard, S.D.; Woodward, J.; Stanley, N.S.; Gizenski, F.; Willaim, M.S.; Clevenger, B.S. *Earth and Earth Rock Dams*; Chapter 2; John Wiley and Sons: New York, NY, USA, 1963.
13. Wan, C.F.; Fell, R. Investigation of Rate of Erosion of Soils in Embankment Dams. *J. Geotech. Geoenviron. Eng.* **2004**, *130*, 373–380. [CrossRef]
14. Yamin, M.; Attom, M.F.; Atabay, S.; Vandanapu, R. The Effect of Compaction Effort on Shear Strength Parameters of Low/High Plasticity Clay Soils. *Geotech. Eng. J. SEAGS AGSSEA* **2021**, *52*, 1–8.
15. Attom, M.F.; Khan, Z.; Vandanapu, R. Efficacy of sand columns to increase the sub-soil moisture content of clay. *SN Appl. Sci.* **2020**, *2*, 1–6. [CrossRef]
16. Vandanapu, R.; Omer, J.R.; Attom, M.F. Laboratory simulation of the influence of groundwater rise and drip irrigation on the settlement of a sample of collapsible desert soil. *Int. J. Geotech. Eng.* **2018**, *15*, 760–766. [CrossRef]
17. Negi, A.S.; Faizan, M.; Siddharth, D.P.; Singh, R. Soil stabilization using lime. International Journal of Innovative Research in Science. *Eng. Technol.* **2013**, *2*, 1–6.
18. Lemaire, K.; Deneele, D.; Bonnet, S.; Legret, M. Effects of lime and cement treatment on the physicochemical, microstructural and mechanical characteristics of a plastic silt. *Eng. Geol.* **2013**, *166*, 255–261. [CrossRef]
19. Karimi, M.; Ghorbani, A.; Daghigh, Y.; Kia, S. Stabilization of Silty Sand Soils with lime and Microsilica Admixture in presence of Sulfates. In Proceedings of the 2011 Pan-Ame CGS Geotechnical Conference–14th Pan-American Conference on Soil Mechanics and Geotechnical Engineering, Toronto, ON, Canada, 2–6 October 2011.
20. Khemissa, M.; Mahamedi, A. Cement and lime mixture stabilization of an expansive overconsolidated clay. *Appl. Clay Sci.* **2014**, *95*, 104–110. [CrossRef]
21. Al-Aghbari, M.; Mohamedzein, Y.E.A.; Taha, R. Stabilisation of desert sands using cement and cement dust. *Proc. Inst. Civ. Eng. Ground Improv.* **2009**, *162*, 145–151. [CrossRef]
22. Mohamedzein, Y.E.-A.; Al-Aghbari, M.Y.; Taha, R.A. Stabilization of desert sands using municipal solid waste incinerator ash. *Geotech. Geol. Eng.* **2006**, *24*, 1767–1780. [CrossRef]
23. Soundarya, N. Effect of fly ash and GGBS on lime stabilized mud block. *Mater. Today Proc.* **2021**, *47*, 4636–4640. [CrossRef]
24. Malkanthi, S.; Balthazaar, N.; Perera, A. Lime stabilization for compressed stabilized earth blocks with reduced clay and silt. *Case Stud. Constr. Mater.* **2019**, *12*, e00326. [CrossRef]
25. Vinod, J.S.; Indraratna, B.; Al Mahamud, M.A. Stabilisation of an erodible soil using a chemical admixture. *Proc. Inst. Civ. Eng. Ground Improv.* **2010**, *163*, 43–51. [CrossRef]
26. Vakili, A.H.; Ghasemi, J.; bin Selamat, M.R.; Salimi, M.; Farhadi, M.S. Internal erosional behaviour of dispersive clay stabilized with lignosulfonate and reinforced with polypropylene fiber. *Constr. Build. Mater.* **2018**, *193*, 405–415. [CrossRef]
27. Indraratna, B.; Mahamud, M.; Vinod, J.S.; Wijeyakulasuriya, V. Stabilization of an erodible soil using a chemical admixtures. In Proceedings of the 2nd International Conference on Geotechnical Engineering, Hammamet City, Tunisia, 25–27 October 2010.
28. Basha, E.A.; Hashim, R.; Mahmud, H.B.; Muntohar, A.S. Stabilization of residual soil with rice husk ash and cement. *Constr. Build. Mater.* **2005**, *19*, 448–453. [CrossRef]
29. Kolias, S.; Kasselouri-Rigpoulou, V.; Karahalios, A. Stabilization of clayey soils with high calcium fly ash and cement. *Cem. Concr. Compos.* **2005**, *27*, 301–313. [CrossRef]

30. Attom, M.; Shatnawi, M.Y. Stabilization of clayey soils using hay material. *J. Solid Waste Technol. Manag.* **2005**, *31*, 84–92.
31. Baldovino, J.d.J.A.; Izzo, R.L.d.S.; Moreira, E.B.; Rose, J.L. Optimizing the evolution of strength for lime-stabilized rammed soil. *J. Rock Mech. Geotech. Eng.* **2019**, *11*, 882–891. [CrossRef]
32. Nagaraj, H.; Sravan, M.; Arun, T.; Jagadish, K. Role of lime with cement in long-term strength of Compressed Stabilized Earth Blocks. *Int. J. Sustain. Built Environ.* **2014**, *3*, 54–61. [CrossRef]
33. Burns, B.; Barker, R.; Ghataora, G.S. Investigating internal erosion using a miniature resistivity array. *NDT E Int.* **2006**, *39*, 169–174. [CrossRef]
34. Herrier, G.; Chevalier, C.; Froumentin, M.; Cuisinier, O.; Bonelli, S.; Fry, J. Lime treated soil as an erosion-resistant material for hydraulic earthen structures. In Proceedings of the 6th International Conference on Scour and Erosion, Paris, France, 27 August–1 September 2012.
35. Rahman, M.; Rashiduzzaman, M.; Akhand, F.Z.; Kabir, K.B. Compressed Stabilized Earth Block: A Green Alternative for Non-load Bearing Building Block in Developing Countries like Bangladesh. *Am. Chem. Sci. J.* **2016**, *12*, 1–10. [CrossRef]
36. Estabragh, A.; Soltannajad, K.; Javadi, A. Improving piping resistance using randomly distributed fibers. *Geotext. Geomembr.* **2014**, *42*, 15–24. [CrossRef]
37. Consoli, N.C.; da Silva Lopes, L., Jr.; Prietto, P.D.M.; Festugato, L.; Cruz, R.C. Variables Controlling Stiffness and Strength of Lime-Stabilized Soils. *J. Geotech. Geoenviron. Eng.* **2011**, *137*, 628–632. [CrossRef]
38. Consoli, N.C.; Bittar Marin, E.J.; Quiñónez Samaniego, R.A.; Heineck, K.S.; Johann, A.D.R. Use of Sustainable Binders in Soil Stabilization. *J. Mater. Civ. Eng.* **2019**, *31*, 06018023. [CrossRef]
39. Consoli, N.C.; da Silva Lopes, L., Jr.; Heineck, K.S. Key Parameters for the Strength Control of Lime Stabilized Soils. *J. Mater. Civ. Eng.* **2009**, *21*, 210–216. [CrossRef]
40. Banu, S.A.; Attom, M.F. Effect of Curing Time on Lime-Stabilized Sandy Soil against Internal Erosion. *Geosciences* **2023**, *13*, 102. [CrossRef]
41. Elandaloussi, R.; Bennabi, A.; Dupla, J.C.; Canou, J.; Benamar, A.; Gotteland, P. Effectiveness of Lime Treatment of Coarse Soils Against Internal Erosion. *Geotech. Geol. Eng.* **2018**, *37*, 139–154. [CrossRef]
42. Benaissa, K.; Angel, P.V.M.; Dolores, R.; Larbi, E.; Abdellatif, K.; Bezzazi, M.; Dubujet, P. Effect of Wall Roughness and Concentration of Clay on Erosion in the Hole Erosion Test. *Am. J. Eng. Appl. Sci.* **2010**, *3*, 734–739.
43. Marine Propellers and Propulsion, "Reynolds' Number", Science Direct. 2019. Available online: https://www.sciencedirect.com/topics/engineering/reynolds-number (accessed on 1 June 2020).
44. Baecque, M.D.; Chevalier, C.; Lopes, S.P.; Feuvre, M.L.; Reiffsteck, P.; Charles, I.; Herrier, G. Resistance to erosion of lime treated soils and perspectives for coastal dikes. In Proceedings of the 25th Meeting of the European Working Group on Internal Erosion, Delft, The Netherlands, 4–7 September 2017; pp. 133–142.
45. Bonelli, S.; Benahmed, N. Piping flow erosion in water retaining structures. *Int. J. Hydropower Dams* **2011**, *18*, 94–98.

water

MDPI

Article

An Ensemble of Weight of Evidence and Logistic Regression for Gully Erosion Susceptibility Mapping in the Kakia-Esamburmbur Catchment, Kenya

Lorraine K. Nkonge [1], John M. Gathenya [2], Jeremiah K. Kiptala [3], Charles K. Cheruiyot [3] and Andrea Petroselli [4,*]

[1] Civil Engineering (Environment, Arid and Semi-Arid Lands), The Pan African University Institute for Basic Sciences, Technology and Innovation (PAUSTI), Jomo Kenyatta University of Agriculture and Technology, Nairobi P.O. Box 62000-00200, Kenya
[2] Soil, Water and Environmental Engineering Department, Jomo Kenyatta University of Agriculture and Technology, Nairobi P.O. Box 62000-00200, Kenya
[3] Department of Civil, Construction and Environmental Engineering, Jomo Kenyatta University of Agriculture and Technology, Nairobi P.O. Box 62000-00200, Kenya
[4] Department of Economics, Engineering, Society and Business Organization (DEIM), Tuscia University, 01100 Viterbo, Italy
* Correspondence: petro@unitus.it

Abstract: Gully erosion is the most intensive type of water erosion and it leads to land degradation across the world. Therefore, analyzing the spatial occurrence of this phenomenon is crucial for land management. The objective of this research was to predict gully erosion susceptibility in the Kakia-Esamburmbur catchment in Narok, Kenya, which is badly affected by gully erosion. GIS and ensemble techniques using weight of evidence (WoE) and logistic regression (LR) models were used to map the susceptibility to gully erosion. First, 130 gullies were detected in the study area and portioned out 70:30 for training and validation, respectively. Nine gully erosion conditioning factors were selected as predictors. The relationships between the gully locations and the factors were identified and quantified using WoE, LR and WoE–LR ensemble models. The results show that land use/cover, distance to road, sediment transport index (STI) and topographic wetness index (TWI) are the factors that have the most influence on gully occurrence in the catchment. Additionally, the WoE–LR model performed better than the WoE and LR models, producing an AUC value of 0.88, which was higher than that of the WoE model, 0.62 and the LR model, 0.63. Therefore, the WoE–LR ensemble model is useful in gully erosion susceptibility mapping and is of help to decision makers in land-use planning.

Keywords: gully susceptibility; weight of evidence (WoE); logistic regression (LR); ensemble; soil loss; Kakia-Esamburmbur; Narok

check for updates

Citation: Nkonge, L.K.; Gathenya, J.M.; Kiptala, J.K.; Cheruiyot, C.K.; Petroselli, A. An Ensemble of Weight of Evidence and Logistic Regression for Gully Erosion Susceptibility Mapping in the Kakia-Esamburmbur Catchment, Kenya. *Water* **2023**, *15*, 1292. https://doi.org/10.3390/w15071292

Academic Editor: Jean-Luc PROBST

Received: 30 January 2023
Revised: 13 March 2023
Accepted: 20 March 2023
Published: 24 March 2023

1. Introduction

Soil erosion caused by water is considered to be one of the primary causes of land degradation worldwide [1]. Soil erosion, which is a natural process that entails detachment, transportation and deposition of soil particles, can be exacerbated by anthropogenic activities [2]. Water induced soil erosion is usually caused by deforestation, climate change, intensification of agriculture, urbanization and tectonic activities, all of which threaten land and water resources sustainability [3]. Soil erosion leads to reservoir siltation [4]; loss of fertile topsoil [5]; flooding and decline in species [6,7]; and eutrophication in streams, reservoirs and lakes [8]. Furthermore, there has been an increase in soil erosion rates over the years [9], for instance soil erosion is responsible for soil loss in most landscapes in Ethiopia and Kenya [10].

Water erosion occurs in various forms including splash, interrill (sheet), rill, stream bank and gully erosion [11]. Gully erosion is the most complex type of erosion amongst these, as it is triggered and accelerated by heavy rainfall and land use change [12]. Studies conducted around the world have shown that gully erosion can contribute between 10% and 94% of the total catchment erosion [13]. Gullies decrease soil productivity through the incision of agricultural lands, and cause restrictions on roads, land use and structures [14].

A gully is defined as an erosion channel with steep side walls and head-cut that is actively eroding due to the surface flows that remove and transport soil particles [15]. It can be defined as having a cross-sectional area greater than 929 cm^2 [13], and is too big to be removed by ordinary tillage practices [16]. Gully erosion is a threshold process and therefore several studies have put emphasis on defining the topographic as well as hydraulic conditions to predict gully erosion susceptibility [16,17].

In contrast to the significant contribution of gullies to total erosion on watersheds, few models have been developed for quantifying the effects of this phenomenon [18]. There are few physically based models that can assess gully erosion. They include CREAMS (chemicals, runoff and erosion from agricultural management systems), EGEM (ephemeral gully erosion mode) and WEPP (Water Erosion Prediction Project), which have been applied in the Loess Plateau [19,20]. However, the above models need a lot of geophysical and geochemical input data. Furthermore, the physical methods cannot assess the spatial distribution of gullies, which is important in visualizing the exposure of an area to gullying [21]. In regions where there is data scarcity, statistical and data mining methods can be successfully used.

Natural hazards have been assessed using a variety of data mining, bivariate and multivariate statistical methods. Some of these have been used to assess the susceptibility to gully erosion. Logistic regression (LR) [14,22]; weight of evidence (WoE) [23]; maximum entropy (ME) [24]; stochastic gradient tree boost (SGT) [16]; frequency ratio (FR) [25]; random forest (RF) [26]; multivariate adaptive regression splines (MARS) [17,27]; index of entropy (IoE) [28]; certainty factor (CF) [29] and support vector machine (SVM) [30].

All the aforementioned techniques have some drawbacks in the effectiveness of their predicted results, despite being able to determine gully erosion susceptibility. These disadvantages can be reduced through ensemble modeling. Recently, ensemble modeling has received significant attention in the spatial prediction of natural hazards such as landslides [31], groundwaters [32], droughts [33] and floods [34] due to improved prediction performance and the ability to deal with complex data [11]. Ensemble methods are techniques in which a prediction model is formed from a combination of various base classifiers [35].

WoE and LR have been used in particular in gully erosion susceptibility mapping, nonetheless, they have their own weaknesses. The main advantage of WoE is that it calculates the weight of the factors through statistical methods and thus avoids subjective weighting. In addition, input maps with missing data can be accommodated and have no significant impact on the result [23]. However, it neglects the correlation between the factors, which is important in gully erosion susceptibility because the conditioning factors should not be correlated. LR is able to evaluate the association between conditioning factors but it cannot analyze the influence of different classes within a conditioning factor [36]. Hence, their drawbacks can be solved and performance improved through their integration.

Starting from such premises, the main objectives of this study were to: (i) determine the level of influence of environmental conditioning factors causing the occurrence of gully erosion in the study area and (ii) assess the capability of WoE–LR to predict gully erosion susceptibility.

2. Materials and Methods

2.1. Study Area

The study area is composed of Kakia and Esamburmbur sub-catchments, which are 30.5 km^2 and 15.7 km^2, respectively (Figure 1), and are located in Narok County, South-West of Kenya. It lies between longitudes 35.83° E and 35.93° E and between latitudes 1.00° S and 1.10° S. The region's elevation ranges from 1828 to 2147 m above sea level [37]. The main drainage channel is the permanent Enkare Narok River, which rises from the Mau Forest and flows through Narok town. The Enkare Narok has two tributaries, which are the seasonal Esamburmbur and Kakia streams that flow through Narok town and converge a few meters before draining into it [38]. The catchment experiences bimodal rainfall in a year [36]. The long rainy season is usually between the months of March and May, while the short rainy season occurs between October and December. The mean annual rainfall for the area is 750 mm while the temperature range for the area is between 8 °C and 28 °C [39]. The main land use/cover in the catchment includes cropland, forests, built-up areas and shrubs. Cropland, which is the main land use, is made up of crops such as maize and wheat [38].

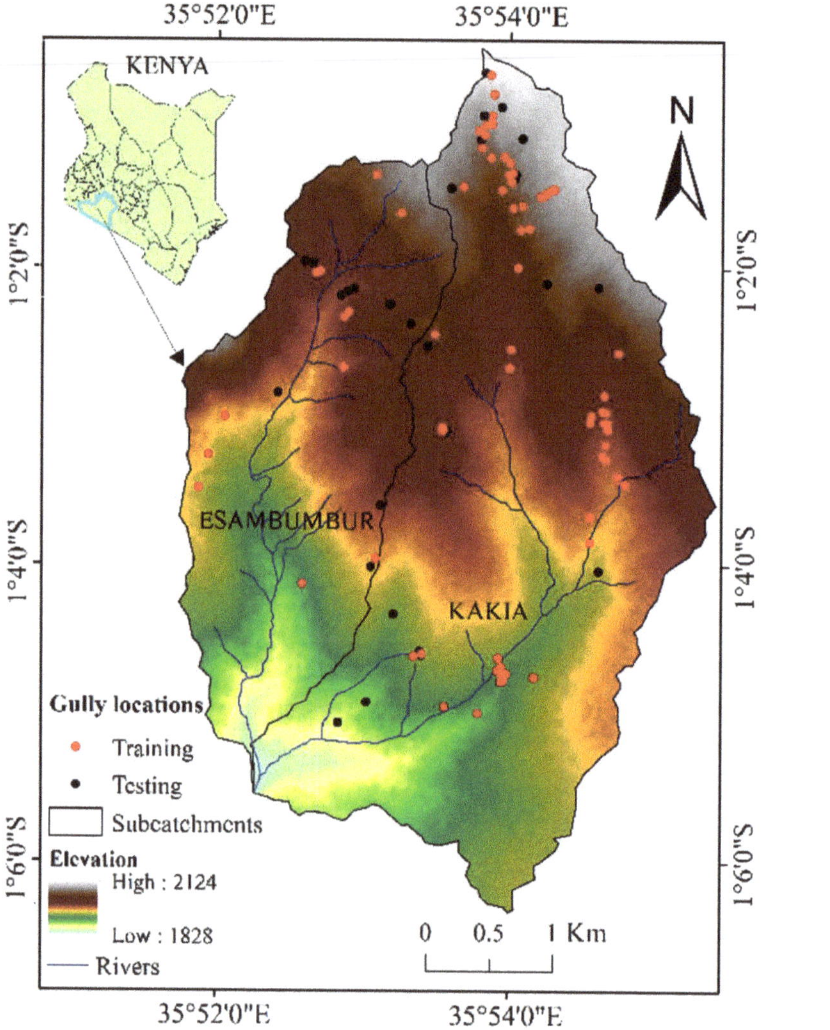

Figure 1. Map of Kakia-Esamburmbur catchment showing the training and testing datasets for modeling.

2.2. Methodology

Gully erosion susceptibility was carried out in four steps including: (1) preparation of a gully inventory which contains locations of gullies in the catchment; (2) preparation of the gully conditioning factors; (3) gully erosion susceptibility modeling using WoE and WoE–LR; and (4) model evaluation by constructing the receiver operating characteristic (ROC) curve and calculating the area under the ROC curve (AUC). Figure 2 is a flow diagram indicating the methodology used in this study.

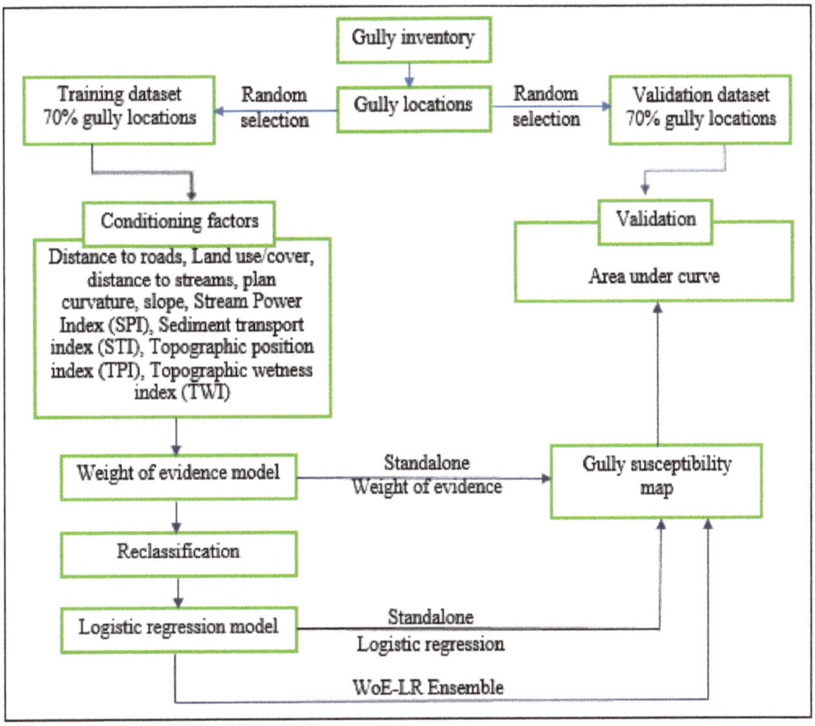

Figure 2. Flowchart for the research methodology on gully erosion susceptibility mapping in Kakia-Esamburmbur catchment, Narok.

2.2.1. Gully Erosion Inventory

Gully erosion mapping in the Kakia-Esamburmbur catchment was performed through field survey using GPS and Google Earth satellite images. Figure 3 shows some of the gully erosion in the study area. Eventually, a reliable and detailed inventory map with a total of 130 gullies was created. The gully locations were randomly divided into two groups which include the training (70%) and the validation (30%) sets, and which were differentiated from one another using a random dividing algorithm [40]. Both the training and validation sets were merged with an equal number of randomly selected locations that represent absence of gully erosion [24]. The absence dataset was created in ArcGIS by using the random-point tool [24].

2.2.2. Conditioning Factors

There are various environmental factors that control the critical conditions for gully development and these are mainly related to rainfall, topography, soil and land use [18]. In order to recognize the susceptible areas, good knowledge of the main gully erosion-related factors is needed. Therefore, conditioning factors were chosen from previous studies [21,41].

In this study, ArcGIS, QGIS and a system for automated geoscientific analyses (SAGA) were used to generate and exhibit such a data grid. The conditioning factors that were used in this study are (1) slope, (2) plan curvature, (3) topographic wetness index (TWI), (4) distance to streams (m), (5) distance to roads (m), (6) topographic position index (TPI), (7) stream power index (SPI), (8) sediment transport index (STI) and (9) land use/land cover.

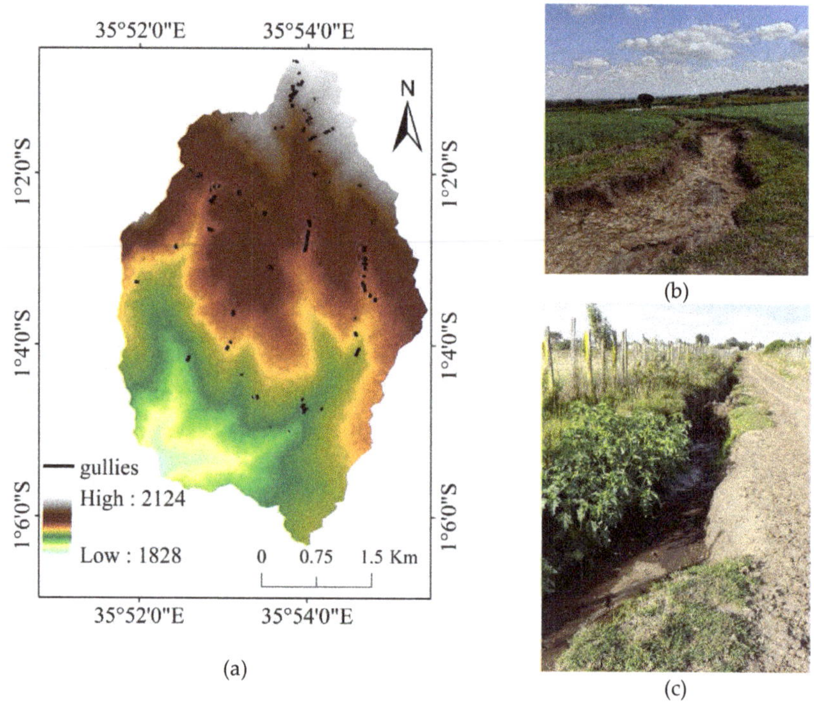

Figure 3. Map of the observed gullies (**a**) and pictures of some of the gullies identified in the Kakia-Esamburmbur catchment (**b**,**c**).

Slope controls the velocity and volume of concentrated flow thus affecting surface erosion and soil erosion [18]. Slope was derived from a 12.5 m resolution DEM from ALOS PALSAR [42] using ArcGIS and ranges from 0% to 42.19% (Figure 4a). Curvature assessment is important for inferring suitable geomorphological data. Plan curvature influences slope erosion processes through the convergence and divergence of water fluxes downhill. In this research, the plan curvature was derived from the DEM with a spatial resolution of 12.5 m in ArcGIS. The plan curvature values range from −3.2 to 3.2 (Figure 4b). Plan curvature can be classified into three classes: concave (positive curvature), flat (zero curvature) and convex (negative curvature) conditions, which were used in this study [36].

Topographic wetness index (TWI) is a secondary topographic factor that is used to evaluate the hydrological features of a region and is a crucial gully erosion determining factor. It represents the spatial distribution of wetness conditions [43]. It was calculated using Equation (1). The ArcGIS software was used for TWI mapping and the values range from 3.6 to 18.6 (Figure 4c).

$$\mathrm{TWI} = \ln\left(\frac{A_c}{\tan\,\beta}\right) \quad (1)$$

where

A_c = upstream contributing area (m^2) and

β = slope gradient (°).

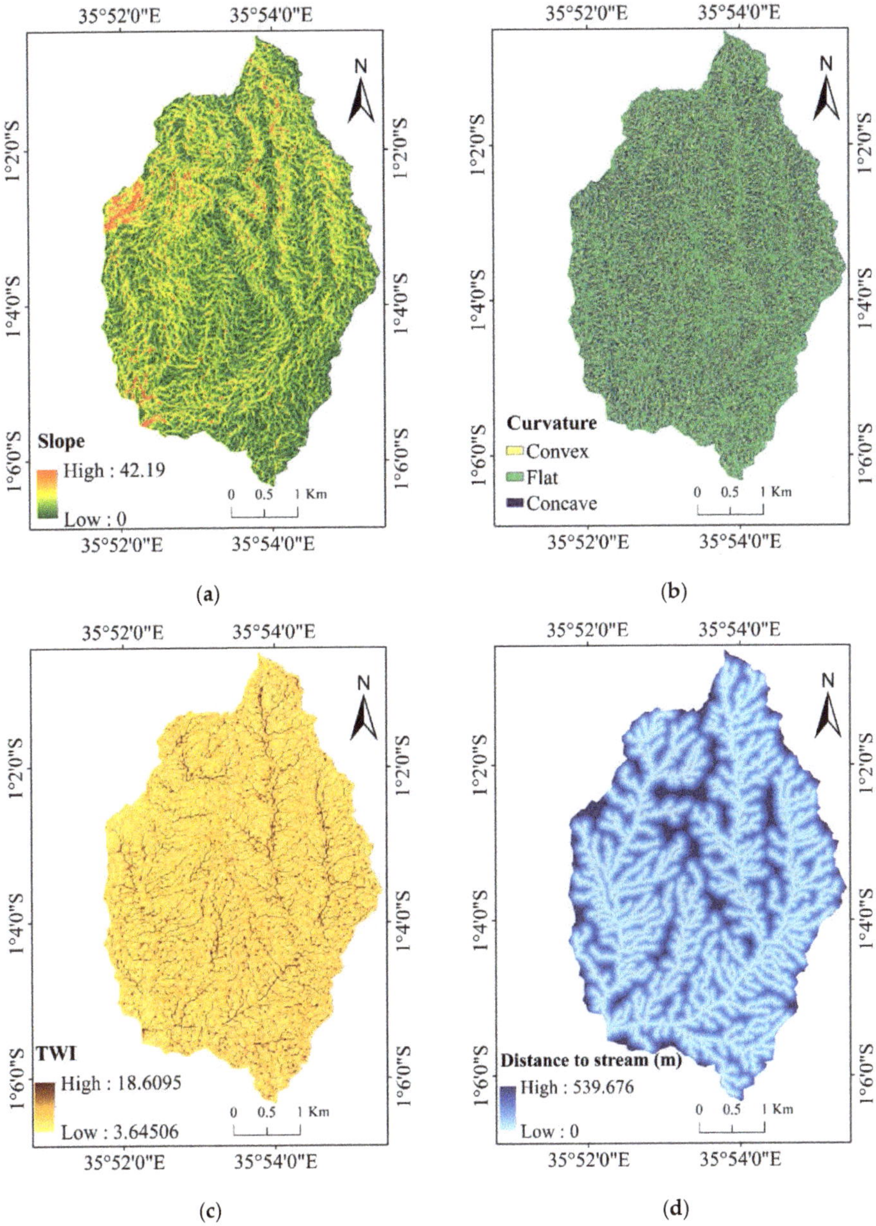

Figure 4. Gully erosion conditioning factors: (**a**) slope (percent), (**b**) curvature, (**c**) topographic wetness index (TWI) and (**d**) distance to stream/channel.

Distance to streams is important in evaluating the role of runoff in gully erosion [44]. Road construction has an adverse influence on hill sustainability at which flow may be suitable for gullies [18]. They were both calculated in ArcGIS using the Euclidean distance tool, which will give the distance (m) from each raster cell to the closest stream/road section [45]. The distance ranged from 0 to 2530 m for streams (Figure 4d) and 0 to 1674 m for roads (Figure 5a).

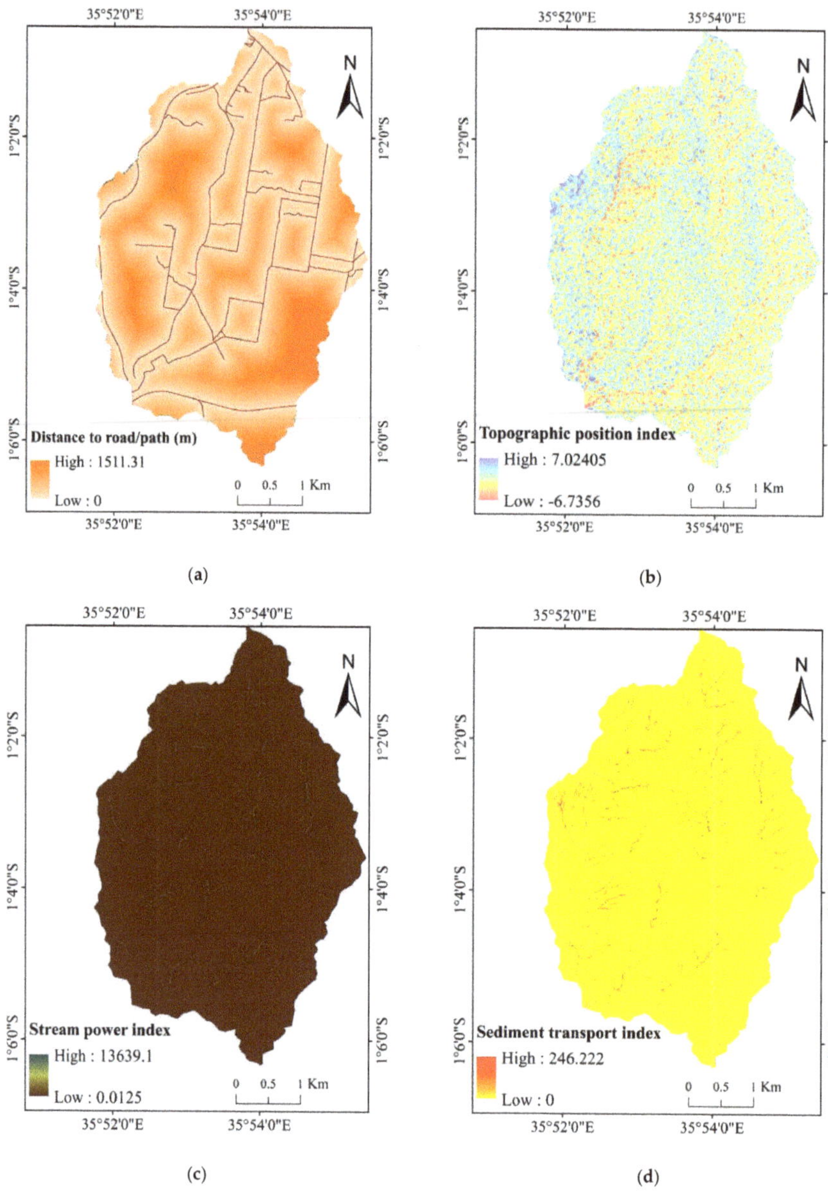

Figure 5. Gully erosion conditioning factors: (**a**) distance to road, (**b**) topographic position index (TPI), (**c**) stream power index (SPI) and (**d**) sediment transport index (STI).

Topographic position index (TPI) is a widely used approach for evaluating topographic slope location. It is the difference between the elevation of a cell and the average of the surrounding cells [17]. It was calculated for each cell using the algorithm [46] denoted by Equation (2) in QGIS using SAGA-GIS. The TPI values for the study area range between −6.7 and 7 and were divided into five classes (Figure 5b).

$$TPI = Z_0 - Z \tag{2}$$

where

Z_0 = elevation at the central point and

Z = average elevation around it within a predetermined radius (R).

The stream power index (SPI) has direct proportionality to stream power, which is the time rate of energy being used, and therefore can be used to estimate the overland flow erosive power. The SPI reflects the discharge and flow erosive power, which influences gully erosion susceptibility [47]. A higher value of SPI indicates that the stream has a much more powerful erosion on the slope surface [35]. The SPI value was calculated from the DEM in ArcGIS using the Equation (3).

$$\mathrm{SPI} = (\mathrm{A_c} \times \tan\beta) \tag{3}$$

where A_c is the upstream contributing area (m^2) and β is slope gradient (°). The SPI values of the study area range between 0 and 13,639 (Figure 5c).

The sediment transport index (STI) has been used to analyze erosion and deposition processes and topographic effects on soil loss. It was calculated using Equation (4) [28] in ArcGIS.

$$\mathrm{STI} = \left(\frac{\mathrm{A_c}}{22.13}\right)^{0.6}\left(\frac{\sin\beta}{0.0896}\right)^{1.3} \tag{4}$$

where A_c is the upstream contributing area (m^2) and β is slope gradient (°). The STI values in the study range from 0 to 246 and were classified into five classes (Figure 5d).

Land use/land cover (LULC) has a significant influence on hydrological and geomorphological pathways due to its direct or indirect effect on infiltration, evapotranspiration, runoff and sediment dynamics [48]. It also influences nutrients, structure and soil properties [45]. Vegetation protects the soil against erosion and surface runoff. Roots offer reinforcement thus increasing shear strength of the soil [45]. The land use/land cover map was acquired from Sentinel-2 [49] and prepared in ArcGIS. It has seven classes, i.e., trees, shrubs, grassland, cropland, vegetation aquatic/regularly flooded, bare and built-up areas (Figure 6).

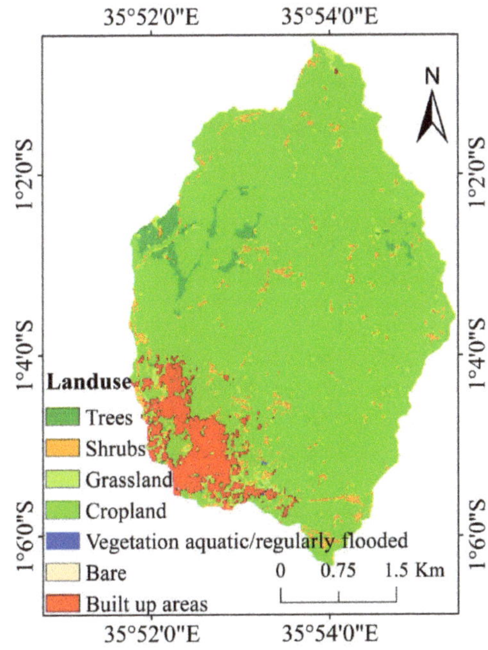

Figure 6. Gully erosion conditioning factors: land use/cover.

2.3. Multi-Collinearity Test

As the independent variable, the above-mentioned conditioning factors were used to examine the effect of correlation among them. It is a problem in the modeling process if both predictor variables are strongly related. The problem is known as collinearity. Tolerance and the VIF (variance inflation factor) are both significant measures of multi-collinearity identification. Tolerance is the inverse of VIF [50]. The test was carried out in the R software.

2.4. Weight of Evidence Model

The weight of evidence (WoE) is a probabilistic approach model based on a log linear form of Bayes' theorem of conditional probability [51]. It has been used for flood susceptibility mapping [37] and landslide susceptibility mapping [52]. By overlaying gully locations with each conditioning factor, their statistical relationship can be identified and assessed to establish whether and how significantly the conditioning factor is responsible for gully occurrence. WoE is expressed as shown in Equation (5)

$$P(A|B) = P(B|A) \times \frac{P(A)}{P(B)} \tag{5}$$

where A represents the presence of a gully and B represents the presence of the gully conditioning factor. Conditional independence is a very important aspect that should be considered in the WoE method. Positive, W^+ and negative, W^- weights are used to calculate the WoE model. The weight for each B is calculated based on the presence or absence of A as shown in Equations (6) and (7).

$$W^+ = \ln \frac{P(B|A)}{P\left(B\middle|\bar{A}\right)} \tag{6}$$

$$W^- = \ln \frac{P\left(\bar{B}\middle|A\right)}{P\left(\bar{B}\middle|\bar{A}\right)} \tag{7}$$

where P denotes the probability and ln denotes the natural log function. A and $\bar{A}$ indicate the presence and absence of a gully, respectively. The presence of a gully conditioning factor is represented by B, and the absence of a gully conditioning factor is represented by $\bar{B}$. A positive (W^+) weight explains the presence of the conditioning factor at the gully location and its value indicates the correlation between the gullies and the gully conditioning factor [53]. A negative (W^-) weight explains the gully conditioning factor's absence and indicates the degree of negative correlation [41]. The final weight, $W_{final} = W^+ - W^-$ will be used to quantify and indicate the spatial relationship between the effective gully conditioning factors and gully occurrence. If W_{final} is positive, it indicates a positive spatial relationship and if it is negative, it indicates a negative spatial relationship [40]. After applying the WoE model, the weights of the factors W_{final} were summed to produce a map of gully erosion susceptibility based on the Equation (8) [54].

$$GESI = \sum W_{final} \tag{8}$$

where GESI is the gully erosion susceptibility index.

2.5. Logistic Regression Model

Logistic regression (LR) is a multivariate statistical technique used in the development of predictive models from either discrete or continuous explanatory variables [18]. LR allows the development of a multivariate regression association linking a dependent variable and explanatory variables, which are important in predicting presence or absence of an outcome based on values of a set of explanatory variables [51]. In this case, the outcome is the presence of gullies and the LR is applied to predict variable (Y) that can be equal to presence of gully (1) or absence of gully (0). The LR uses a logistic function to assess the association linking the dependent variable and the explanatory variables. The logistic function is expressed as Equation (9) [52].

$$P = \frac{1}{1 + e^{-l}} \tag{9}$$

where P is the probability of gully erosion occurrence, which varies from 0 to 1 and l is the linear logistic factor, whose value varies from $-\infty$ to $+\infty$ and is defined by Equation (10).

$$l = \text{intercept} + b_1x_1 + b_2x_2 + \ldots + b_kx_k \tag{10}$$

where x_1, x_2 and x_k are the explanatory variables and b_1, b_2 and b_k are the coefficients of the LR model. In this study, LR was performed using the R software and ArcGIS, adopting forward stepwise regression to select the explanatory variables.

For the LR–WoE ensemble model, the determined class values of the WoE model were used to create the input for the LR model. The LR model requires a numerical rather than a categorical dependent variable. In order to address this issue, the weights of the classes of the conditioning factors were determined using the WoE model. The weights were then employed to transform the categorical variables into numerical variables. Therefore, all the classes of the conditioning factors were converted to numerical variables as shown in Tables 1 and 2. Finally, using the natural break method, the gully susceptibility map was reclassified into very low, low, moderate, high and very high classes after applying the LR–WoE model [22].

Table 1. Multicollinearity test statistics of gully conditioning factors.

Parameter	Collinearity statistics	
	Tolerance	VIF
Curvature	0.85	1.17
Distance to road	0.83	1.20
Distance to stream	0.92	1.09
Landcover	0.94	1.06
Slope	0.88	1.14
SPI	0.96	1.04
STI	0.41	2.41
TPI	0.88	1.14
TWI	0.40	2.49

Table 2. Weights of conditioning factors Slope, Curvature, TWI, Distance to stream and Distance to road as analyzed using WoE method.

Factors	Class	W^+	W^-	W_{final}
Slope (%)	0–4.4	−0.8852	0.0099	−0.8951
	4.5–8.1	−0.5559	0.0441	−0.6000
	8.2–11.2	−0.3376	0.0359	−0.3735
	11.3–16.0	−0.2115	0.1523	−0.3638
	16.1–42.1	0.4138	−0.3441	0.7579
Curvature	Concave	−0.1319	0.0420	−0.1739
	flat	0.1961	−0.2273	0.4233
	convex	−0.3327	0.0938	−0.4266
TWI	3–6.1	−0.4293	0.2090	−0.6383
	6.1–7.7	−0.1982	0.0832	−0.2814
	7.7–9.6	0.0501	−0.0105	0.0606
	9.6–11.8	0.4188	−0.0441	0.4629
	11.8–18.7	1.8308	−0.1672	1.9979

Table 2. *Cont.*

Factors	Class	W^+	W^-	W_{final}
Distance to stream (m)	0–38	−0.2149	0.2410	−0.4559
	38–89	−0.6371	0.1579	−0.7949
	89–150	1.3807	−0.3503	1.7310
	150–224	−0.7861	0.0261	−0.8122
	224–539	−0.5384	0.0051	−0.5434
Distance to road (m)	0–204	3.8957	−0.2429	4.1387
	204–434	0.5204	−0.0521	0.5725
	434–683	−0.7537	0.1270	−0.8807
	683–959	−0.9306	0.2262	−1.1568
	959–1511	0.0563	−0.0442	0.1005

Note: W^+ is positive weight, W^- is negative weight, W_{final} is $W^+ - W^-$.

2.6. Model Evaluation

The model performance was evaluated by drawing the receiver operating characteristic (ROC) and calculating the area under the curve (AUC). It was generated by plotting the sensitivity, also known as true positive rate (TPR), on the *y*-axis, Equation (11) against the 1-specificity, which is the false positive rate (FPR), on the *x*-axis using Equation (12) [52]. A ROC of 1 indicates perfect prediction. The model sensitivity is the percentage of existing gully pixels correctly predicted using the model, while the 1-specificity is the percentage of predicted gully pixels over the total study area.

$$\text{y-axis} = \text{TPR} = \left(\frac{\text{TP}}{\text{TP} + \text{FN}}\right) \quad (11)$$

$$\text{x-axis} = \text{FPR} = 1 - \left(\frac{\text{TN}}{\text{TN} + \text{FP}}\right) \quad (12)$$

where TP stands for true positive, TN stands for true negative, FP stands for false positive and FN stands for false negative. The TP and TN are the proportions of gully cells that are correctly classified as gully and non-gully, while the FP and FN are the proportions of gully cells that are incorrectly classified as gully and non-gully [11]. The AUC is a common method for assessing models used in prediction of natural hazards [55,56]. The AUC is calculated using the Equation (13).

$$\text{AUC} = \left(\sum \text{TP} + \sum \text{TN}\right) / (\text{P} + \text{N}) \quad (13)$$

where P denotes the total number of gullies and N denotes the total number of non-gullies.

3. Results

3.1. Conditioning Factor's Effect

3.1.1. Multicollinearity Test

The VIF ranges from 1.04 to 2.49, according to the multicollinearity test, while the TOL, which is the reciprocal of the VIF, ranges from 0.40 to 0.96. (Table 1). VIF values greater than five, with corresponding TOL values less than 0.2, indicate significant multicollinearity among factors [57]. The results indicate that there is no multi-collinearity between the conditioning factors.

3.1.2. Weight of Evidence Model

It is observed that there is a higher chance of gully occurrence at the high slope range of 16.1–42.1 while the lower slope classes have a low possibility of gully occurrence, which means the relationship is positive as seen from Table 2. The convex curvature had the highest W_{final} value, indicating that such areas in the study area are most vulnerable to gully erosion. From the results it can be seen that higher values of TWI indicate a higher chance of gully erosion occurrence while lower values indicate lower chances of gully erosion occurrence, thus there is a positive association. This can be attributed to the fact that at low TWI, less runoff is generated and at higher TWI, gully development is encouraged [54].

The results in Table 2 show that there is a higher possibility of gully occurrence near a road network. This can be seen through the high positive value of the final weight for the least distance from the roads (W_{final} = 4.1). This can be explained by the fact that the development of roads leads to soil disturbance which exacerbates gully formation [47].

TPI has a negative relationship with gully erosion. The lowest values of TPI had the highest W_{final} values while the highest TPI values had the lowest W_{final} values. Table 3 shows that higher values of SPI are associated with higher chances of experiencing gully erosion and vice versa.

Table 3. Weights of conditioning factors TPI, SPI, STI, LULC as analyzed using WoE method.

Factors	Class	W^+	W^-	W_{final}
TPI	−6.8–−1.5	0.4019	−0.0427	0.4446
	−1.5–−0.5	0.2975	−0.1021	0.3996
	−0.5–0.5	0.0118	−0.0067	0.0185
	0.5–1.6	−0.3065	0.0890	−0.3955
	1.6–7.1	−0.8978	0.0529	−0.9507
SPI	0–160	−0.1122	1.3792	−1.4915
	160–750	−1.3833	0.0208	−1.4041
	750–1925	2.1774	−0.0437	2.2211
	1925–4120	3.9965	−0.0769	4.0735
	4120–13640	2.9584	−0.0065	2.9649
STI	0–4	−0.1340	0.6750	−0.8090
	4–11	0.2549	−0.0254	0.2803
	11–27	1.0322	−0.0511	1.0834
	27–60	1.7170	−0.0346	1.7516
	60–247	1.7125	−0.0056	1.7181
LULC	Trees	−1.3375	0.0196	−1.3570
	Shrubs	−0.9576	0.0343	−0.9918
	Grassland	−0.8827	0.0098	−0.8925
	Cropland	0.1143	−1.0517	1.1660
	Vegetation aquatic/regularly flooded	3.1360	−0.0066	3.1425
	Bare	4.0024	−0.0067	4.0091
	Built-up areas	−2.1744	0.0553	−2.2296

Note: W^+ is positive weight, and W^- is negative weight, W_{final} is $W^+ - W^-$.

This shows that gully occurrence increases with an increase in SPI. This can be attributed to the fact that SPI is a product of slope and catchment area. It is used in showing areas of concentrated runoff and thus higher values of SPI indicate higher chances of experiencing gully erosion. The results also agree with a study carried out by [58].

Lower STI values have lower W_{final} values while higher STI values have higher W_{final} values. There is a positive association between STI and gully erosion. It indicates that gully occurrence increases with an increase in STI. This can be attributed to the fact that STI accounts for topographic effect on erosion. Similar results were also found in research carried out by [23].

The main land coverages which were identified in the catchment are: trees, shrubs, grassland, cropland, vegetation aquatic/regularly flooded, bare and built-up areas. The built-up area is mostly around Narok town at the outlet of the catchment. The bare land cover class has a high positive final weight (W_{final} = 4.0) while the built-up area has the highest negative final weight (W_{final} = −2.2).

These results show that there is a positive association between gully erosion occurrence and bare land while there is a strong positive association between the occurrence of gullies and built-up areas. From the above results, there is an implication that change in land cover/use from forest to agriculture, which increases the presence of bare soil, leads to gully development in the study area.

These findings agree with [59] who found that there is an association between gully erosion formation with the change in land cover/use from forest to cropland. Land use change has occurred in the Kakia-Esamburmbur catchment where forest cover has declined by 39% and agriculture has increased by 55.4% [39]. These findings are in agreement with the knowledge that forested areas experience less erosion in the form of gullies than barren land or wastelands [60].

3.1.3. Logistic Regression Model

The result for the LR model is given by Equation (14).

$$\begin{aligned} Z = -23.4199 + (5.519 \times \text{slope}) + (-1.890 \times \text{curvature}) + (7.590 \times \text{SPI}) + \\ (1.183 \times \text{Distance to road}) + (-5.323 \times \text{STI}) + \\ (0.639 \times \text{Distance to stream}) + (2.211 \times \text{TPI}) + \\ (4.154 \times \text{LULC}) + (5.994 \times \text{TWI}) \end{aligned} \quad (14)$$

3.1.4. LR–WoE Ensemble Model

The LR–WoE ensemble model developed for prediction of gully erosion susceptibility in the study area at 10% significance level is given by Equation (15). The significance of the conditioning factors is shown in Table 4.

$$\begin{aligned} Z = -7.026 + (0.061 \times \text{TWI}) + (0.03 \times \text{Distance to stream}) + (0.007 \times \text{Distance to road}) - \\ (0.023 \times \text{STI}) + (0.008 \times \text{LULC}) \end{aligned} \quad (15)$$

From Equation (15) it is obvious that TWI, distance to stream, distance to road and LULC have positive coefficients.

This means that these factors are related to the occurrence of gullies in a positive way. The coefficient of STI, on the other hand, is negative, indicating a negative relationship with the occurrence of gullies in the study area.

Table 4. Significance levels of the conditioning factors.

	Estimate	Standard Error	Z Value	Pr (>|z|)	
(Intercept)	−7.025994	741.6681	−0.009	0.99244	
Curvature	−0.003545	0.002803	−1.265	0.20593	
Distance to road	0.00677	0.001605	4.218	2.46×10^{-5}	***
Distance to stream	0.002653	0.001544	1.719	0.08569	.
Land use/cover	0.007804	0.003642	2.143	0.03214	*
Slope	0.004048	0.00634	0.639	0.52313	
SPI	0.008114	0.030244	0.268	0.78847	
STI	−0.023267	0.006674	−3.486	0.00049	***
TPI	−0.00194	0.002445	−0.793	0.42765	
TWI	0.061095	0.014213	4.299	1.72×10^{-5}	***

Note: Significance codes: '***' 0.001 '*' 0.05 '.' 0.1.

3.2. *Gully Erosion Hazard*

3.2.1. Weight of Evidence Model

The total weight, which is a result of summing the final weight (W_{final}) values presented in Tables 2 and 3, is shown in Figure 7. The total weighted map was converted into five classes: very low, low, moderate, high and very high hazard. It can be seen from the map that the areas around the forests have very low susceptibility levels. This implies that increasing forest cover in the Kakia-Esamburmbur catchment can be carried out in order to reduce soil erosion. The majority of the catchment that has moderate to high susceptibility levels is agricultural land.

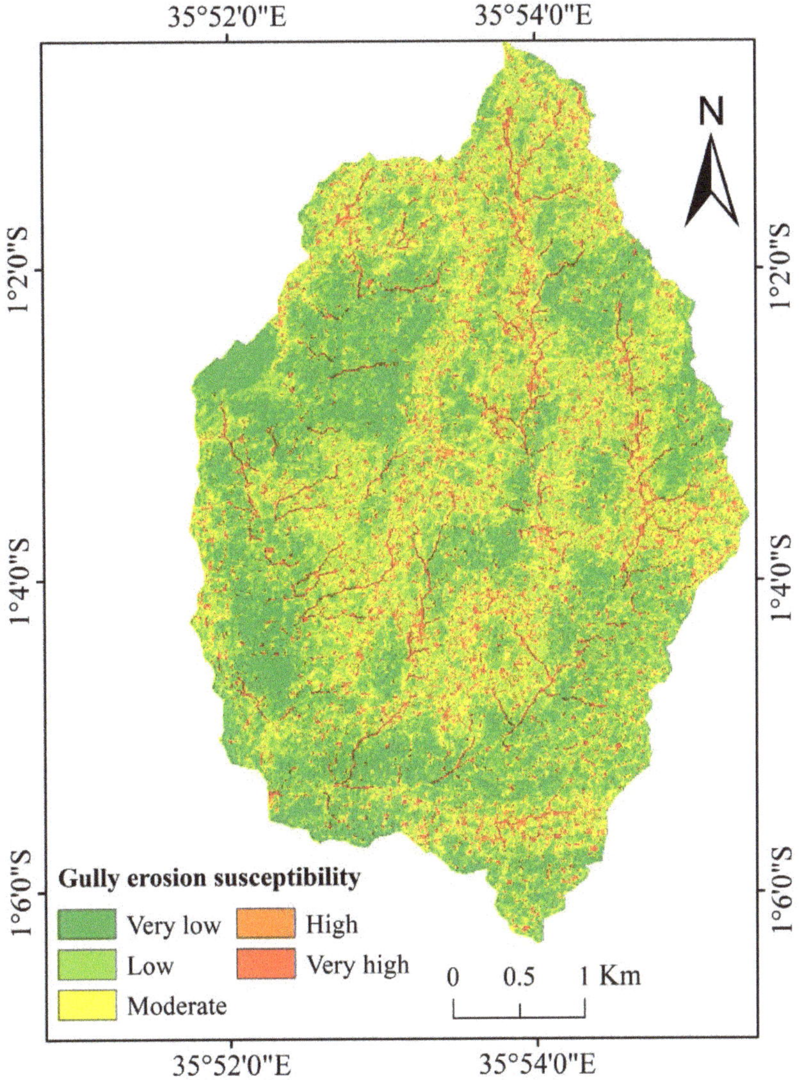

Figure 7. Map of gully erosion susceptibility of Kakia-Esamburmbur catchment using WoE.

3.2.2. LR Model and WoE–LR Ensemble Model

A gully erosion susceptibility map was created with the LR (Figure 8) and WoE–LR ensemble (Figure 9) using Equations (14) and (15), respectively, in ArcGIS interface and using the Jenks natural breaks classification method, reclassified into five classes (very low, low, moderate, high and very high) for visual interpretation.

3.3. Model Validation

The model performance was evaluated by drawing the receiver operating characteristic (ROC) curve. A ROC of 1 indicates perfect prediction. The following is a classification of the relationship between prediction accuracy and AUC: 0.9–1 equals excellent; 0.8–0.9 equals very good; 0.7–0.8 equals good; 0.6–0.7 equals average; and 0.5–0.6 equals poor [55]. It was found that the WoE model has an accuracy of 62% during training and a validation accuracy of 67% as shown in Figure 10. This accuracy level is average but lower than that obtained by [5] of 79.5% and by [23] of 67.8%. The LR model has an accuracy of 63% and 67% during training and validation, respectively, as shown in Figure 11. The WoE–LR ensemble performed much better and had an accuracy of 88% during training and a validation accuracy of 78% as shown in Figure 12.

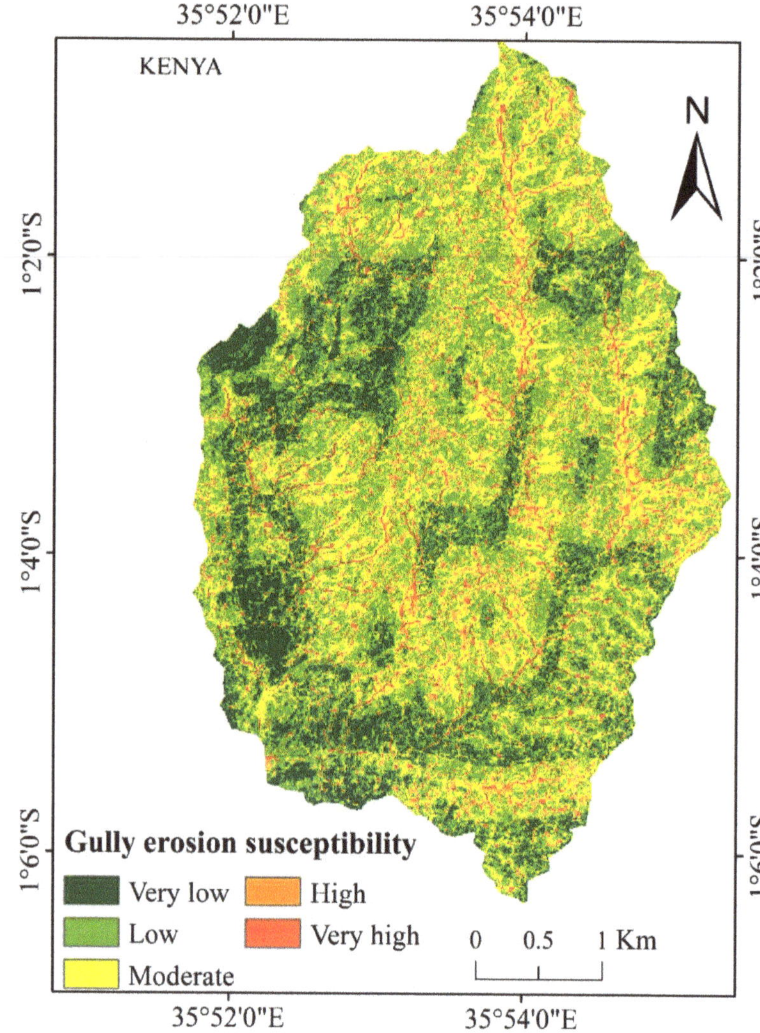

Figure 8. Map of gully erosion susceptibility of Kakia-Esamburmbur catchment using LR.

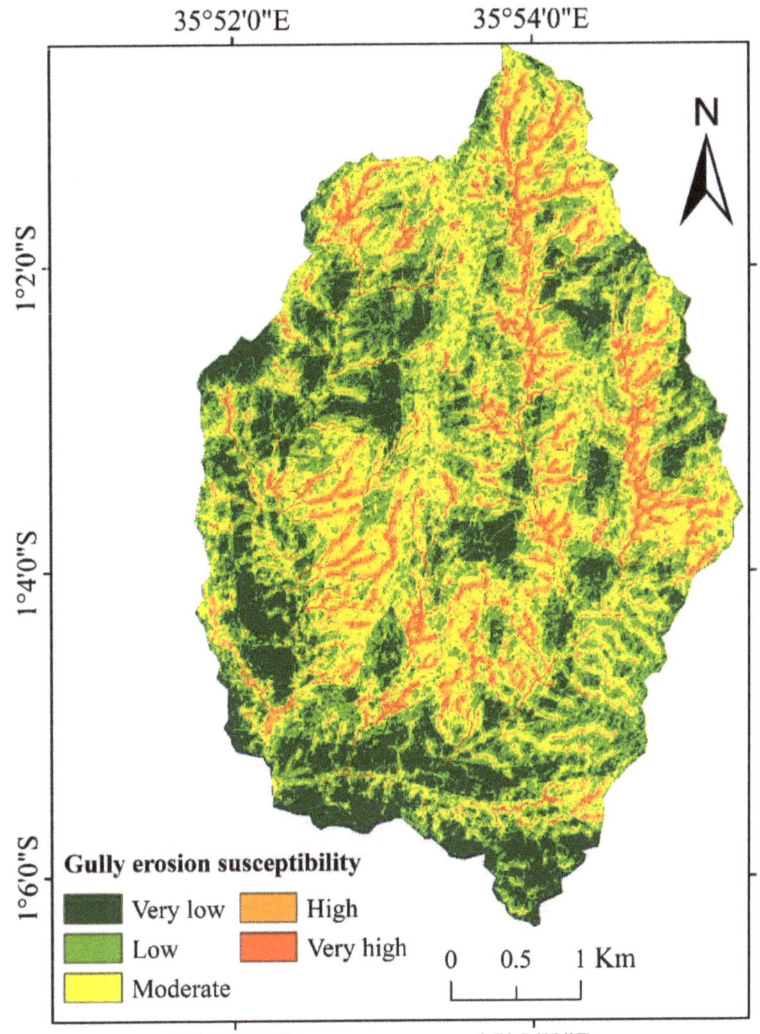

Figure 9. Map of gully erosion susceptibility of Kakia-Esamburmbur catchment using WoE–LR.

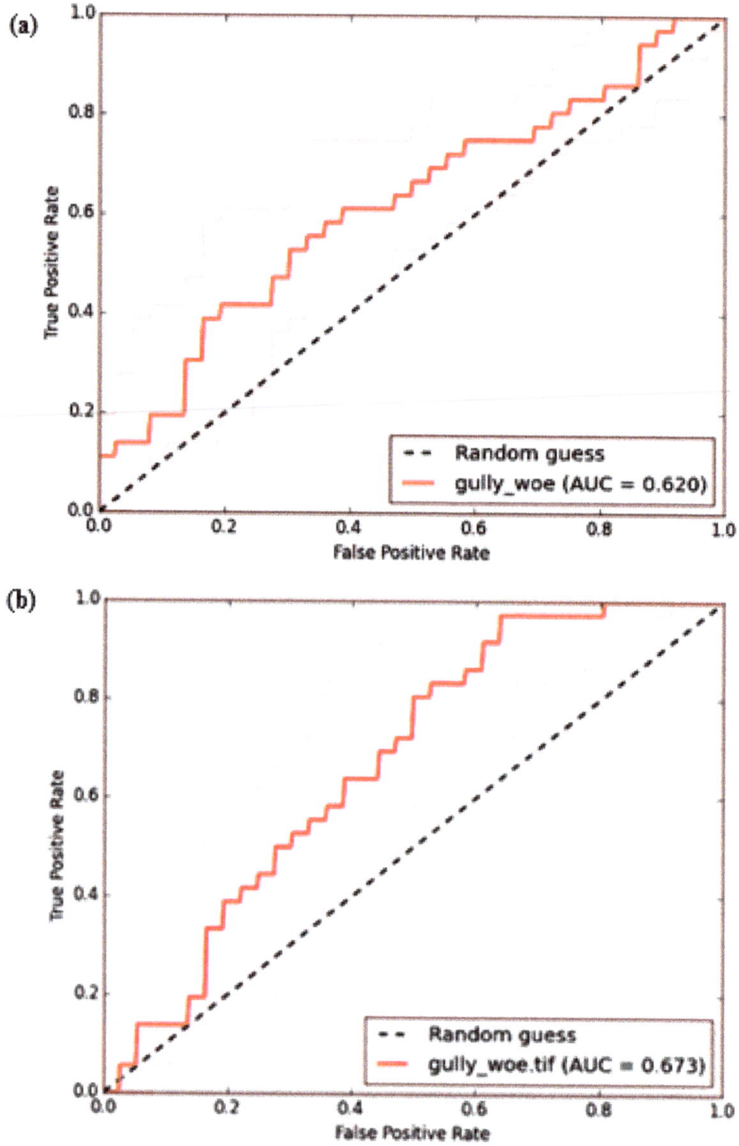

Figure 10. Prediction rate curve for gully erosion susceptibility map using WoE in Kakia-Esamburmbur catchment: (**a**) training and (**b**) validation.

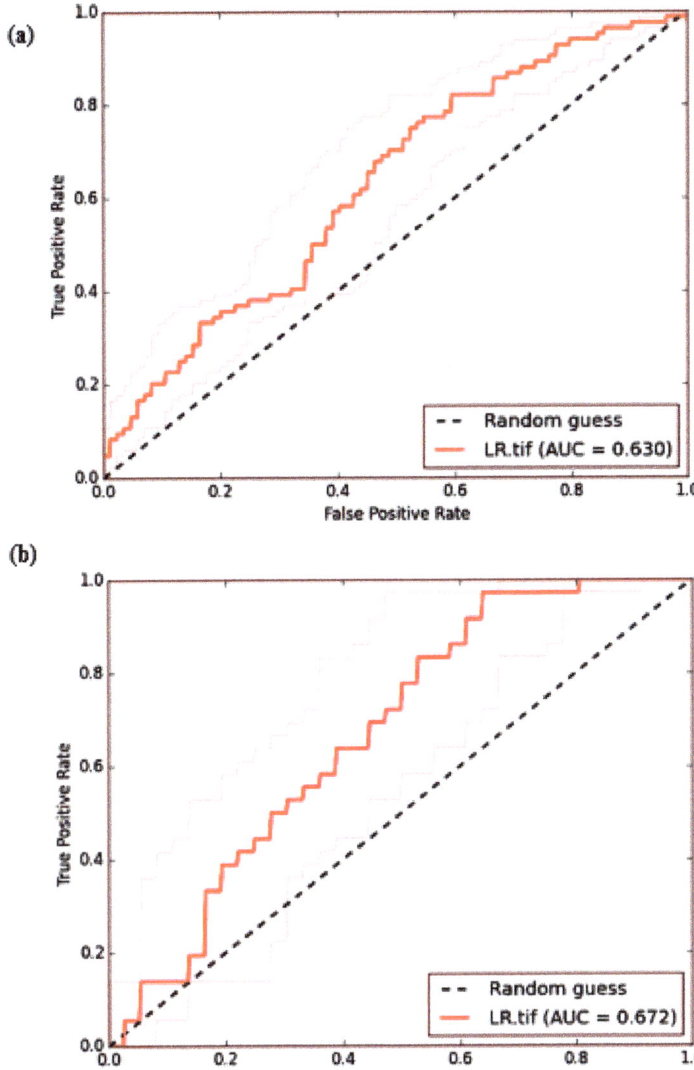

Figure 11. Prediction rate curve for gully erosion susceptibility map using LR in Kakia-Esamburmbur catchment: (**a**) training and (**b**) validation.

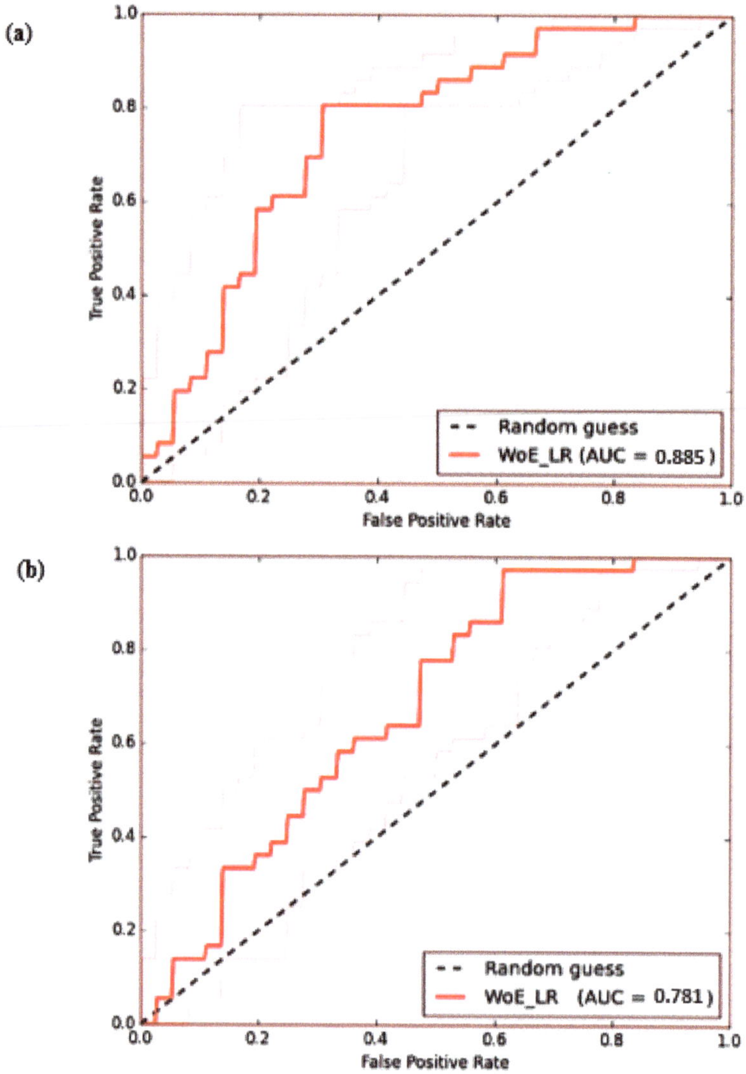

Figure 12. Prediction rate curve for gully erosion susceptibility map using WoE–LR in Kakia-Esamburmbur catchment: (**a**) training and (**b**) validation.

4. Discussion

Gully erosion has a two-fold impact on the environment: first, it degrades the surface and subsurface horizons of the soil, increasing the production of sediment, and second, it increases surface runoff and decreases groundwater recharge [61,62]. The first step in managing gully erosion is through gully erosion risk assessment that is conducted using gully erosion susceptibility models. Susceptibility maps can be helpful to government institutions, particularly since they can facilitate mitigation decisions for gully erosion [63]. In this research, three approaches along with the GIS technique were used for gully erosion susceptibility mapping: (1) using weight of evidence, (2) logistic regression and (3) their ensemble model. Using ROC curves and AUC values, the overall accuracy of the gully erosion susceptibility models was assessed.

In gully erosion investigations, it is critical to assess the significance of conditioning factors in gully susceptibility mapping. The results of the current study also showed that distance to roads, distance to streams, land use, TWI and STI are the most influential factors on gully occurrence. The output of the models reveals different gullying susceptibility value ranges. The results show that gully formation is more likely to develop in regions that are closer to roads and streams, have sparser vegetation and have higher drainage densities than other areas. These results are consistent with [17].

Maps showing gully erosion susceptibility classified areas of the basin with low slopes and close proximity to roads/paths as having high and extremely high susceptibility. In contrast, forest regions with steep slopes showed low susceptibility to gully erosion. In the forest, roughness due to vegetation cover may result in medium runoff factors in this area as well, resulting in a low concentrated flow force of degradation [63]. The outcomes further showed that the LR model's prediction accuracy was improved by combining the WoE method with the LR approach. Therefore, rather than employing a single model that is tailored to each study area, it is crucial to assess the efficacy of a combination/ensemble of multiple models.

5. Conclusions and Recommendations

Gully erosion can lead to damage to the environment and agriculture lands thus leading to migration and less agricultural productivity. Many methods have recently been used to assess gully erosion susceptibility. In this paper, the WoE model and a proposed WoE–LR model were used to produce a gully erosion susceptibility map.

Nine conditioning factors were used including: slope, plan curvature, distance to streams (m), distance to roads (m), topographic position index (TPI), topographic wetness index (TWI), stream power index (SPI), sediment transport index (STI) and land use/land cover. Using the natural break method, the susceptibility maps were classified into five susceptibility classes (very low, low, moderate, high and very high). The WoE technique produced a map of areas susceptible to gully erosion with a prediction accuracy of 62% while the LR model and the WoE–LR ensemble had an accuracy of 63% and 88%, respectively.

These results show that an ensemble model performs better than an individual model. Amongst the conditioning factors, it was found that STI, land use/cover, TWI, distance to stream and distance to road had the most influence on the occurrence of gullies. The areas that were found to have low susceptibility include the forested areas and also near the already developed Narok town. The areas where agriculture is practiced and next to the roads were found to more susceptible to gully erosion.

The Kakia-Esamburmbur faces major land degradation due to loss of top fertile soil caused by water erosion. The susceptibility map can be used as a basic rule of thumb reference for land management in the Kakia-Esamburmbur catchment. The proposed ensemble model and gully erosion susceptibility map should be helpful to environmentalists and engineers for future planning and management.

Author Contributions: L.K.N.: Data curation, investigation, formal analysis, and original draft preparation; J.M.G.: funding acquisition, supervision, validation, resources, review and editing; J.K.K.: supervision, review and editing; C.K.C.: supervision, review and editing; A.P.: supervision, review and editing. All authors have read and agreed to the published version of the manuscript.

Funding: This research was funded by the Pan African University of Basic Sciences, Technology and Innovation (PAUSTI) and National Research Fund (NRF), Kenya (Flash Floods and Erosion in Enkare Narok Basin: Causes and Management).

Data Availability Statement: The data presented in this study are available on request from the corresponding author. The data are not publicly available due to privacy restrictions.

Acknowledgments: The authors acknowledge and appreciate the administrative and technical support from the Pan African University of Basic Sciences, Technology and Innovation (PAUSTI) and support of NRF project partners Water Resources Authority (WRA) and Ewaso Ng'iro South Development Authority (ENSDA).

Conflicts of Interest: The authors declare no conflict of interest. The funders had no role in the design of the study; in the collection, analyses, or interpretation of data; in the writing of the manuscript; or in the decision to publish the results.

References

1. Arabameri, A.; Pradhan, B.; Pourghasemi, H.R.; Rezaei, K. Identification of erosion-prone areas using different multi-criteria decision-making techniques and gis. *Geomat. Nat. Hazards Risk* **2018**, *9*, 1129–1155. [CrossRef]
2. Apollonio, C.; Petroselli, A.; Tauro, F.; Cecconi, M.; Biscarini, C.; Zarotti, C.; Grimaldi, S. Hillslope erosion mitigation: An experimental proof of a nature-based solution. *Sustainability* **2021**, *13*, 6058. [CrossRef]
3. Sholagberu, A.T.; Mustafa, M.R.; Yusof, K.W.; Hashim, A.M. Geo-statistical based susceptibility mapping of soil erosion and optimization of its causative factors: A conceptual framework. *J. Eng. Sci. Technol.* **2017**, *12*, 2880–2895.
4. Fox, G.A.; Sheshukov, A.; Cruse, R.; Kolar, R.L.; Guertault, L.; Gesch, K.R.; Dutnell, R.C. Reservoir Sedimentation and Upstream Sediment Sources: Perspectives and Future Research Needs on Streambank and Gully Erosion. *Environ. Manag.* **2016**, *57*, 945–955. [CrossRef]
5. Zabihi, M.; Mirchooli, F.; Motevalli, A.; Darvishan, A.K.; Pourghasemi, H.R.; Zakeri, M.A.; Sadighi, F. Spatial modelling of gully erosion in Mazandaran Province, northern Iran. *Catena* **2018**, *161*, 1–13. [CrossRef]
6. Odunuga, S.; Ajijola, A.; Igwetu, N.; Adegun, O. Land susceptibility to soil erosion in Orashi Catchment, Nnewi South, Anambra State, Nigeria. *Proc. Int. Assoc. Hydrol. Sci.* **2018**, *376*, 87–95. [CrossRef]
7. Slimane, A.B.; Raclot, D.; Evrard, O.; Sanaa, M.; Lefevre, I.; Bissonnais, Y.L. Relative contribution of rill/interrill and gully/channel erosion to small reservoir siltation in mediterranean environments. *Land Degrad. Dev.* **2016**, *27*, 785–797.
8. Issaka, S.; Ashraf, M.A. Impact of soil erosion and degradation on water quality: A review. *Geol. Ecol. Landsc.* **2017**, *1*, 1301053. [CrossRef]
9. Borrelli, P.; Robinson, D.A.; Fleischer, L.R.; Lugato, E.; Ballabio, C.; Alewell, C.; Meusburger, K.; Modugno, S.; Schütt, B.; Ferro, V.; et al. An assessment of the global impact of 21st century land use change on soil erosion. *Nat. Commun.* **2017**, *8*, 1–13. [CrossRef]
10. Hassen, G.; Bantider, A. Assessment of drivers and dynamics of gully erosion in case of Tabota Koromo and Koromo Danshe watersheds, South Central Ethiopia. *Geoenviron. Disasters* **2020**, *7*, 5. [CrossRef]
11. Bui, D.T.; Shirzadi, A.; Shahabi, H.; Chapi, K.; Omidavr, E.; Pham, B.T.; Asl, D.T.; Khaledian, H.; Pradhan, B.; Panahi, M.; et al. A novel ensemble artificial intelligence approach for gully erosion mapping in a semi-arid watershed (Iran). *Sensors* **2019**, *19*, 2444. [CrossRef]
12. Arabameri, A.; Pourghasemi, H.R. Spatial Modeling of Gully Erosion Using Linear and Quadratic Discriminant Analyses in GIS and R. In *Spatial Modeling in GIS and R for Earth and Environmental Sciences*; Elsevier: Amsterdam, The Netherlands, 2019; pp. 299–321. [CrossRef]
13. Poesen, J.; Nachtergaele, J.; Verstraeten, G.; Valentin, C. Gully erosion and environmental change: Importance and research needs. *Catena* **2003**, *50*, 91–133. [CrossRef]
14. Zakerinejad, R.; Maerker, M. An integrated assessment of soil erosion dynamics with special emphasis on gully erosion in the Mazayjan basin, southwestern Iran. *Nat. Hazards* **2015**, *79*, 25–50. [CrossRef]
15. Arabameri, A.; Cerda, A.; Tiefenbacher, J.P. Spatial pattern analysis and prediction of gully erosion using novel hybrid model of entropy-weight of evidence. *Water* **2019**, *11*, 1129. [CrossRef]
16. Angileri, S.E.; Conoscenti, C.; Hochschild, V.; Märker, M.; Rotigliano, E.; Agnesi, V. Water erosion susceptibility mapping by applying Stochastic Gradient Treeboost to the Imera Meridionale River Basin (Sicily, Italy). *Geomorphology* **2016**, *262*, 61–76. [CrossRef]
17. Javidan, N.; Kavian, A.; Pourghasemi, H.R.; Conoscenti, C.; Jafarian, Z. Gully erosion susceptibility mapping using multivariate adaptive regression splines-replications and sample size scenarios. *Water* **2019**, *11*, 2319. [CrossRef]
18. Conoscenti, C.; Angileri, S.; Cappadonia, C.; Rotigliano, E.; Agnesi, V.; Märker, M. Gully erosion susceptibility assessment by means of GIS-based logistic regression: A case of Sicily (Italy). *Geomorphology* **2014**, *204*, 399–411. [CrossRef]
19. Han, F.; Ren, L.; Zhang, X.; Li, Z. The WEPP Model Application in a Small Watershed in the Loess Plateau. *PLoS ONE* **2016**, *11*, e0148445. [CrossRef]
20. Langendoen, E.J.; Wells, R.R.; Ursic, M.E.; Vieira, D.A.N.; Dabney, S.M. Evaluating sediment transport capacity relationships for use in ephemeral gully erosion models. *IAHS-AISH Proc. Rep.* **2014**, *367*, 128–133. [CrossRef]
21. Pourghasemi, H.R.; Yousefi, S.; Kornejady, A.; Cerdà, A. Performance assessment of individual and ensemble data-mining techniques for gully erosion modeling. *Sci. Total Environ.* **2017**, *609*, 764–775. [CrossRef]
22. Wu, Z.; Wu, Y.; Yang, Y.; Chen, F.; Zhang, N.; Ke, Y.; Li, W. A comparative study on the landslide susceptibility mapping using logistic regression and statistical index models. *Arab. J. Geosci.* **2017**, *10*, 187. [CrossRef]
23. Dube, F.; Nhapi, I.; Murwira, A.; Gumindoga, W.; Goldin, J.; Mashauri, D.A. Potential of weight of evidence modelling for gully erosion hazard assessment in Mbire District–Zimbabwe. *Phys. Chem. Earth* **2014**, *67–69*, 145–152. [CrossRef]
24. Kornejady, A.; Ownegh, M.; Bahremand, A. Landslide susceptibility assessment using maximum entropy model with two different data sampling methods. *Catena* **2017**, *152*, 144–162. [CrossRef]
25. Rahmati, O.; Haghizadeh, A.; Pourghasemi, H.R.; Noormohamadi, F. Gully erosion susceptibility mapping: The role of GIS-based bivariate statistical models and their comparison. *Nat. Hazards* **2016**, *82*, 1231–1258. [CrossRef]

26. Rahmati, O.; Pourghasemi, H.R.; Melesse, A.M. Application of GIS-based data driven random forest and maximum entropy models for groundwater potential mapping: A case study at Mehran Region, Iran. *Catena* **2016**, *137*, 360–372. [CrossRef]
27. Gómez-Gutiérrez, Á.; Conoscenti, C.; Angileri, S.E.; Rotigliano, E.; Schnabel, S. Using topographical attributes to evaluate gully erosion proneness (susceptibility) in two mediterranean basins: Advantages and limitations. *Nat. Hazards* **2015**, *79*, 291–314. [CrossRef]
28. Jaafari, A.; Najafi, A.; Pourghasemi, H.R.; Rezaeian, J.; Sattarian, A. GIS-based frequency ratio and index of entropy models for landslide susceptibility assessment in the Caspian forest, northern Iran. *Int. J. Environ. Sci. Technol.* **2014**, *11*, 909–926. [CrossRef]
29. Hong, H.; Chen, W.; Xu, C.; Youssef, A.M.; Pradhan, B.; Tien Bui, D. Rainfall-induced landslide susceptibility assessment at the Chongren area (China) using frequency ratio, certainty factor, and index of entropy. *Geocarto Int.* **2017**, *32*, 139–154. [CrossRef]
30. Chen, W.; Pourghasemi, H.R.; Kornejady, A.; Zhang, N. Landslide spatial modeling: Introducing new ensembles of ANN, MaxEnt, and SVM machine learning techniques. *Geoderma* **2017**, *305*, 314–327. [CrossRef]
31. Nhu, V.-H.; Mohammadi, A.; Shahabi, H.; Bin Ahmad, B.; Al-Ansari, N.; Shirzadi, A.; Clague, J.J.; Jaafari, A.; Chen, W.; Nguyen, H. Landslide Susceptibility Mapping Using Machine Learning Algorithms and Remote Sensing Data in a Tropical Environment. *Int. J. Environ. Res. Public Health* **2020**, *17*, 4933.
32. Chen, W.; Li, H.; Hou, E.; Wang, S.; Wang, G.; Panahi, M.; Li, T.; Peng, T.; Guo, C.; Niu, C.; et al. GIS-based groundwater potential analysis using novel ensemble weights-of-evidence with logistic regression and functional tree models. *Sci. Total Environ.* **2018**, *634*, 853–867. [CrossRef]
33. Roodposhti, M.S.; Safarrad, T.; Shahabi, H. Drought sensitivity mapping using two one-class support vector machine algorithms. *Atmos. Res.* **2017**, *193*, 73–82. [CrossRef]
34. Shafizadeh-Moghadam, H.; Valavi, R.; Shahabi, H.; Chapi, K.; Shirzadi, A. Novel forecasting approaches using combination of machine learning and statistical models for flood susceptibility mapping. *J. Environ. Manag.* **2018**, *217*, 1–11. [CrossRef]
35. Kadavi, P.R.; Lee, C.W.; Lee, S. Application of ensemble-based machine learning models to landslide susceptibility mapping. *Remote Sens.* **2018**, *10*, 1252. [CrossRef]
36. Tehrany, M.S.; Pradhan, B.; Jebur, M.N. Flood susceptibility mapping using a novel ensemble weights-of-evidence and support vector machine models in GIS. *J. Hydrol.* **2014**, *512*, 332–343. [CrossRef]
37. Tchouateu, O.T.; Mwangi, H.M.; Sang, J.K.; Ngugi, H.N. Impacts of climate change on peak streamflow in Kakia-Esamburmbur Sub-catchment of Enkare Narok River catchment, Kenya. *J. Sustain. Res. Eng.* **2020**, *5*, 194–205.
38. Mireille, N.M.; Mwangi, H.M.; Mwangi, J.K.; Gathenya, J.M. Analysis of Land Use Change and Its Impact on the Hydrology of Kakia and Esamburmbur of the. *Hydrology* **2019**, *6*, 86. [CrossRef]
39. Umukiza, E.; Raude, J.M.; Wandera, S.M.; Petroselli, A.; Gathenya, J.M. Impacts of Land Use and Land Cover Changes on Peak Discharge and Flow Volume in Kakia and Esamburmbur Sub-Catchments of Narok Town, Kenya. *Hydrology* **2021**, *8*, 82. [CrossRef]
40. Rahmati, O.; Tahmasebipour, N.; Haghizadeh, A.; Pourghasemi, H.R.; Feizizadeh, B. Evaluation of different machine learning models for predicting and mapping the susceptibility of gully erosion. *Geomorphology* **2017**, *298*, 118–137. [CrossRef]
41. Lombardo, L.; Cama, M.; Maerker, M.; Rotigliano, E. A test of transferability for landslides susceptibility models under extreme climatic events: Application to the Messina 2009 disaster. *Nat. Hazards* **2014**, *74*, 1951–1989. [CrossRef]
42. Alaska Satellite Facility Data Search Vertex (2020). Available online: https://search.asf.alaska.edu/#/?dataset=ALOS (accessed on 29 January 2023).
43. Arabameri, A.; Chen, W.; Loche, M.; Zhao, X.; Li, Y.; Lombardo, L.; Cerda, A.; Pradhan, B.; Bui, D.T. Comparison of machine learning models for gully erosion susceptibility mapping. *Geosci. Front.* **2019**, *11*, 1609–1620. [CrossRef]
44. Hu, C.; Wright, A.L.; Lian, G. Estimating the Spatial Distribution of Soil Properties Using Environmental Variables at a Catchment Scale in the Loess Hilly Area, China. *Int. J. Environ. Res. Public Heal.* **2019**, *16*, 491. [CrossRef]
45. Trigila, A.; Iadanza, C.; Esposito, C.; Scarascia-Mugnozza, G. Comparison of Logistic Regression and Random Forests techniques for shallow landslide susceptibility assessment in Giampilieri (NE Sicily, Italy). *Geomorphology* **2015**, *249*, 119–136. [CrossRef]
46. Jenness, J. Topographic Position Index (tpi_jen.avx) Extension for ArcView 3.x, v. 1.2. Jenness Enterprises. 2006. Available online: http://www.jennessent.com/arcview/tpi.htm (accessed on 1 March 2023).
47. Chen, W.; Pourghasemi, H.R.; Naghibi, S.A. Prioritization of landslide conditioning factors and its spatial modeling in Shangnan County, China using GIS-based data mining algorithms. *Bull. Eng. Geol. Environ.* **2018**, *77*, 611–629. [CrossRef]
48. Meinhardt, M.; Fink, M.; Tünschel, H. Landslide susceptibility analysis in central Vietnam based on an incomplete landslide inventory: Comparison of a new method to calculate weighting factors by means of bivariate statistics. *Geomorphology* **2015**, *234*, 80–97. [CrossRef]
49. Copernicus. Available online: https://scihub.copernicus.eu/ (accessed on 31 August 2020).
50. Ozdemir, A. Using a binary logistic regression method and GIS for evaluating and mapping the groundwater spring potential in the Sultan Mountains (Aksehir, Turkey). *J. Hydrol.* **2011**, *405*, 123–136. [CrossRef]
51. Chen, W.; Sun, Z.; Han, J. Landslide susceptibility modeling using integrated ensemble weights of evidence with logistic regression and random forest models. *Appl. Sci.* **2019**, *9*, 171. [CrossRef]
52. Pourghasemi, H.R.; Moradi, H.R.; Fatemi Aghda, S.M. Landslide susceptibility mapping by binary logistic regression, analytical hierarchy process, and statistical index models and assessment of their performances. *Nat. Hazards* **2013**, *69*, 749–779. [CrossRef]

53. Xu, C.; Xu, X.; Dai, F.; Xiao, J.; Tan, X.; Yuan, R. Landslide hazard mapping using GIS and weight of evidence model in Qingshui River watershed of 2008 Wenchuan earthquake struck region. *J. Earth Sci.* **2012**, *23*, 97–120. [CrossRef]
54. Shit, P.K.; Bhunia, G.S.; Pourghasemi, H.R. Gully Erosion Susceptibility Mapping Based on Bayesian Weight of Evidence. *Adv. Sci. Technol. Innov.* **2020**, *8*, 133–146. [CrossRef]
55. Chen, W.; Reza, H.; Seyed, P.; Naghibi, A. A comparative study of landslide susceptibility maps produced using support vector machine with different kernel functions and entropy data mining models in China. *Bull. Eng. Geol. Environ.* **2018**, *77*, 647–664. [CrossRef]
56. Youssef, A.M.; Pourghasemi, H.R.; Pourtaghi, Z.S.; Al-Katheeri, M.M. Landslide susceptibility mapping using random forest, boosted regression tree, classification and regression tree, and general linear models and comparison of their performance at Wadi Tayyah Basin, Asir Region, Saudi Arabia. *Landslides Landslides* **2015**, *13*, 839–856. [CrossRef]
57. Pourtaghi, Z.S.; Pourghasemi, H.R.; Rossi, M. Erratum to: Forest fire susceptibility mapping in the Minudasht forests, Golestan province, Iran. *Environ. Earth Sci.* **2015**, *73*, 1535. [CrossRef]
58. Meliho, M.; Khattabi, A.; Mhammdi, N. A GIS-based approach for gully erosion susceptibility modelling using bivariate statistics methods in the Ourika watershed, Morocco. *Environ. Earth Sci.* **2018**, *77*, 655. [CrossRef]
59. Mukai, S. Gully Erosion Rates and Analysis of Determining Factors: A Case Study from the Semi-arid Main Ethiopian Rift Valley. *Land Degrad. Dev.* **2016**, *28*, 602–615. [CrossRef]
60. Chuma, G.B.; Mondo, J.M.; Ndeko, A.B.; Mugumaarhahama, Y.; Bagula, E.M.; Blaise, M.; Valérie, M.; Jacques, K.; Karume, K.; Mushagalusa, G.N. Forest cover affects gully expansion at the tropical watershed scale: Case study of Luzinzi in Eastern DR Congo. *Trees For. People* **2021**, *4*, 100083. [CrossRef]
61. Kou, M.; Jiao, J.; Yin, Q.; Wang, N.; Wang, Z.; Li, Y.; Yu, W.; Wei, Y.; Yan, F.; Cao, B. Successional Trajectory Over 10Years of Vegetation Restoration of Abandoned Slope Croplands in the Hill-Gully Region of the Loess Plateau. *Land Degrad. Dev.* **2016**, *27*, 919–932. [CrossRef]
62. Azareh, A.; Rahmati, O.; Rafiei-Sardooi, E.; Sankey, J.B.; Lee, S.; Shahabi, H.; Ahmad, B. Modelling gully-erosion susceptibility in a semi-arid region, Iran: Investigation of applicability of certainty factor and maximum entropy models. *Sci. Total Environ.* **2019**, *655*, 684–696. [CrossRef]
63. Arabameri, A.; Pradhan, B.; Bui, D.T. Spatial modelling of gully erosion in the Ardib River Watershed using three statistical-based techniques. *Catena* **2020**, *190*, 104545. [CrossRef]

water

MDPI

Article

Soil Loss Estimation Coupling a Modified USLE Model with a Runoff Correction Factor Based on Rainfall and Satellite Soil Moisture Data

Francesca Todisco [1,*], Lorenzo Vergni [1], Sofia Ortenzi [2] and Lucio Di Matteo [2]

1 Department of Agricultural, Food and Environmental Science, University of Perugia, 06124 Perugia, Italy; lorenzo.vergni@unipg.it
2 Department of Physics and Geology, University of Perugia, 06123 Perugia, Italy; sofia.ortenzi@studenti.unipg.it (S.O.); lucio.dimatteo@unipg.it (L.D.M.)
* Correspondence: francesca.todisco@unipg.it; Tel.: +39-075-5856052

Abstract: Satellite observations (Copernicus Sentinell-1) can supply antecedent soil moisture data, which helps to predict thresholds triggering runoff and runoff volume. In the paper, we developed a runoff correction factor to the USLE, using rainfall and satellite antecedent soil moisture data, following the approach of the modified USLE models such as the USLE-M and USLE-MM. The runoff and soil loss estimations accuracy are validated by plot-scale measurements (2008–2020 period) provided by SERLAB (Soil Erosion Laboratory) of the University of Perugia. The results show that the event rainfall depth added to the antecedent soil moisture is a fairly suitable predictor of the runoff. Using the simulated runoff in a USLE-MM model, the capability to predict event soil losses is enhanced with an RMSE = 0.57 Mg/ha lower than the RMSE ≈ 3.1 Mg/ha obtained by the USLE model. Using a modified USLE model, albeit with remote estimated runoff data, is still more advantageous at the event scale than the USLE model, which does not consider the runoff. These results are particularly significant for the estimation of runoff and soil losses. Satellite data shows the potential of applying the modified USLE models for large-scale monitoring and quantification of event soil erosion and runoff.

Keywords: USLE; remote sensing; soil water erosion; Copernicus Sentinel-1; runoff thresholds; runoff generation; rainfall runoff erosivity factor; runoff models; erosion models; hydrological processes seasonality

Citation: Todisco, F.; Vergni, L.; Ortenzi, S.; Di Matteo, L. Soil Loss Estimation Coupling a Modified USLE Model with a Runoff Correction Factor Based on Rainfall and Satellite Soil Moisture Data. *Water* **2022**, *14*, 2081. https://doi.org/10.3390/w14132081

Academic Editor: Nufang Fang

Received: 29 April 2022
Accepted: 27 June 2022
Published: 29 June 2022

1. Introduction

Several models (empirical, conceptual, physical, and process-oriented) have been proposed to estimate soil loss at different spatial and temporal scales [1,2]. The empirical approaches, among which the Universal Soil Loss Equation (USLE) and its subsequent updates (RUSLE and RUSLE2) [3] dominate, continue to find a wide application at all the spatial scales [4,5]. The reason for this diffusion of USLE models is to be found in the relative ease of application and reliability of the estimates [6]. The growing availability of remote sensing data, useful for quantifying USLE factors [7], has increased its popularity even more. On the other hand, the original USLE model [8], developed to predict the long-term average annual soil loss, cannot be considered the most effective tool for planning soil conservation measures since a large fraction of the total soil loss in a given area is due to a relatively few large storms [9,10]. The most common way to improve the reliability of the USLE model at the event scale is to modify the rainfall erosivity factor by multiplying it by the runoff coefficient (i.e., the ratio of the event runoff to the event rainfall depth) as in the USLE-M [11] and USLE-MM models [12,13]. Runoff, however, is not just a function of certain constant characteristics of the soil and the space-time characteristics of the rainfalls. On the contrary, using runoff means knowing how to predict it with reasonable precision through sufficiently accurate models, applicable in most cases. Todisco et al. [14] explained that although the modification of the erosivity with the runoff coefficient makes the USLE/RUSLE model more accurate at the event scale, it also makes its application more difficult, given the poor availability of runoff measurements. This problem can be solved by employing rainfall runoff simulation models, but in the literature, there are only a few examples of the application of USLE/RUSLE models in combination with rainfall runoff models [14–16]. The reason lies in the complexity of the rainfall runoff processes, which depend on several interacting factors, such as topographic characteristics, meteo-climatic conditions (including rainfall intensity and duration), lithological and hydraulic properties of soils, vegetation cover types, agricultural practices, etc. [17–20]. A recent approach to this problem is investigating the threshold patterns in stormflow resulting from changes in antecedent moisture conditions such as antecedent volumetric water content (θ), Antecedent Soil Moisture Index (ASI), degree of saturation (Sr), or a combination of these with cumulated rainfall, P, (e.g., ASI + P). The ASI index was introduced by [21]; it is expressed in mm, and it is defined by the product of the average volumetric water content (θ) and the depth of investigation (D). Some of these parameters are widely available all over the Earth as satellite data. Detty et al. [19] highlighted a high correlation between ASI + P and quick flow volumes for 13 runoff events, investigating runoff generation mechanisms in till-mantled headwater catchment (NH, USA). Penna et al. [22] proposed a threshold between θ and streamflow, analyzing a small alpine headwater catchment located in the Italian Dolomites. In detail, soil moisture and streamflow started to rise concurrently during wet conditions, while soil moisture peaked earlier than streamflow during dry conditions. In addition, runoff resulting from changes in antecedent moisture conditions was investigated. Ali et al. [23] reported different nonlinear input-output relationship shapes, such as the hockey stick shape [24], step function (or Heaviside) [25], Dirac function [26], and sigmoid function [27]. Threshold behavior in different ecohydrological processes (e.g., runoff generation) is strictly related to the spatial and temporal scales, climatic conditions, and physical controls. Following [23], other studies focused on the hydrologic response threshold values, trying to explain what might drive the differences among the sites. [28], noticed differences in runoff coefficients and hydrological dynamics between summer and fall/spring rainfall events, which were related to ASI + P. The study on a small, forested catchment in the Italian Pre-Alps equipped with soil moisture probes, weirs, and piezometers suggested a mix of intensity-dependent and wetness-dependent processes (ASI + P) controlling the runoff production mechanism. As reported by [29], seasonal variations in the threshold behaviors depend on vegetation canopy interception and wet/dry conditions. Scaife et al. [30] obtained different ASI + P thresholds on three forested catchments in North Carolina (USA). In these cases, maximum hourly rainfall intensity had a relatively small effect on quick flow generation compared to ASI + P. The authors presented straight lines above and below the thresholds, which are useful to compute the runoff based on ASI + P values. The role of P and ASI + P on the surface, subsurface, and epikarst seepage runoff (SR, SSR, and ESR) have been recently investigated in twelve 5 × 20 m karst plots in southwest China [31]. Based on the various shapes of the relationship between P or ASI + P and runoff [23], different models for karst slopes, such as hockey stick, step function, and the sigmoid function, have been detected [31]. Therefore, using P or ASI + P to derive generation thresholds (triggering stormflow), rise thresholds (rapid responses of stormflow), and predict runoff is challenging. The cited studies are focused on equipped small catchments or plots, where the volumetric water content is monitored continuously to capture the moisture content before the stormflow event. Unfortunately, the water content is not monitored in most of the small catchments worldwide, and when done, the data acquired may not represent the whole basin. Moreover, soil moisture equipment for continuous monitoring, especially those working at low frequency, needs to be calibrated before their use in the field to improve the reliability of measurements [32–36]. Nowadays, the increased detail of remote sensing satellite data products can provide information on moisture conditions in the first centimeters of soils, making available information helpful for defining runoff thresholds.

In this framework, the general objective of the present work is to investigate the possibility of using satellite soil moisture data to predict runoff thresholds, runoff and soil loss on an experimental plot (8 × 22 m) at the SERLAB experimental station (central Italy), characterized by fine fluvial-lacustrine deposits. The specific objectives are (i) to investigate the hydrologic response threshold values and dynamics with satellite soil moisture and rainfall data and (ii) to predict soil loss at the event scale with a USLE-MM model coupled with runoff models based on rainfall and satellite soil moisture data. Soil moisture data from Copernicus Sentinel-1 are integrated with data acquired on the ground (rainfall and runoff) at SERLAB. Runoff thresholds are presented by considering the seasonality and intensity of the events. Similar to the Scaife et al. [30] approach, the empirical runoff models below and above the thresholds are defined and coupled with the USLE-MM model to simulate the event of soil loss.

2. Materials and Methods

2.1. The SERLAB Experimental Station and the Runoff/Soil Loss Database

The Masse experimental station for soil erosion measurement of the Perugia University (SERLAB, Soil Erosion LABoratory, Figure 1) was established in 2007. It is located 20 km south of Perugia (42°59′34″ N 12°17′27″ E) in the Umbria region, central Italy. The station includes twelve plots:

- Four plots of 8 × 22 m^2;
- Two plots of 4 × 22 m^2;
- Two plots of 4 × 11 m^2;
- Two plots of 2 × 11 m^2;
- Two microplots of about 1 × 0.92 m^2.

Figure 1. View of the Masse experimental station, SERLAB (Central Italy). (42°59′34″ N 12°17′27″ E).

All plots are oriented parallel to a 16% slope. The area has a characteristic Mediterranean climate with an average annual rainfall of about 900 mm. The soil texture is silty-clay-loam (clay = 34%, silt = 59%, and sand = 7%) and the gravel content is negligible. The soil structure, massive when the soil is wet, becomes weak, fine, subangular, and blocky upon drying.

Each plot is equipped with a runoff collection system whose capacity (0.86 to 2.72 m^3) varies with the plot size. When a rainfall event producing runoff occurs, the collected volume is analyzed to measure the sediment concentration in the tanks and to derive the soil loss. All the plots (except for the first two on the right in Figure 1) are maintained in cultivated fallow through frequent tillage operations to remove any spontaneous vegetation and obliterate the presence of any rills formed during erosive events. The consistency of the data set is over 600 measures of event runoff and soil loss. Furthermore, the climatic station (tipping-bucket rain gauge, anemometer, thermometer, hygrometer) located within the experimental site provides the complete time series of climatic data since 2007 at a 5 min resolution. More details about the experimental schemes, technical features, and procedures of SERLAB can be found in previous papers [6,12–14,37,38]. For the study, only the data of one plot (8×22 m^2) were considered. This plot was equipped with a sonic distance sensor that enabled a detailed analysis of runoff formation and accumulation dynamics in the collecting tanks.

2.2. USLE Models for Soil Loss Estimation at the Event Scale

The original USLE equation for the estimation of the mean annual soil loss [8] has the following expression:

$$A = RKLSCP \tag{1}$$

where A (Mg ha^{-1} $year^{-1}$) is the mean annual soil loss per unit area, R = EI_{30} (MJ mm ha^{-1} h^{-1} $year^{-1}$) is the mean annual rainfall erosivity calculated as the product of the rainfall kinetic energy E (MJ ha^{-1}) by the maximum 30 min rainfall intensity observed during the event, I_{30} (mm h^{-1}), K (Mg ha h ha^{-1} MJ^{-1} mm^{-1}) is the soil erodibility factor, L (dimensionless) is the slope length factor, S (dimensionless) is the slope steepness factor, C (dimensionless) is the cover-management factor, and P (dimensionless) is the support practices factor.

Equation (1) can also be used to estimate the soil loss at event scale A_e (Mg ha^{-1}) using the erosivity factor of the single event R_e. A normalized value of the event soil loss $A_{e,N}$ can be obtained as:

$$A_{e,N} = \frac{A_e}{LSCP} = R_e K \tag{2}$$

Equation (2), however, tends to overestimate the lowest and underestimate the highest values [11]. The reason for this systematic error is the lack of explicit consideration of runoff. Various authors have proposed modifications to the USLE model to consider this aspect. The best-known models are the USLE-M, proposed by Kinnell and Risse [11], and the USLE-MM, proposed by Bagarello et al. [12]; both models include the runoff to correct the erosivity factor that becomes, actually, a rainfall runoff factor. The following expression can describe the two models:

$$A_{e,N} = (Q_R R_e)^{\alpha} K_u \tag{3}$$

where Q_R (mm/mm) is the event runoff coefficient, defined as the ratio of runoff volume Q (mm) and rainfall volume P (mm), with $\alpha = 1$ in the USLE-M and $\alpha > 1$ in the USLE-MM and where K_u is a modified erodibility factor that varies with the selected model.

The use of the USLE-MM model proved to be more efficient than the USLE-M model in the Italian experimental sites of Sparacia (Sicily) and Masse SERLAB [6]. In [6], based on the 532 measures of the SERLAB data set from 2008 to 2015, model (3) was parametrized with $K_u = 0.0896$ and $\alpha = 1.0479$. Other attempts were made in [14] that modified Equation (3) using soil moisture in place of the runoff coefficient in the rainfall runoff factor, deriving the model named Soil Moisture For Erosion, SM4E.

In this work, the possibility of using in the USLE-MM model the Q_R estimated by models based on the integrated use of satellite soil moisture data and event rainfall depth (i.e., $Q_R = f[(ADSI + P)]/P$) is evaluated and discussed. To focus the analysis on this objective, the USLE-MM model will be parametrized using the data selected in this study and deriving from the intersection of the SERLAB and Sentinel-1 data sets. The efficacy of the methodology is tested by comparing the estimated runoff, runoff coefficient, and soil loss with the corresponding measures available in the SERLAB data set.

2.3. Monitoring Soil Moisture from Satellite Data

The European Copernicus Project (Copernicus Global Land Service—CGLS), in service since 2013, aims to provide timely and high-quality essential bio-geophysical information and climate variables on the Earth's surface on a global scale. The satellite product data (e.g., land surface temperature, water bodies area, vegetation index, soil moisture, etc.) make it possible to monitor the status and evolution of vegetation, energy cycles, and water on a global scale [39,40]. The Sentinel-1 mission comprises a constellation of two identical radar imagery satellites in the same orbit; Sentinel-1A was launched in April 2014 and Sentinel-1B in April 2016. Specifically, the Sentinel-1 SAR satellite sensor can derive the Copernicus Global Land SSM (Surface Soil Moisture) product that describes the soil moisture of the first 50 mm of soil on a 1 km (0.1°) spatial sampling. SSM1 km data are derived from observations using the model TU-Wien-Change-Detection [41–43]. Since January 2015, the Copernicus SSM 1 km product has been available for the European continent per individual location every 3–8 days, and since October 2016, every 1.5–4 days (thanks to Sentinel-1B satellite) in general. The SSM1 km products can be requested and downloaded free of charge from the website https://land.copernicus.eu/global/products/ssm (accessed on 10 January 2022), with the following naming standard: c_gls_SSM1 km_<YYYYMMDDHHMM>_CEURO_S1CSAR_<VX.Y.Z>.nc., where <YYYMMDDHHMM> gives the temporal location of the file. The SSM1 km products include a netCDF4 file containing full resolution bands (ssm and ssm_noise), an XML file containing the metadata, and a colored geo-tiff file.

Sentinel-1 derives SSM(t) (Surface Soil Moisture at t time) directly from the observed radar backscatter; the changes in backscatter are interpreted as changes in soil moisture, while other surface properties such as geometry and roughness, and vegetation structure are interpreted as static parameters. The incidence angle dependency of the backscatter is modeled by the linear slope parameter, which allows normalization to a common reference observation angle ($\theta r = 40°$). The model parameters describe maximum dry and wet conditions and average signal contributions from vegetation and surface geometry. The long-term backscatter measurements are used to derive dry and wet soil conditions (parameters $\sigma^0{}_{dry(40)}$ and $\sigma^0{}_{wet(40)}$) from the radar backscatter signal [44]. The relative surface soil moisture is given by:

$$\mathrm{SSM(t)} = \frac{\sigma_{40}(t) - \sigma^0_{dry(40)}}{\sigma^0_{wet} - \sigma^0_{dry}} \tag{4}$$

SSM(t) represents a degree of saturation and ranges between 0% and 100% [40]. The volumetric water content $\mathrm{SSM_a(t)}$ can be obtained by introducing the soil porosity (n) as follows:

$$\mathrm{SSM_a(t)} = n\frac{\mathrm{SSM(t)}}{100} \tag{5}$$

As reported in the introduction, identifying runoff thresholds by empirical approaches requires the knowledge of the antecedent soil moisture content, i.e., $\mathrm{SSM_a(t)}$ at the time (t) just before the rainfall event triggering runoff. Since the first centimeters of soil layers are often tilled and, therefore, the porosity is very variable in space and time [45–47], obtaining $\mathrm{SSM_a(t)}$ values from SSM(t) is not possible. In other words, $\mathrm{SSM_a(t)}$ values obtained by Equation (5) may not be reliable. A possible way to overcome this problem is using SSM(t) satellite data to calculate the Antecedent Degree of Saturation Index (ADSI) similarly to ASI in [21]:

$$\mathrm{ADSI} = h\frac{\mathrm{SSM(t)}}{100} \tag{6}$$

where h is the thickness of the investigated soil, which is the first 50 mm for satellite observations, and ADSI is expressed in mm.

For this investigation, the SERLAB data set overlapped with the SSM1 km data by Copernicus Sentinel-1 available from January 2015. Considering that SSM(t) data are not continuous, only rainfall runoff events that occurred concurrently or just after the satellite SSM(t) measures were selected for the analysis.

3. Results

3.1. SERLAB Data Set

Forty-seven measurements were captured by overlapping the SSM1km Copernicus Sentinel-1 and the SERLAB data set; eighteen correspond to rainfall events that did not produce runoff, and three lacked soil loss measures. Table S1 shows the information available for the crossing set of data. From the SERLAB data set, the rainfall depth (P), runoff event (Q), and the soil loss, A_e, for the erosive events and other rainfall characteristics along with the corresponding date of occurrence are extracted. From Copernicus Sentinel-1, the corresponding SSM values with the date of satellite data acquisitions are extracted.

Most runoff values are lower than 2.0 mm (about 83% of the observations), while the others are higher than 5 mm, with a maximum value of 15.5 mm. The no runoff events (about 38% of the observation) have rainfall depth, P, lower than 10.0 mm, while the runoff, along with the soil loss, occurs for rainfall higher than 18.8 mm. Overall, the maximum P is about 87.4 mm, with a mean value of 30.4 mm. Six runoff events occurred with antecedent SSM(t) higher than 80%, one of which contributed to producing the highest runoff and erosive event during the autumn–winter season coupled with a cumulated rainfall of about 68.8 mm. The maximum rainfall intensity in 30 min, I_{30}, varies between 2.0 and 54.0 mm/h, with three events over 40.0 mm/h and soil loss values higher than 17.0 t/ha. These events in the data set are surveyed as rill events while the others are classified as interrill.

3.2. Definition of the Relationship between the Runoff and the ADSI + P

The rainfall data compared with the runoff values measured at SERLAB since 2008 consist of about a hundred measures (Figure 2a). Data points for which Copernicus SSM1km data are available (Table S1) are highlighted in Figure 2a. As expected, P values do not explain the runoff processes alone (i.e., for the same rainfall event, different runoff values correspond). Hence it is impossible to derive an accurate mathematical relationship estimating runoff from rainfall only. Therefore, when available, we tried to improve the runoff prediction model by correcting the rainfall depth with the antecedent soil moisture conditions. Although the consistency of the database is drastically reduced (47 measurements), a first attempt to improve the runoff estimation was carried out. Similar to the widely used approach that relates runoff and (ASI + P), the runoff values have been associated with ADSI + P (Figure 2b). Multiple thresholds are highlighted under the hypothesis that the data set can be classified on the basis of the season of occurrence. By separating the data set into the autumn–winter and the spring–summer events and according to [29], it is possible to identify the generation threshold (occurring at ADSI + P = 38.0 mm), the rise threshold for spring–summer events (ADSI + P = 63.2 mm), and the rise threshold for autumn–winter events (ADSI + P = 93.2 mm). In this analysis generation threshold corresponds to the ADSI + P value triggering runoff, while rise thresholds identify the ADSI + P values when the runoff behavior changes mainly due to the specific characteristics of the rainfall graph. Therefore, the rise thresholds separate events in high from the low hydrological response. A single linear model has been fitted to the events with low runoff (lower than 2.0 mm) to obtain an empirical relationship between ADSI + P and the runoff. Although runoff events higher than 2.0 mm are few, straight lines are used to describe the high runoff events maintaining the seasonal classification. This procedure seems to be viable even if, in this case, the few data available do not provide certainty that the hypothesis is reliable. The regressions models that are also drawn in Figure 2b, are as follows:

- For autumn–winter events

$$Q = 0.0188\,(ADSI + P) - 0.503 \quad 38.0 < ADSI + P < 93.2 \quad R^2 = 0.390 \tag{7a}$$

$$Q = 0.814\,(ADSI + P) - 75.826 \quad 93.2 \leq ADSI + P < 109.0 \quad R^2 = 0.934 \tag{7b}$$

- For spring–summer events

$$Q = 0.0188\,(ADSI + P) - 0.503 \quad 38.0 < ADSI + P < 63.2 \quad R^2 = 0.39 \tag{8a}$$

$$Q = 0.425\,(ADSI + P) - 26.823 \quad 63.2 \leq ADSI + P < 84.0 \quad R^2 = 0.824 \tag{8b}$$

The analysis indicates that some spring–summer events cannot be explained only by the ADSI + P approach (the three events indicated with triangles in Figure 2b). These three events are classified as rill in the SERLAB data set (Table S1) and have not been included in the regression analysis to determine the models given in Equations (7a,b) and (8a,b). This choice is discussed later in the paper. Moreover, few events behave differently, deviating from the linear regression models proposed. For example, point no. 15 in Figure 2b (occurring in September 2016, Table S1) provides a low runoff than that predicted by the spring–summer rise threshold model (Equation (8b)); the point no. 22 (occurring in May 2018, Table S1) is among the spring–summers but behaves like the autumn–winter ones. This latter event is characterized by one of the most prolonged duration (12.33 h), and it is preceded by very wet soil conditions (SSM(t) = 76%).

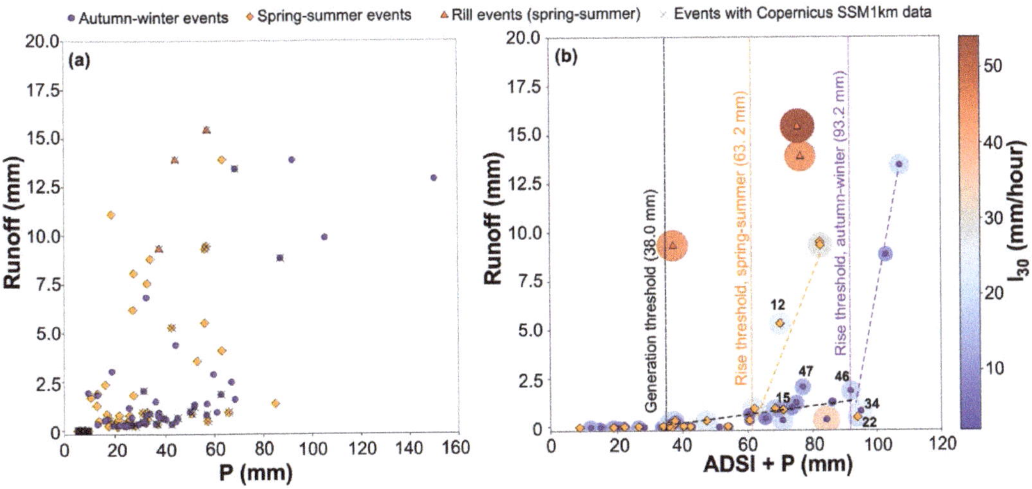

Figure 2. Runoff events separated into spring–summer and autumn–winter periods. (**a**) Runoff vs. event rainfall depth (P) for the 2008–2020 period; (**b**) Runoff vs. the sum of Antecedent Degree of Saturation Index and P (ADSI + P), with generation threshold and rise thresholds for the two periods (only events with Copernicus SSM1km data available are considered, Table S1). The numbers associated with some selected events in Figure 2b refer to Table S1.

3.3. Estimation of the Runoff, Runoff Coefficient, and the Soil Loss

The functions Q = f (ADSI + P) given in Figure 2b (Equations (7a,b) and (8a,b)) have been used to estimate the runoff (Q) from ADSI data and the gauging station P records. The comparison between the estimated and measured runoff (Table S1) is given in Figure 3a. In Figure 3b, the same comparison is presented in terms of runoff coefficients (Q_R).

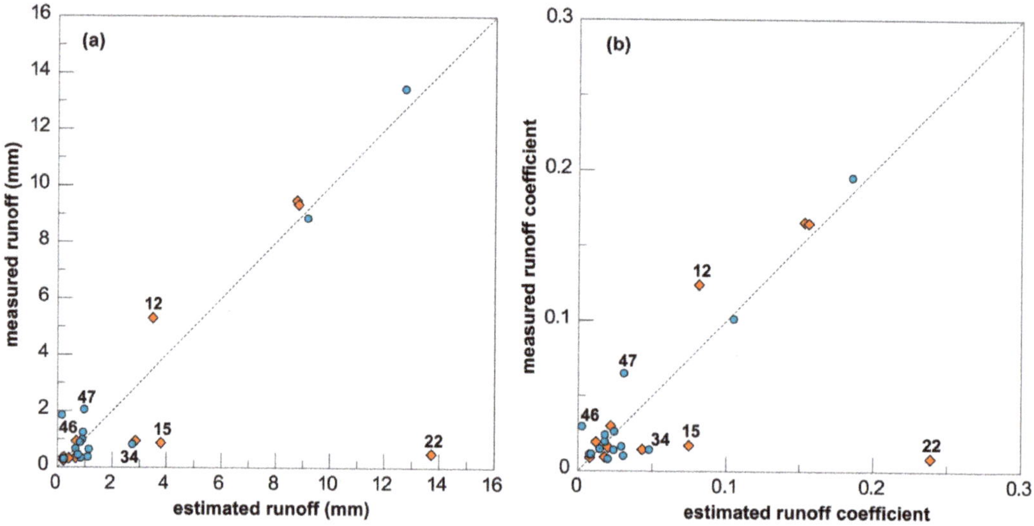

Figure 3. Comparison between the measured runoff and the runoff estimated by the models Q = f (ADSI + P), Equations (7a,b) and (8a,b) (**a**); same comparison based on the runoff coefficient (**b**). Blue dots and orange diamonds indicate autumn–winter and spring–summer events, respectively. The numbers associated with some selected events referred to in Table S1.

The highest estimation errors derive from the events nos. 15 and 22 (Table S1). Although these events have ADSI + P values higher than the rise threshold ones, the measured runoff is lower than those estimated by steeper linear models. The root mean square errors (RMSE) associated with the estimated runoff (Figure 3a) are 0.8, 4.17, and 2.78 mm, for the autumn–winter, spring–summer, and overall data set, respectively. The RMSE obtained for the spring–summer, and overall data sets are very negatively affected by events 15 and 22 (Figure 3) already mentioned above in the paper. However, even excluding these events, the RMSE for spring–summer events becomes 2.0 mm, remaining higher than autumn–winter events. Similar considerations can be made for the runoff coefficient (Figure 3b). In this case, the RMSEs for the different data sets are 0.02, 0.07, and 0.05 for the autumn–winter, spring–summer, and overall data sets, respectively (average 2.14 mm).

The 27 events for which the soil loss measures are available (Table S1) are used to parametrize a USLE-MM model. This model is optimized for this study but is not to be intended as representative of the full SERLAB data set (2008–2020).

Referring to the parametrized USLE-MM model (Equation (3)), the regression analysis in Figure 4 for the data set in Table S1 gives Equation (9).

$$A_{e,N} = 0.0317(Q_R R_e)^{1.1038} \quad (9)$$

Using Equation (9), with the Q_R values calculated as the ratio between the runoff calculated from the satellite and rainfall data (Q = f (ADSI + P), Equations (7a,b) and (8a,b)), and the corresponding P, the normalized soil loss $A_{e,N}$ was estimated. Figure 5 compares these $A_{e,N}$ values and those estimated by Equation (9) using the measured Q_R (Table S1). Most events are distributed around the 1:1 line, thus indicating a moderate effect on the USLE-MM model prediction by using a Q_R estimated by satellite and rainfall data instead of the measured one. The role of the season does not appear particularly marked. In addition, in this case, the estimation errors are very negatively affected by the events commented on above in the paper. In particular, for the event no. 22, there is a relevant overestimation of the soil loss deriving from the estimated runoff. Events nos. 15 and 34 provide the same kind of error, although less. Opposite consideration can be done for the events nos.

12, 47, and 46. The graph in Figure 5 is bi-logarithmic to visualize better the low events otherwise grouped around the origin of the axes. Meanwhile, this representation enables one to appreciate the goodness of fit of the higher values of soil loss, which greatly affects the quantification of the erosion process over a long period.

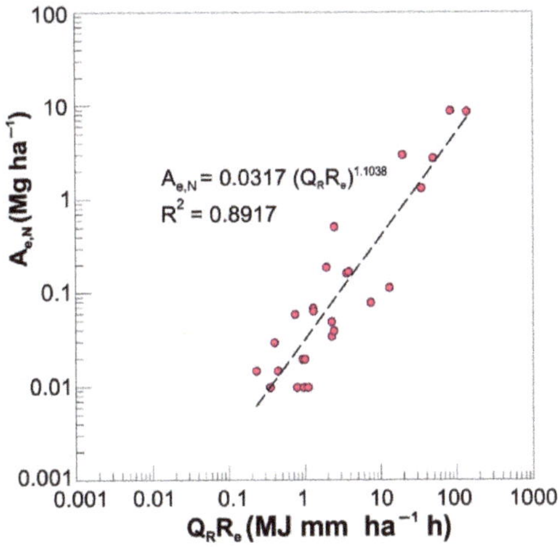

Figure 4. Scatter plots of the measured pairs ($Q_R R_e$, $A_{e,N}$) and USLE-MM model (equation in the graph) for the data set given in Table S1.

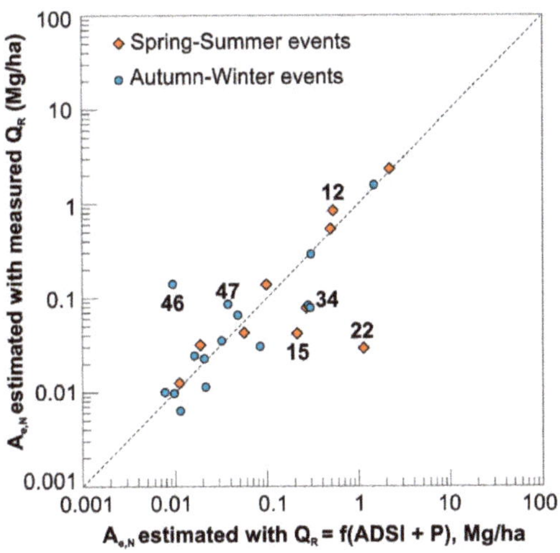

Figure 5. Comparison between the soil loss estimated by Equation (4) with the Q_R measured and estimated as the ratio between the runoff, Q = f (ADSI + P), Equations (7a,b) and (8a,b), and the corresponding measured rainfall P. Blue dots and orange diamonds indicate autumn–winter and spring–summer events, respectively. The numbers associated with some selected events are referred to in Table S1.

The estimation error of the runoff and runoff coefficient results in a noticeable worsening of soil loss estimation. In Figure 6, the observed values $A_{e,N}$ are compared with those estimated by the USLE-MM model (Equation (9)) in two conditions: with Q_R measured and with Q_R derived from satellite and rainfall data (Q = f (ADSI + P) by using Equation (7a,b) for autumn–winter events, Equation (8a,b), for spring–summer events.

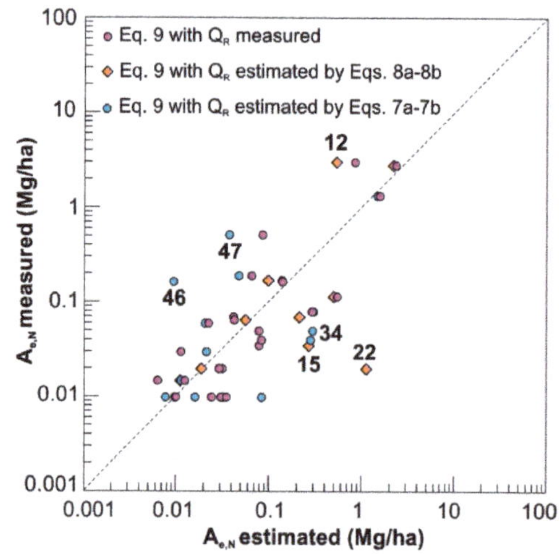

Figure 6. Goodness of fit of the soil loss measured and estimated by the USLE-MM model (Equation (9)) with Q_R measured and estimated by Equations (7a,b) and (8a,b). Blue dots and orange diamonds indicate autumn–winter and spring–summer events, respectively. The numbers associated with some selected events are referred to in Table S1.

In terms of RMSE, the estimation errors are similar, 0.46 and 0.57 t/ha, respectively, but as expected, are higher when the ADSI + P data are used to calculate the runoff instead of Q_R measured. In addition, in this case, the results are very negatively affected by the events with nos. 12, 15, 22, 34, 46, and 47 (Table S1).

4. Discussion

The data set used for the analysis included events with records of satellite soil water content (expressed by the index ADSI), and rainfall (P), runoff (Q), rainfall erosivity (R_e), soil loss (A_e) measured at plot scale (SERLAB experimental field). Four different behaviors can be identified when plotting the points in a cartesian plane (Q vs. ADSI + P, Figure 2b). First of all, the three events classified as rill in the SERLAB data set show behavior on their own, with a very high hydrological response (runoff and soil loss) at a medium ADSI + P value. A specific model describing rill events has not been built due to the small data set. Classifying of the events in rill/interrill is also possible according to specific activation thresholds [48,49] based only on widely available rainfall graph characteristics. In particular, the three recognized rill events are characterized by R_e values (Table S1) higher than or very close to the value R_e = 359 mm ha^{-1} h^{-1} proposed in [49] as the threshold for the rill erosion process activation at the SERLAB. In addition, in terms of I_{30}, these events differentiate from the others (Table S1, Figure 2b).

Another group of five events (nos. 1, 4, 12, 26, and 45 in Table S1) shows an intermediate hydrological response (runoff and soil loss) at a medium ADSI + P. Three of them enable the determination of the summer-spring relationship above the rise threshold (Equation (8b)) and the other two the analogous line for the winter-autumn period (Equation (7b)). In addition, in this case, it is possible to explain the specific behavior of this group in terms of hydrological response, mainly based on the characteristics of the rainfall graph. The two events intercepted by the autumn–winter rise (nos. 26 and 45) are characterized by mean intensity (IM), generally much lower than the events belonging to the spring–summer rise (nos. 1, 4, and 12). An exception is represented by event no. 4, which has an I_M only slightly higher than those of events 26 and 45. However, a detailed analysis of the rainfall graphs of the spring–summer events (nos. 1, 4, and 12) revealed that these consist almost entirely of a single shower ("burst") to which more than 90% of rain belongs.

Autumn–winter events, instead, have their total precipitation distributed in numerous showers. Evidently, events with higher IM and more concentrated volumes over time (as typically occurs for summer events) can lead to a more rapid decay of the infiltration rate and an earlier occurrence of the runoff with the same ADSI + P. The last two groups show, respectively, null runoff response at low ADSI + P and low hydrological response (runoff and soil loss) at a medium ADSI + P. The models predicting Q from ADSI + P data were derived based on this classification. The segmented linear regression models with a low slope for low hydrological response and a high slope for high hydrological response, in the literature, are named "hockey stick" behavior [23] and seem the most appropriate to describe the relationship between the ADSI + P and Q in the database analyzed. A different hockey stick model was associated with "spring–summer" and "autumn–winter" groups, respectively, with a higher rise threshold for the latter.

The performance of the Q = f (ADSI + P) models is fairly satisfactory, estimating quite accurately both the low and high runoff (Figure 3). This result is very important as the most intense events, although rare, are those that produce the greatest runoff and soil loss. It is widely recognized [6,10] that the annual soil loss is produced almost entirely by a few very intense events that modify the hydrological connectivity and drag downstream the detached material only partially displaced by less intense events.

This fact results in a pretty suitable estimate of the event soil loss when the runoff simulated with the inferred models is used to input the USLE-MM (Equation (9)). As expected, the USLE-MM model provides a less accurate estimate with a slightly higher RMSE (0.57) if compared with the performance obtained using the measured Q_R (RMSE = 0.49) with a more scattered point cloud (Figure 6). However, the benchmark for the correct assessment of the USLE-MM (Equation (9)) with runoff coefficient derived from satellite and rainfall data ADSI + P is the performance of USLE-derived models that include predicted runoff coefficient [14,15,50]. Todisco et al. [14], analyzing a SERLAB database of about 60 plot soil loss values (average 3.5 Mg/ha) and the estimated soil loss by a USLE-MM model coupled with MISDc model [51], found RMSE = 2.96 Mg/ha. We obtain a moderately worse result by using the Q values obtained by the presented models (Equations (7a,b) and (8a,b)) in Equation (9): RMSE 0.57 Mg/ha compared to an average of 0.35 Mg/ha.

Furthermore, the differentiation of the rainfall runoff models according to the season (autumn–winter and spring–summer) agrees with the results obtained by [14] when stating that the hydrological behavior is different for summer and winter rainfalls due to the particular characteristics of summer rainfall events in central Europe [48]. As a consequence, the performance in the estimation of the soil loss differs whether evaluated for the entire data set or separately for the two seasons, being much better for the autumn–winter period. The small database considered and the resulting uncertainty about the "hockey stick" relationships identified may certainly have negatively affected the results.

Due to the small data set available, we used all data to create the predictive relationships, representing the main weak point of the analysis. Hence, it was impossible to validate the relationships using an independent data set. Unfortunately, in recent years, in which satellite data are available, few rainy events were able to produce runoff and soil loss at SERLAB. Therefore, the relationships derived (Equations (7a,b) and (8a,b)) and the actual contribution of the antecedent soil water content in the accuracy of runoff prediction has to be considered as a preliminary result. The study aims mainly to show the concept, which will have to be confirmed with further measures. On the other hand, runoff is a very complex process depending on several components such as locally specific rainfall, crop and soil use, soil conditions, and slope [52]. In this context, the satellite moisture data, even at low resolutions, can represent an explanatory variable helpful in improving the accuracy of the runoff estimate, especially when in-field monitoring data are unavailable. This is demonstrated by the growing use of Copernicus SSM1km data for hydrological applications, even at the slope scale [14,53–55].

Furthermore, it is necessary to underline that the performance obtained is better than that provided by the original USLE model, which is still the most widely used model for estimating soil loss, given the difficulty in measuring runoff. In fact, by applying the USLE model (Equation (2)) with the soil erodibility factor K = 0.02 estimated from the SERLAB database (2008–2020), we obtain for the data set of Table S1 an RMSE $\approx$ 3.1 Mg/ha that is much higher than the value (RMSE 0.57 Mg/ha) derived using Q = f (ADSI + P) in Equation (9). Therefore, using a USLE-MM model (albeit with remote estimated runoff data) is still advantageous at the event scale compared to the USLE model, which does not consider the runoff.

5. Conclusions

The work analyzed the Masse SERLAB data set of erosive and non-erosive events, jointly with that derived from Copernicus Sentinel-1 of surface soil moisture (from 2015 onward). The events simultaneously present in the two data sets were extracted and analyzed. For these events, we have all the erosion data: soil loss (A_e), runoff event (Q), runoff coefficient (Q_R), and rainfall data (P, I_{30}, R_e), and at the same time, the surface soil moisture (SSM) and the corresponding Antecedent Degree of Saturation Index (ADSI) in the first 50 mm of the soil. The data set was analyzed with the dual purpose of verifying the possibility of identifying a relationship between ADSI + P and Q and evaluating the possibility of using these models to estimate the runoff Q and the corresponding runoff coefficient QR. The estimated runoff coefficient has also been used in a USLE-MM model to estimate the soil loss event. The following findings were achieved:

1. As expected, the event rainfall depth (P), if corrected with the corresponding ADSI value, does explain the runoff process reasonably well, at least with the available data;
2. By separating the data set into the autumn–winter and the spring–summer events, it is possible to identify the generation threshold (ADSI + P value triggering runoff) and the rise thresholds (ADSI + P values at which the runoff behavior changes, Table S1);
3. The relationship Q = f (ADSI + P) was derived for spring–summer events and autumn–winter events (Equations (7a,b) and (8a,b));
4. Using a USLE-MM model (albeit with remote estimated runoff data) is still advantageous at the event scale compared to the USLE model.

The small database considered and the resulting uncertainty about the relationships identified may certainly have negatively affected the result. The database used is not large enough to interpret the results correctly. In particular, there is a double open question: the events that it was possible to analyze (as they fall within the temporal intersection of the two databases) are few in absolute terms; furthermore, the number of rill data is low in the Masse SERLAB data set, and this prevented an overall assessment of events with high runoff. Future developments could use other erosion databases to make the analysis more robust.

We highlight that the obtained results open interesting scenarios in the overview of the studies aimed at defining USLE-derived models that could improve the unit soil loss estimation at the event scale. In particular, the choice of using SSM data to estimate the runoff is very important for the practice. Moreover, remote sensing soil moisture data are widely available globally. Satellite data shows the potential of applying the developed USLE-derived model for large-scale monitoring and quantification of the soil erosion process.

Supplementary Materials: The following supporting information can be downloaded at: https://www.mdpi.com/article/10.3390/w14132081/s1, Table S1. SERLAB and Sentinel-1 databases of rainfall, runoff and antecedent soil moisture observations.

Author Contributions: Conceptualization, F.T. and L.D.M.; methodology, F.T. and L.D.M.; validation, L.V. and S.O.; investigation, F.T. and L.V.; formal analysis, resources, data curation, writing, F.T., L.V., L.D.M., and S.O.; funding acquisition, F.T., L.V. All authors have read and agreed to the published version of the manuscript.

Funding: This work was funded by the 'Ricerca di Base, 2019' project, University of Perugia - Italy (principal investigator Francesca Todisco) and 'Ricerca di Base, 2017' project, University of Perugia - Italy (principal investigator Lorenzo Vergni).

Data Availability Statement: All the data used in the work are available in Table S1.

Conflicts of Interest: The authors declare no conflict of interest.

References

1. Bisantino, T.; Bingner, R.; Chouaib, W.; Gentile, F.; Trisorio, L.G. Estimation of runoff, peak discharge and sediment load at the event scale in a medium-size Mediterranean watershed using the ANNAGNPS model. *Land Degrad. Dev.* **2015**, *26*, 340–355. [CrossRef]
2. Borrelli, P.; Märker, M.; Schütt, B. Modelling post-tree-harvesting soil erosion and sediment deposition potential in the Turano River basin (Italian Central Apennine). *Land Degrad. Dev.* **2015**, *26*, 356–366. [CrossRef]
3. Foster, G.R.; Toy, T.E.; Renard, K.G. Comparison of the USLE, RUSLE1.06 and RUSLE2 for application to highly disturbed lands. *First Interag. Conf. Res. Watersheds* **2003**, *27*, 154–160.
4. Borrelli, P.; Robinson, D.A.; Fleischer, L.R.; Lugato, E.; Ballobio, C.; Alewell, A.; Meusburger, K.; Modugno, S.; Schutt, B.; Ferro, V.; et al. An assessment of the global impact of 21st century land use change on soil erosion. *Nat. Commun.* **2017**, *8*, 2013. [CrossRef]
5. Panagos, P.; Ballabio, C.; Borrelli, P.; Meusburger, K.; Klik, A.; Rousseva, S.; Tadić, M.P.; Michaelides, S.; Hrabalíková, M.; Olsen, P.; et al. Rainfall erosivity in Europe. *Sci. Total Environ.* **2015**, *511*, 801–814. [CrossRef]
6. Bagarello, V.; Ferro, V.; Pampalone, V.; Porto, P.; Todisco, F.; Vergni, L. Predicting soil loss in central and south Italy with a single USLE-MM model. *J. Soils Sediments* **2018**, *18*, 3365–3377. [CrossRef]
7. Ganasri, B.P.; Ramesh, H. Assessment of soil erosion by RUSLE model using remote sensing and GIS-a case study of Nethravathi Basin. *Geosci. Front.* **2016**, *7*, 953–961. [CrossRef]
8. Wischmeier, W.H.; Smith, D.D. *Predicting Rainfall-Erosion Losses: A Guide to Conservation Farming*; U.S. Department of Agriculture, Science and Education Administration: Washington, DC, USA, 1978; p. 537.
9. Larson, W.E.; Lindstrom, M.J.; Schumacher, T.E. The role of severe storms in soil erosion: A problem needing consideration. *J. Soil Water Conserv.* **1997**, *52*, 90–95.
10. Di Stefano, C.; Pampalone, V.; Todisco, F.; Vergni, L.; Ferro, V. Testing the Universal Soil Loss Equation-MB equation in plots in Central and South Italy. *Hydrol. Processes* **2019**, *33*, 2422–2433. [CrossRef]
11. Kinnell, P.I.A.; Risse, L.M. USLE-M: Soil Empirical modeling rainfall erosion through runoff and sediment concentration. *Sci. Soc. Am. J.* **1998**, *62*, 1667–1672. [CrossRef]
12. Bagarello, V.; Di Piazza, G.V.; Ferro, V.; Giordano, G. Predicting unit plot soil loss in Sicily, south Italy. *Hydrol. Processes* **2008**, *22*, 586–595. [CrossRef]
13. Bagarello, V.; Ferro, V.; Pampalone, V.; Porto, P.; Todisco, F.; Vergni, L. Statistical check of USLE-M and USLE-MM to predict bare plot soil loss in two Italian environments. *Land Degrad. Dev.* **2018**, *18*, 3365–3377. [CrossRef]
14. Todisco, F.; Brocca, L.; Termite, L.F.; Wagner, W. Use of satellite and modeled soil moisture data for predicting event soil loss at plot scale. *Hydrol. Earth Syst. Sci.* **2015**, *19*, 3845–3856. [CrossRef]
15. Gao, G.Y.; Fu, B.J.; Lü, Y.H.; Liu, Y.; Wang, S.; Zhou, J. Coupling the modified SCS-CN and RUSLE models to simulate hydrological effects of restoring vegetation in the Loess Plateau of China. *Hydrol. Earth Syst. Sci.* **2012**, *16*, 2347–2364. [CrossRef]
16. Kinnell, P.I.A. Modelling event soil losses using the QREI30 index within RUSLE2. *Hydrol. Processes* **2014**, *28*, 2761–2771. [CrossRef]
17. McGlynn, B.L.; McDonnell, J.J. Quantifying the relative contributions of riparian and hillslope zones to catchment runoff. *Water Resour. Res.* **2003**, *39*, 1310. [CrossRef]
18. Thomaz, E.L. The influence of traditional steep land agricultural practices on runoff and soil loss. *Agric. Ecosyst. Environ.* **2009**, *130*, 23–30. [CrossRef]
19. Detty, J.M.; McGuire, K.J. Threshold changes in storm runoff generation at a till-mantled headwater catchment. *Water Resour. Res.* **2010**, *46*, W07525. [CrossRef]
20. Scaife, C.I.; Band, L.E. Nonstationarity in threshold response of stormflow in southern Appalachian headwater catchments. *Water Resour. Res.* **2017**, *53*, 6579–6596. [CrossRef]
21. Haga, H.; Matsumoto, Y.; Matsutani, J.; Fujita, M.; Nishida, K.; Sakamoto, Y. Flow paths, rainfall properties, and antecedent soil moisture controlling lags to peak discharge in a granitic unchanneled catchment. *Water Resour. Res.* **2005**, *41*, W12410. [CrossRef]
22. Penna, D.; Tromp-van Meerveld, H.J.; Gobbi, A.; Borga, M.; Dalla Fontana, G. The influence of soil moisture on threshold runoff generation processes in an alpine headwater catchment. *Hydrol. Earth Syst. Sci.* **2011**, *15*, 689–702. [CrossRef]
23. Ali, G.; Oswald, C.J.; Spence, C.; Cammeraat, E.L.; McGuire, K.J.; Meixner, T.; Reaney, S.M. Towards a unified threshold-based hydrological theory: Necessary components and recurring challenges. *Hydrol. Processes* **2013**, *27*, 313–318. [CrossRef]
24. Weiler, M.; McDonnell, J.J.; Tromp-van Meerveld, I.; Uchida, T. Subsurface stormflow. *Encycl. Hydrol. Sci.* **2006**, 1–14. [CrossRef]
25. Tromp-van Meerveld, H.J.; McDonnell, J.J. Threshold relations in subsurface stormflow: 1. A 147-storm analysis of the Panola hillslope. *Water Resour. Res.* **2006**, *42*, 1–11. [CrossRef]

26. James, A.L.; Roulet, N. Investigating hydrologic connectivity and its association with threshold change in runoff response in a temperate forested watershed. *Hydrol. Processes* **2007**, *21*, 3391–3408. [CrossRef]
27. Zehe, E.; Elsenbeer, H.; Lindenmaier, F.; Schulz, K.; Blöschl, G. Patterns of predictability in hydrological threshold systems. *Water Resour. Res.* **2007**, *43*, W07434. [CrossRef]
28. Penna, D.; van Meerveld, H.J.; Oliviero, O.; Zuecco, G.; Assendelft, R.S.; Dalla Fontana, G.; Borga, M.A.R.C.O. Seasonal changes in runoff generation in a small forested mountain catchment. *Hydrol. Processes* **2015**, *29*, 2027–2042. [CrossRef]
29. Wei, L.; Qiu, Z.; Zhou, G.; Kinouchi, T.; Liu, Y. Stormflow threshold behaviour in a subtropical mountainous headwater catchment during forest recovery period. *Hydrol. Processes* **2020**, *34*, 1728–1740. [CrossRef]
30. Scaife, C.I.; Singh, N.K.; Emanuel, R.E.; Miniat, C.F.; Band, L.E. Non-linear quickflow response as indicators of runoff generation mechanisms. *Hydrol. Processes* **2020**, *34*, 2949–2964. [CrossRef]
31. Wang, S.; Yan, Y.; Fu, Z.; Chen, H. Rainfall-runoff characteristics and their threshold behaviors on a karst hillslope in a peak-cluster depression region. *J. Hydrol.* **2021**, *605*, 127370. [CrossRef]
32. Vaz, C.M.; Jones, S.; Meding, M.; Tuller, M. Evaluation of standard calibration functions for eight electromagnetic soil moisture sensors. *Vadose Zone J.* **2013**, *12*, 1–16. [CrossRef]
33. Di Matteo, L.; Pauselli, C.; Valigi, D.; Ercoli, M.; Rossi, M.; Guerra, G.; Cambi, C.; Ricco, R.; Vinti, G. Reliability of water content estimation by profile probe and its effect on slope stability. *Landslides* **2018**, *15*, 173–180. [CrossRef]
34. Kargas, G.; Londra, P.; Anastasatou, M.; Moustakas, N. The effect of soil iron on the estimation of soil water content using dielectric sensors. *Water* **2020**, *12*, 598. [CrossRef]
35. Di Matteo, L.; Spigarelli, A.; Ortenzi, S. Processes in the Unsaturated Zone by Reliable Soil Water Content Estimation: Indications for Soil Water Management from a Sandy Soil Experimental Field in Central Italy. *Sustainability* **2021**, *13*, 227. [CrossRef]
36. Ortenzi, S.; Mangoni, M.; Di Matteo, L. Estimating moisture content and hydraulic properties of unsaturated sandy soils of Tiber River (Central Italy): Integrating data from calibrated PR2/6 probe and hydraulic property estimator. *Acque Sotter.—Ital. J. Groundw.* **2022**, *11*, 17–25. [CrossRef]
37. Bagarello, V.; Ferro, V.; Giordano, G.; Mannocchi, F.; Pampalone, V.; Todisco, F.; Vergni, L. Effect of plot size on measured soil loss for two Italian experimental sites. *Biosyst. Eng.* **2011**, *108*, 18–27. [CrossRef]
38. Todisco, F.; Vergni, L.; Mannocchi, F.; Bomba, C. Calibration of the soil loss measurement method at the Masse experimental station. *Catena* **2012**, *91*, 4–9. [CrossRef]
39. Doubkova, M.; Wagner, W.; Naeimi, V.; Cao, S.; Bauer-Marschallinger, B.; Kidd, R.; Hasenauer, S.; Dostalova, A.; Paulik, C. The use of Sentinel-1 for monitoring of soil moisture within the copernicus global land service. *Living Planet Symp.* **2016**, *740*, 105.
40. Copernicus Global Land Operations. Available online: https://land.copernicus.eu/global/sites/cgls.vito.be/files/products/CGLOPS1_PUM_SSM1km-V1_I1.30.pdf (accessed on 16 February 2022).
41. Wagner, W. Soil Moisture Retrieval from ERS Scatterometer Data. Institut für Photogrammetrie u. Fernerkundung d. Technologie University: Karlsruhe, Germany, 1998; p. 49.
42. Wagner, W.; Lemoine, G.; Rott, H. A method for estimating soil moisture from ERS scatterometer and soil data. *Remote Sens. Environ.* **1999**, *70*, 191–207. [CrossRef]
43. Hornacek, M.; Wagner, W.; Sabel, D.; Truong, H.L.; Snoeij, P.; Hahmann, T.; Diedrich, E.; Doubková, M. Potential for high resolution systematic global surface soil moisture retrieval via change detection using Sentinel-1. *IEEE J. Sel. Top. Appl. Earth Obs. Remote Sens.* **2012**, *5*, 1303–1311. [CrossRef]
44. Bauer-Marschallinger, B.; Freeman, V.; Cao, S.; Paulik, C.; Schaufler, S.; Stachl, T.; Modanesi, S.; Massari, C.; Ciabatta, L.; Brocca, L.; et al. Toward global soil moisture monitoring with Sentinel-1: Harnessing assets and overcoming obstacles. *IEEE Trans. Geosci. Remote Sens.* **2018**, *57*, 520–539. [CrossRef]
45. Richard, G.; Cousin, I.; Sillon, J.F.; Bruand, A.; Guérif, J. Effect of compaction on the porosity of a silty soil: Influence on unsaturated hydraulic properties. *Eur. J. Soil Sci.* **2001**, *52*, 49–58. [CrossRef]
46. Fu, Y.; Tian, Z.; Amoozegar, A.; Heitman, J. Measuring dynamic changes of soil porosity during compaction. *Soil Tillage Res.* **2016**, *193*, 114–121. [CrossRef]
47. de Oliveira, J.A.; Cássaro, F.A.; Pires, L.F. Estimating soil porosity and pore size distribution changes due to wetting-drying cycles by morphometric image analysis. *Soil Tillage Res.* **2021**, *205*, 104814. [CrossRef]
48. Todisco, F. The internal structure of erosive and non-erosive storm events for interpretation of erosive processes and rainfall simulation. *J. Hydrol.* **2014**, *519*, 3651–3663. [CrossRef]
49. Todisco, F.; Vergni, L.; Vinci, A.; Pampalone, V. Practical thresholds to distinguish erosive and rill rainfall events. *J. Hydrol.* **2019**, *579*, 124173. [CrossRef]
50. Todisco, F.; Brocca, L.; Mannocchi, F.; Melone, F.; Moramarco, T. Using continuous hydrological modeling coupled to a modified USLE model for predicting plot soil loss in Umbria. In *Quaderni di Idronomia Montana*; Edibios: Cosenza, Italy, 2012; Volume 30, pp. 353–362.
51. Brocca, L.; Melone, F.; Moramarco, T. Distributed rainfall-runoff modelling for flood frequency estimation and flood forecasting. *Hydrol. Processes* **2011**, *25*, 2801–2813. [CrossRef]
52. Mitchell, J.; Banasik, K.; Hirschi, M.; Cooke, R.; Kalita, P. There is not always surface runoff and sediment transport. In *Proceedings of the Soil Erosion Research for the 21st Century: Proceedings of the International Symposium, Honolulu, HI, USA, 3–5 January 2001*; pp. 575–578.

53. Bazzi, H.; Baghdadi, N.; El Hajj, M.; Zribi, M.; Belhouchette, H. A comparison of two soil moisture products S 2 MP and copernicus-SSM over southern France. *IEEE J. Sel. Top. Appl. Earth Obs. Remote Sens.* **2019**, *12*, 3366–3375. [CrossRef]
54. Rabiei, S.; Jalilvand, E.; Tajrishy, M. A Method to Estimate Surface Soil Moisture and Map the Irrigated Cropland Area Using Sentinel-1 and Sentinel-2 Data. *Sustainability* **2021**, *13*, 11355. [CrossRef]
55. Attarzadeh, R.; Amini, J.; Notarnicola, C.; Greifeneder, F. Synergetic use of Sentinel-1 and Sentinel-2 data for soil moisture mapping at plot scale. *Remote Sens.* **2021**, *10*, 1285. [CrossRef]

water

MDPI

Review

Measurement of Water Soil Erosion at Sparacia Experimental Area (Southern Italy): A Summary of More than Twenty Years of Scientific Activity

Vincenzo Pampalone *, Francesco Giuseppe Carollo, Alessio Nicosia, Vincenzo Palmeri, Costanza Di Stefano, Vincenzo Bagarello and Vito Ferro

Department of Agricultural, Food and Forestry Sciences, University of Palermo, Viale delle Scienze, Building 4, 90128 Palermo, Italy; francescogiuseppe.carollo@unipa.it (F.G.C.); alessio.nicosia@unipa.it (A.N.); vincenzo.palmeri02@unipa.it (V.P.); costanza.distefano@unipa.it (C.D.S.); vincenzo.bagarello@unipa.it (V.B.); vito.ferro@unipa.it (V.F.)

* Correspondence: vincenzo.pampalone@unipa.it

Abstract: The main purpose of this article is to give a general idea of the scientific activity that was carried out starting from the 2000s on the basis of the data collected in the plots installed at the Sparacia experimental station for soil erosion measurement in Sicily, South Italy. The paper includes a presentation of the experimental site, a description of the methods and procedures for measuring soil erosion processes both available in the literature and applied at the Sparacia station (sediment sampling and water level reading in the storage tanks for total erosion measurements; profilometer, and Structure from Motion technique for rill erosion measurements), and the main results obtained in the monitoring period in the experimental site. The latter concern the effects of plot size and steepness on soil loss, the measurement variability, the frequency analysis of soil loss, the rill erosion characterization, and the comparison between rill and interrill erosion rates. Each of these topics is addressed with multi-temporal analyses performed with increasing size of the available database, which allowed to draw robust conclusions. Soil loss did not vary appreciably with plot length in contrast with the assumption made in the USLE/RUSLE. The variability of the measurements of soil loss, runoff volume, and sediment concentration at the event scale in replicated plots decreased as the mean measured value increased. The normalized event soil loss was distributed according to a two-component distribution. A power relationship between rill volumes and lengths was established. The measurements also confirmed the morphological similarity between the channels of the rills and ephemeral gullies described by a power dimensionless relationship. Rill erodibility of the sampled clay soil varied over time, maintaining relatively low values. Finally, rill erosion was dominant relative to interrill erosion, and a more efficient sediment transport system through the rill network occurred as plot steepness increased.

Keywords: soil loss; plot measurements; interrill; rill; physical model concept; SfM technique

check for updates

Citation: Pampalone, V.; Carollo, F.G.; Nicosia, A.; Palmeri, V.; Di Stefano, C.; Bagarello, V.; Ferro, V. Measurement of Water Soil Erosion at Sparacia Experimental Area (Southern Italy): A Summary of More than Twenty Years of Scientific Activity. *Water* **2022**, *14*, 1881. https://doi.org/10.3390/w14121881

Academic Editor: Yeshuang Xu

Received: 12 May 2022
Accepted: 7 June 2022
Published: 11 June 2022

1. Introduction

In the research field of water soil erosion, the measurement of a representative variable of the phenomenon (e.g., plot soil loss, basin sediment yield) and its controlling factors (e.g., rainfall energetic characteristics, soil grain size distribution) allows for directly evaluating the magnitude of the phenomenon in the area of interest, extending and improving the knowledge of the erosive processes, and establishing the effectiveness of soil conservation practices. Moreover, the simulation of the erosive phenomenon by mathematical models needs measurement of the parameters and/or variables included in the equations used to describe the different subprocesses. The measurements are also necessary to calibrate and validate the models through the comparison between the estimated and measured values of the variable expressive of the erosive phenomenon.

Laboratory erosion research can be carried out in highly controlled experimental conditions and is ideal for fast and repeatable experiments, but it usually requires using scale models and disturbed soil samples whose representativeness of field conditions is rather uncertain. Instead, field erosion research allows for gaining data at the proper scale and with undisturbed soils. Moreover, field data are generally representative of the climatic, soil, morphologic, and land use characteristics of the sampled site. However, these types of data are less accurate than the laboratory ones due to the practical problems of experimental field activities. In addition, to obtain representative data of average or extreme conditions for the sampled site, a multiyear monitoring period is needed, with noticeable costs and practical problems.

This paper summarizes the scientific activity that has been going on for more than twenty years in the Sparacia experimental station for soil erosion measurement in Sicily, South Italy. The Sparacia station of the Department of Agricultural, Food, and Forest Sciences—Palermo University is located approximately 100 km south of Palermo. The climate is of Mediterranean semiarid type with an average annual rainfall of 700 mm. The soil is a Vertic Xerocrept and has a clay texture (62% clay, 33% silt, and 5% sand) and negligible gravel content. The experimental station includes 20 plots with different lengths λ, widths wi, and slope steepness s. Six plots of 22 m × 8 m ($\lambda \times wi$), two plots of 33 m × 8 m, two plots of 44 m × 8 m, two plots of 11 m × 4 m, two plots of 11 m × 2 m and two plots of 22 m × 2 m are installed on a 14.9% slope (Figure 1a), where a recording rain gauge operates with a 1 min resolution. Two plots of 22 m × 6 m are established on a 22% slope, and two plots of 22 m × 6 m are established on a 26% slope (Figure 1b). Rainfall data are measured with a recording rain gauge located near the four plots at 1 min intervals. The oldest plots (four plots of 22 m × 8 m) were constructed in 1999, whereas the most recent plots (two plots of 22 m × 2 m) were constructed in 2007. Plots are maintained in cultivated fallow, with up and downhill tillage, by a power cultivator. Tillage is conducted 3–4 times per year when the soil is relatively dry.

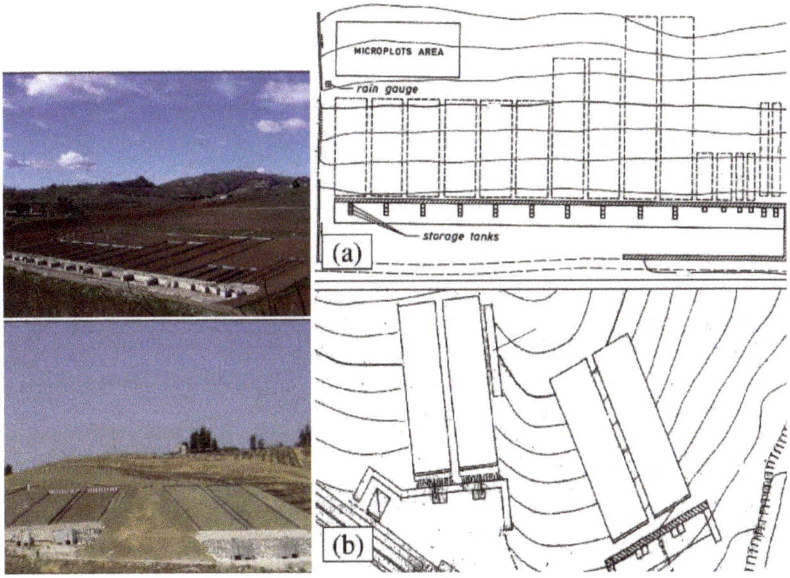

Figure 1. Plan and view of the experimental plots installed on the (**a**) 14.9% slope and (**b**) 22% and 26% slopes (modified from [1]).

Runoff and associated sediments are intercepted by a gutter placed at the lower end of each plot and collected into a storage system consisting of tanks of known geometric characteristics that are arranged in series (Figure 2a). Total runoff and soil loss are measured after each erosive event (i.e., an event producing measurable runoff and sediment) or after a sequence of events if they are separated by a short time interval. Microplots of 0.25 m × 0.25 m and 0.40 m × 0.40 m were also installed on the 14.9 and 22% slopes, and two experimental basins of 3.67 ha and 30 ha are also located in the experimental area. The present paper focuses only on the soil erosion measurement at the plot scale. These measurements allowed parameterizing and validating Universal Soil Loss Equation (USLE)-based models for estimating event soil loss where the runoff coefficient is included in the erosivity factor [2,3]. Among these models, the so-called USLE-MM was developed through main steps starting from 2008 [4], while the USLE-MB [5] was proposed more recently and resulted in similar estimate performances as the former.

The paper reports both the methods to measure rill, interrill erosion, and total, i.e., rill + interrill, erosion at the Sparacia station and the main results concerning the effects of plot size and steepness on soil loss, the measurement variability, the frequency analysis of soil loss, the rill erosion characterization, and the comparison between rill and interrill erosion rates. A section concerning a review on soil loss measurement is first presented.

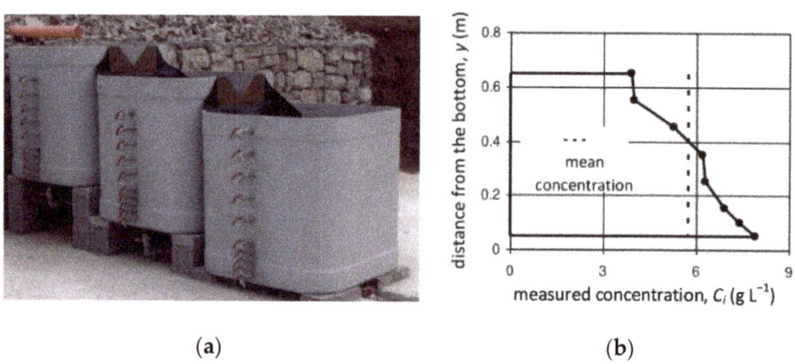

Figure 2. View of the (**a**) storage tanks and (**b**) concentration profile (modified from [6]).

2. A Brief Summary on Soil Loss Measurement in the Literature

The spatial scale at which the measurement is performed (splash cup, microplot, plot, hillslope, or watershed) and the applied measurement technique have to be chosen in the light of the considered erosion subprocesses (e.g., impact, interrill, rill, and gully). Field plots with different sizes and steepness are often used to measure interrill and rill erosion. Microplots, with a small surface area varying approximately between 0.04 m^2 and 10 m^2 [6,7], are used to measure interrill erosion because rill development is not usual. Plots approximately 10–200 m long and 2–50 m wide [8] can be used to measure total, i.e., rill + interrill, erosion. Soil erosion and runoff were recognized to be dependent on the spatial scale [9,10], which makes questionable upscaling of plot erosion measurement to the basin scale. This scale dependency is attributed to spatial variability in infiltration, temporal variations in rainfall intensity during a storm event, hillslope sediment sinks, and sediment delivery along flow paths within a basin [11] that are not generally represented at the plot scale. As a result, measured erosion rates per unit area are higher on the plot than on the basin scale. On the other hand, larger plots allow for a better representation of the hydrological and erosive response of a given ecosystem [9,12], and a plot length of 20 m is sufficient to determine runoff and sediment export rates representative of a basin [13].

The measurement of the total soil loss from the experimental plots is performed either by collecting a portion of the total surface runoff coming from a conveyance system (e.g., H-flume or pipe) [14] in a sampling unit or the whole runoff inside a storage system. In the former case, once calibrated in the field, the sampling unit directly allows for measuring runoff volume and sediment concentration (e.g., Coshocton wheel [15], Fagna-type unit [16]) or storing runoff and sediments for large runoff events (multislot divisors [8,17]). In the second case, the water–sediment mixture is usually collected into tanks of known geometry, and its volume is determined by measuring the water depth. The plot soil loss is equal to the sediment mean concentration by volume. Sediment concentration can be measured by either catching the whole sediment amount after siphoning the supernatant-cleared water or collecting a sample of the mixed suspension [18]. The good accuracy of the first technique is paid by the necessity to remove high mud volumes to oven-dry in the laboratory, which makes this technique more and more cumbersome and time-consuming as stored runoff and plots' number and size increase. In order to obtain reliable soil loss measurements by the second technique, the sample has to be representative of the whole suspension; that is, the sampled suspended sediment concentration has to be equal to the actual one. This occurs in the case of homogeneity of the suspension, i.e., when the sample concentration in any measurement point of the tank is equal to the actual concentration. This occurrence can be hindered by the incomplete mixing of the suspension [6,19].

For instance, Lang [20] detected that sampling with a bottle at the bottom of the tank, where the soil–water mixture was vigorously agitated to suspend sediment, determined underestimations of sediment concentration ranging from −17 to −74%. Zobisch et al. [21] measured total soil loss by adding the suspended sediment stored in a large tank and the sediment settled in a smaller metal bucket placed within the tank. The suspension was stirred, and its concentration was determined on a sample taken by an immersed laboratory beaker. The dry weight of the total settled sediments was also determined. Zobisch et al. [21] verified a poor accuracy of soil loss measurement and its strong dependence on the field workers, which negatively affected the repeatability of the sampling procedure. They concluded that the differences were due to the level of mixing thoroughness of the suspended sediment and the timing and plunging depth of the sample beaker. The sampling procedure of Ciesiolka et al. [22] consisted of stirring the mixture in the tank and taking samples simultaneously at three depths. The average of three sediment concentrations was used as a representative value for the entire depth of the mixture. The measured sediment concentration systematically underestimated the actual one. The authors proposed an equation based on settling theory to remove the error due to the delay time between stirring and sampling, which reduced the underestimation to values generally ranging from −66 to −6%. The residual underestimation could be due to other sources of error, such as a not uniform sediment distribution within the tank and the sampling technique. A cylindrical sampler was used by Nikkami [23] after stirring the suspension in the storage tank to withdraw a column of suspension, and the results were compared with those obtained by bottle and pipette samplings. The latter underestimated the actual sediment concentration and were unreliable, while the measurements by the cylindrical sampler were affected by errors ranging from −18 to −4%. A similar sampling cylinder with a closing valve at the bottom was proposed by Carollo et al. [24] (Figure 3) and tested to be applied for soil loss measurement at the Sparacia experimental area. The details regarding the cylinder and the applied sampling procedure are reported below. Bagarello and Ferro [19], Bagarello et al. [6], and Todisco et al. [25] evaluated factors affecting the measured sediment concentration and derived calibration curves for the sediment storage tanks at the Sparacia and Masse (Umbria, central Italy) stations. The experimental stations are characterized by the same type of plots, runoff storage system, and different soils.

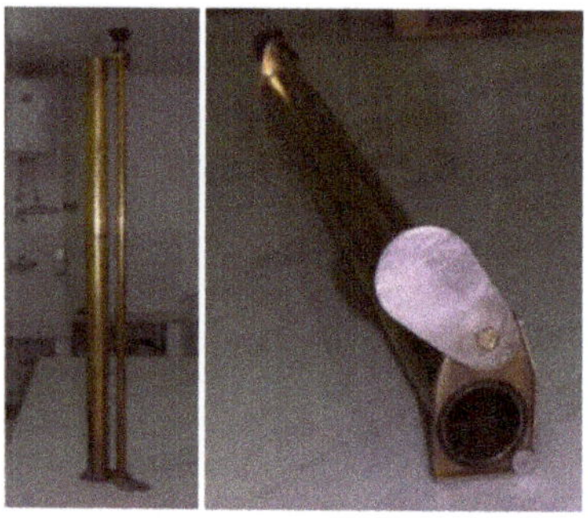

Figure 3. View of the sampler (modified from [24]).

3. Measurement of Total Erosion at Sparacia

3.1. Measurement Methods

The total weight of the solid particles in the suspension collected into each tank is calculated as mean concentration by volume. The latter is easily determined by the water level reading. In order to measure mean concentration, two different procedures were applied over the monitoring period. By using the former procedure, after mixing the suspension for a given time, the sediment is sampled from 10 taps deployed along the middle vertical of the tank wall to determine the sediment concentration profile and, by integration, the mean concentration (Figure 2b). The samples are oven-dried at 105 °C for 48 h, and both the sediment weight and the associated water volume are determined. The measured mean sediment concentration differs from the actual concentration C_e as the sampling procedure determines an incomplete mixing condition, and sedimentation phenomena occur over the sampling time [19,20,25]. The measured mean concentration is then converted into the actual concentration using the calibration curve of the storage system [6,19]. According to the calibration curve, the actual concentration is equal to fourfold the measured mean concentration. When considering the high number of equipped plots in the experimental area, the above procedure was highly time-consuming for both the field sampling phase and the oven-drying one in the laboratory. Therefore, since September 2013, a quick sampling procedure using an efficient sampler has been applied [24]. The sampler is a brass cylinder with a height of 120 cm and an inner diameter of 4.75 cm. It is equipped with a closing valve 'guillotine' fitted with a sealing gasket (Figure 3). The valve is controlled from above using a knob joined to a drive pin. The sampler allows the extraction of a whole column, i.e., bounded by the free surface and the tank bottom of the suspension. The sampling procedure adopted in the field is based on the analysis of laboratory measurements performed by Carollo et al. [24]. Specifically, this analysis allowed us to establish that the mean value of the measured concentration in different verticals can be considered equal to the actual concentration, i.e., this sampling procedure does not need a calibration curve of the storage system, and that five samples are sufficient to obtain a relatively low margin of error. The margin of error decreases sharply as the actual concentration increases.

3.2. Experimental Results

3.2.1. Effects of Plot Size and Steepness

Plot size is expected to affect soil loss due to spatial variability in infiltration, the potential for sediment to be trapped by roughness components and other sinks, and changes in erosion rates and processes with increasing amounts of runoff [26,27].

Bagarello and Ferro [28] selected 19 events monitored during the period November 1999–October 2008 to assess the effects of plot length, λ, on event soil loss per unit plot area A_e, runoff volume per unit plot area V_e, and sediment concentration C_e. The most frequent results were that soil loss did not vary significantly with λ, runoff decreased as plot length increased, and sediment concentration generally did not vary significantly with λ. Therefore, shortening the plot length was not an effective practice to control event soil loss for the Sparacia area. The analysis of the measurements of the interrill and rill erosion components available for two events supported the idea that shortening λ produces a moderate decrease in rill erosion rates but also an appreciable increase in interrill erosion rates and that shortening λ becomes effective when rill erosion is dominant (λ ≥ 33–44 m) (Figure 4). Bagarello et al. [29] investigated the plot width and length effects on A_e, V_e, and C_e by using data collected on plots differing in width (2–8 m) and length (11–22 m). A reduced data set in terms of plot length was considered to compare the results with those available at the Masse station [25]. Plot width did not have a statistically significant effect on long-term mean values of runoff, soil loss, and sediment concentration, while the plot width effects at the event temporal scale varied from negligible to significant for low erosivity events and tended to vanish for highly erosive events. Plot length did not have a statistically significant effect on mean runoff, soil loss, and sediment concentration. Moreover, in this case, for low erosivity events, the plot length effect varied from negligible to significant, while it greatly decreased for highly erosive events (Figure 5). Similar results were obtained for Masse station, characterized by soil with a prevailing silt fraction. Accordingly, both long-term and severe storm-related data collected on narrow and short plots are expected to be representative of wider and longer plots. For low erosivity events, local soil conditions at the beginning of the rainfall and more relevant measurement errors likely produced varying plot width and length effects. For highly erosive events, the importance of local soil conditions and the measurement error is reduced, and the effects of plot size on the measured variables tended to vanish. The analysis of the measurements of the interrill and rill erosion components available for an additional event to those analyzed by Bagarello and Ferro [28] highlighted that shortening λ increased total erosion since the rill erosion rate decreased, but more deposition of sediment coming from the rill channels occurred within the 22 m long plots relative to the 11 m long plots. In other words, rill interruption within the plot, which prevents the sediment from contributing to total soil loss measured at the plot outlet, was more frequent on the longer plot.

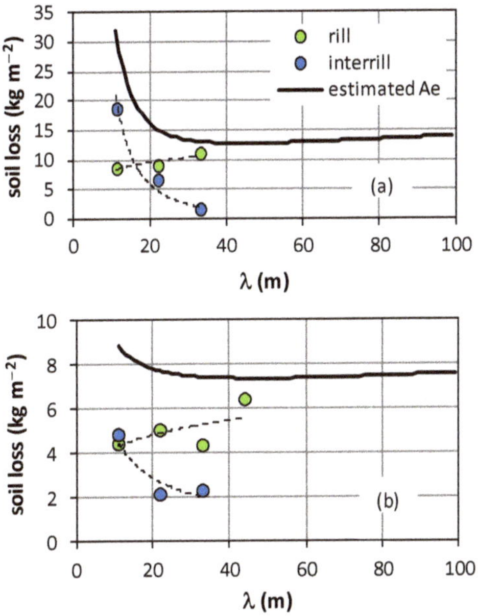

Figure 4. Relationship of total, A_e, rill, and interrill soil loss vs. plot length, λ, for the (**a**) 1 September 2005 event and (**b**) 28 June 2008 event (modified from [28]).

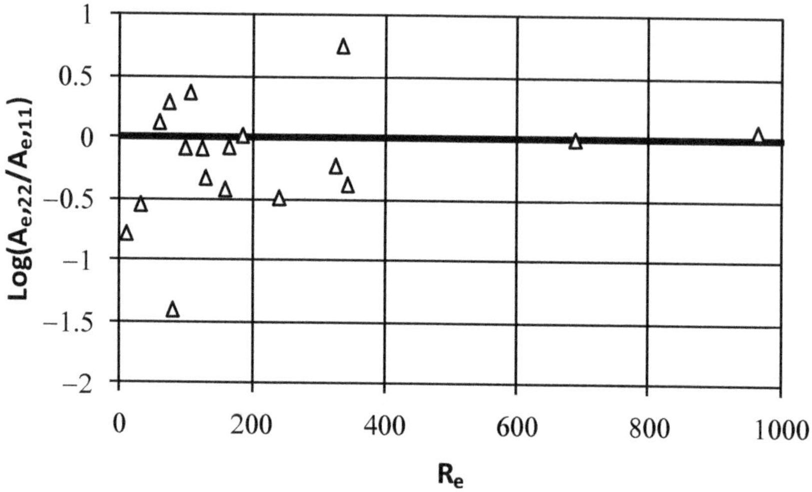

Figure 5. Log of the ratio between soil losses, A_e, measured in the plots with length λ = 22 m and plots with λ = 11 m vs. the erosivity index, R_e (MJ mm ha^{-1} h^{-1}) (modified from [29]).

Bagarello et al. [30] established plot length and steepness effects on A_e, C_e, and the event runoff coefficient Q_{Re}, i.e., V_e per unit rainfall depth, using measurements collected until February 2012. The database was developed by considering, for a given event, the individual plots yielding a simultaneous measurement of both A_e and V_e. A total of 317 data were collected for the plots with λ = 11, 22, 33, and 44 m and s = 14.9% were considered for testing plot length effects. The selected events were those with the simultaneous functioning of at least a plot for each length to obtain reasonably homogeneous and comparable data. The 278 data collected for the 22 m long plots with s = 14.9, 22, and 26% were considered for testing plot steepness effects. The selected events were characterized by the simultaneous functioning of at least a plot for each steepness value. The mean and median values of the erosive and hydrological variables, which were expressive of the long-term average responses, confirmed that, generally, A_e did not vary appreciably with λ because the decrease in runoff coefficient was offset by an increase in sediment concentration (Figure 6). Therefore, how the physical erosion process occurred at the sampled site was not consistent with the assumption made in the USLE/RUSLE that soil loss per unit area increases with plot length. The authors verified that the short duration of individual rain showers, which determines a flow discontinuity along the hillslope [31], can favor the occurrence of more noticeable sediment deposition on longer plots, determining practically constant A_e values on plots differing in length. The moderately increasing relationship of C_e against λ was attributed to the advanced sediment transport efficiency on the longest plots, where concentrated flow, which notoriously has a higher capacity to transport sediment as compared to the overland flow, is expected to be more pronounced.

Moreover, for the event of October 2011, for λ varying from 22 to 44 m A_e decreased because of increasing rill erosion rates and increasing deposition phenomena in the interrill areas. These behaviors agree with the preceding ones obtained by Bagarello and Ferro [28] and Bagarello et al. [29] with reduced databases.

Slope steepness s had a positive effect on A_e because of a runoff coefficient Q_{Re} that did not vary appreciably with s and a sediment concentration C_e increasing with s (Figure 7). The detected steepness effects were well reproduced by the relationships used in the USLE/RUSLE models to predict the slope steepness factor. The availability of the rill and interrill erosion rates for three sampled events on 14.9 and 22% sloped plots (λ = 22 m) did not reveal clear slope effects on the rill, interrill, and total soil loss rates.

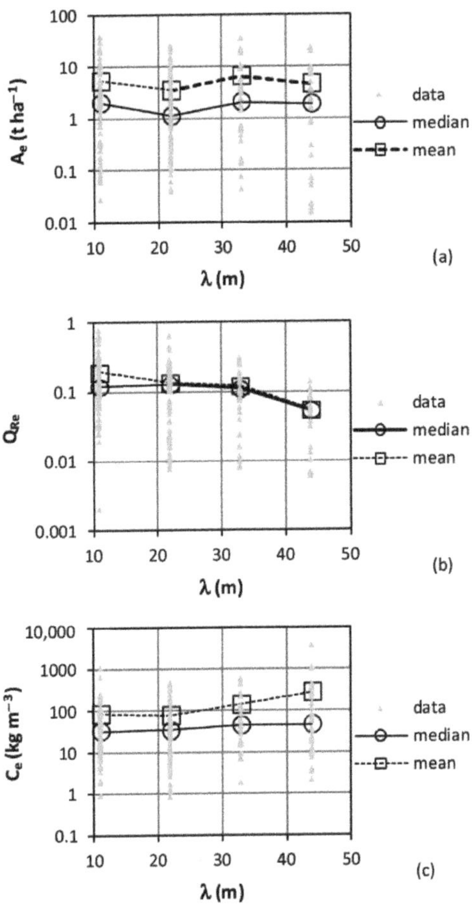

Figure 6. Plot length, λ, effect on event (**a**) soil loss per unit area, A_e, (**b**) runoff coefficient, Q_{Re}, and (**c**) sediment concentration, C_e, for the plots with slope steepness s = 14.9% (modified from [30]).

Bagarello and Ferro [32] checked plot-scale effects on hydrological and erosion variables using a larger dataset as compared with that considered by Bagarello et al. [30]. Similar information on the scale effects was obtained with four different scenarios combining dependent variables, i.e., individual values of A_e, V_e, and C_e for each plot or the mean of their replicated values, and scale indicators, i.e., plot length, λ, or plot area, *A*. Following Chen et al. [33], Bagarello and Ferro [32] checked scale effects using common regression analysis techniques to fit a power relationship of A_e, V_e, and C_e against λ or the plot area *A* to the data. The power exponent, b_1, which is the scaling exponent, represented a measure of the scale effect on the considered variable. Statistically significant relationships were detected for a minor part of the sampled events, regardless of the considered variables and scenarios (19–43%).

Therefore, the lack of any scaling effect was the most common result while, when scale effects were detected, runoff always decreased, sediment concentration generally increased, and soil loss per unit area most frequently decreased in longer or larger plots. For the scenario with the largest number of scale effects, i.e., individual values of the considered variable against plot length, further analysis confirmed that lack of scale effects was the prevailing result and, when scale effects were detected, runoff decreased, sediment concentration generally increased, and soil loss per unit area generally decreased with

λ. For the latter two variables, scale effects were highly dependent on the event (high variability of b_1 with the event), while they showed a reduced variability with the event for runoff (reduced variability of b_1 with the event). Rainfall characteristics were not able to discriminate between significant and nonsignificant scale effects. For events with a significant correlation, for the $V_e - \lambda$ relationship, the exponent b_1 became less negative with an increase in both rainfall amount and erosivity. In other words, scale effects for runoff were less evident when heavy rainfall occurred. The rainfall characteristics were not able to describe b_1 variations for C_e and A_e.

An alternative approach was finally developed for testing scale effects with reference to the whole historical sequences of events, regardless of the single events. A statistical check was performed to establish if the probability distribution of A_e for $\lambda = 11$ m was able to reproduce the empirical frequency distribution of A_e for $\lambda = 44$ m and vice versa. The discrepancy between the probability and frequency distributions indicated the occurrence of scale effects.

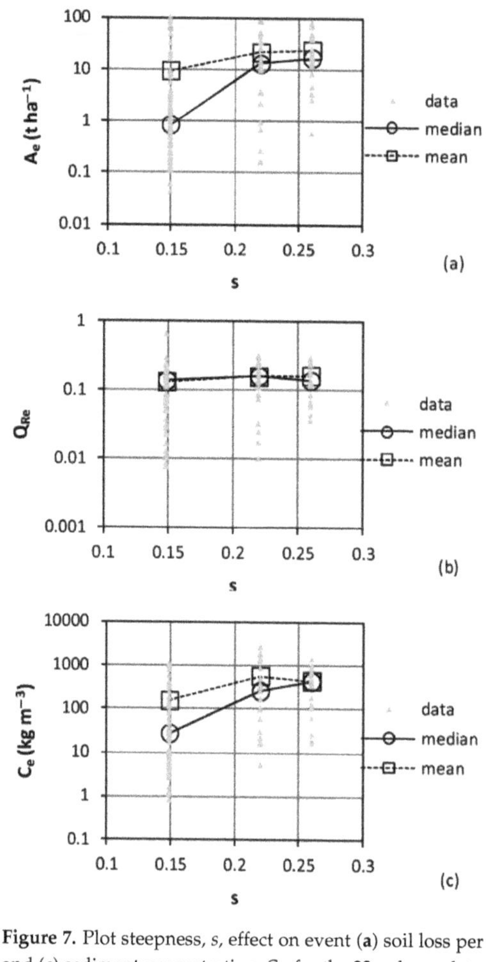

Figure 7. Plot steepness, s, effect on event (**a**) soil loss per unit area, A_e, (**b**) runoff coefficient, Q_{Re}, and (**c**) sediment concentration, C_e, for the 22 m long plots (modified from [30]).

3.2.2. Measurement Variability and Physical Model Concept

Soil loss measurements are affected by a measurement variability due to the applied measurement technique [19] and a natural variability [34] that produces an unexplained variance of runoff and soil loss measurements in identical plots subjected to the same rainfall events. As a consequence of the natural variability, the single or few plot measurements are not coincident with the expected mean value for a particular treatment.

The information derived by Bagarello and Ferro [35] from measurements collected in the period 2000–2001 on 8 m × 22 m plots suggested that the coefficient of variation CV of A_e decreases as the mean measured value increases. Then, Bagarello et al. [36], by using 5 years of data from plots differing in length, highlighted that the variability of both runoff and soil loss measurements decreases as the mean value increases. This trend was detected above a threshold mean value of 3 mm for V_e and 0.1 t ha^{-1} for A_e, while a relatively low variability was generally detected for mean values lower than the threshold one. The analysis of the events that occurred until October 2008 [28] on 8 m × 22 m plots, jointly with the literature plot measurements, clarified that the relative variability of soil loss, runoff volume, and sediment concentration decreased as the mean measured value increased all over the experimental range. This analysis also indicated that runoff and sediment concentration were generally less variable than soil loss measurements and allowed to obtain, for each variable, a decreasing CV vs. mean relationship useful to estimate the optimum number of replicated plots for erosion studies [34,37], i.e., allowing to determine a representative value of the variable for given experimental conditions.

The natural variability of plot soil erosion measurements raises the problem of their usability to test the applicability of erosion prediction models, which is performed by comparing the measurements to the model estimations for a given temporal interval. Indeed, the difference between the measured and predicted value is due to both model error and unexplained variance of the measurements. Nearing [38] suggested a method for evaluating the performances of the model by accounting for the natural variability of plot data. According to the author, the best possible model to predict the erosion from an area of land is a physical model of the area that has similar soil type, land use, size, shape, slope (i.e., a replicated plot), and erosive inputs. Nearing [39], using a large USA soil loss database, suggested that the model prediction has to be considered acceptable if the departure from the measured value lies within the population of differences between pairs of measured values.

Bagarello and Ferro [40] tested the physical model concept by Nearing [39] using erosion data collected on the 33 m and 44 m long plots and the 22 m long plots established on 14.9%, 22%, and 26% slopes from November 1999 to April 2011. The relative difference, R_{diff}, was calculated as follows [39]:

$$R_{diff} = \frac{P - M}{P + M} \quad (1)$$

where P is the predicted value from the physical model and M is the measured value. In order to define the physical model, differences in width between the plot and its replicate were neglected. The highest measured values showed fewer relative differences between replicates (Figure 8), in accordance with the results obtained by Nearing [39]. The R_{diff} data falling within the empirically calculated 95% occurrence interval (Equation (3) in [39]) were equal to 88% of the total. When considering the relatively low discrepancy between these percentages, the investigation supported the applicability of the analysis by Nearing [39] for model evaluations. Bagarello et al. [41] used bare plot soil loss data (November 1999–May 2012) from Sparacia and Masse experimental stations. A total of 819 A_e values were available, varying from 0.00012 to 21.7 kg m^{-2}. The Italian (P, M) database was compiled by considering four plot lengths (λ = 11, 22, 33, and 44 m), four slope steepness values (s = 14.9, 16.0, 22.0, and 26.0%), and defining a physical model as a plot with the same length and width to the sampled one.

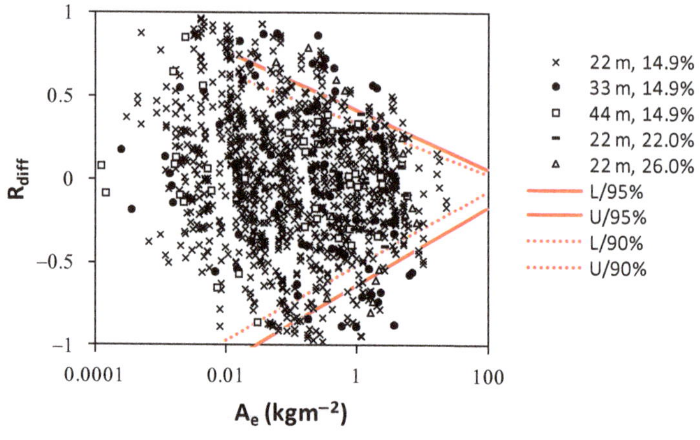

Figure 8. Relative differences in measurements of soil loss between replicated plots, R_{diff}, vs. the measured soil loss value, *M*, for all plots established at the Sparacia station and 90 and 95% occurrence intervals for the data (L = lower limit; U = upper limit) calculated according to [39] (modified from [40]).

Indeed, the preliminary linear regression analysis between the *M* and *P* values suggested that the physical model has to be defined in terms of perfect planimetric equivalence. The percentage of R_{diff} values falling within the 95% occurrence interval (Equation (3) in [39]) was equal to 88.9%, which was in line with the previous result by Bagarello and Ferro [40]. However, R_{diff} values decreasing with an increase in *M*, which is the premise of the analysis by Nearing [39], were only detected for $M > 1$ kg m^{-2}, and the corresponding R_{diff} values falling within the 95% occurrence interval were equal to only 85.1% of the total (Figure 9). Accordingly, a new applicative criterion of the physical model concept was developed and expressed by the following relationship between the absolute difference between the physical model prediction and measurement (Figure 10):

$$|P - M| = 0.356M^{0.91} \quad (2)$$

where $R^2 = 0.72$. For a given soil loss value (*M*), Equation (2) predicts the mean absolute difference associated with the sampling of a replicated plot. A model prediction is acceptable if the absolute difference with the measured soil loss is lower than or equal to the value calculated by Equation (2).

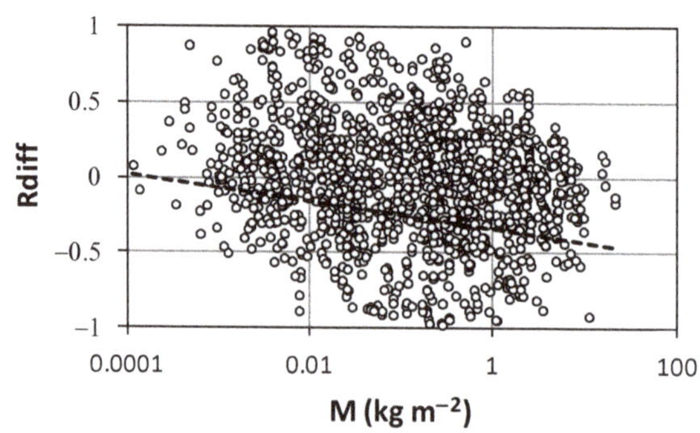

Figure 9. Relative differences in the measurement of soil loss between replicated plots, R_{diff}, vs. the measured soil loss value, *M*, for all plots established at the Sparacia and Masse stations (modified from [41]).

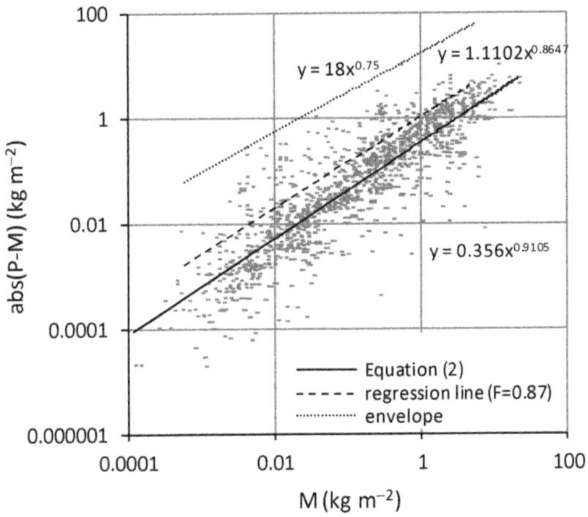

Figure 10. Plot of the absolute difference in measurement of soil loss between replicated plots, $|P - M|$, vs. the measured value, M, and of Equation (2), the regression line associated with a frequency $F = 0.87$ and the data enveloping line (modified from [42]).

The general objective of the investigation by Bagarello et al. [42] was to compare the two applicative criteria of the physical model concept by Nearing [39] (original criterion) and Bagarello et al. [41] (new criterion) through three soil erosion empirical models and plot soil loss data from Sparacia. The criterion by Bagarello et al. [41], which was expressed in terms of R_{diff} to be directly comparable with that of Nearing [39], was analytically demonstrated to be characterized by a greater selectivity. Indeed, the fraction of the acceptable model predictions relative to the total varied from 29% for the former to 54% for the latter.

A less restrictive new criterion was developed by carrying out a frequency analysis of all the data used to obtain Equation (2). In particular, the data were divided into half log-cycle intervals, and within each division, the value associated with a frequency higher than 0.5, which practically yields Equation (2), was determined. The power regression equation between the values associated with a frequency F of 0.87 and M values (Figure 10) featured a similar selectivity level as that of the original criterion. Both the original and new criteria accept predictions characterized by relative maximum errors that decrease as M increases. For example, Equation (2) accepts predictions with a maximum error decreasing from 80% to 27% as M increases from the minimum (0.00012 kg m^{-2}) to the maximum (21.7 kg m^{-2}) value. Based on all available data, the models' accuracy ranking depended on the evaluation criterion, while for the highest soil losses, both criteria gave the same accuracy ranking. In order to increase the significance of the comparison between the two criteria, they should be calibrated by merging the American and Italian databases.

3.2.3. Statistical Analysis of Soil Loss Measurements

Field investigations (e.g., [43]) have clearly shown that rare and severe events control total soil erosion over a long period. Consequently, soil conservation strategies should be developed considering soil loss of a given return period rather than long-term average erosion.

Bagarello et al. [44] developed a frequency analysis of relatively long-term (2000–2008) soil loss data collected in a high number of simultaneously operating plots and microplots of different lengths (λ = 0.25, 0.4, 11, 22, 33, and 44 m) and slope of 14.9%. This analysis allowed establishing both the theoretical distribution reproducing the frequency distribution and the soil loss of a given return period. Each event's soil loss value, A_e, measured on a plot of given length λ, was normalized using the mean soil loss, $\mu(A_e)$, calculated for that event and all the plots with the same length. Normalization allowed assuming that the probability distribution $P(x)$ of the variable $x = A_e/\mu(A_e)$ is unique. For both the microplots and plots, the frequency distributions of x were perfectly overlapping, and thus $P(x)$ was independent of both the scale length λ and the temporal scale, which were completely represented by the mean value, $\mu(A_e)$.

For a given plot length, the values (334 data) of the annual maximum event soil loss were then extracted from the complete dataset (1403 data). As the x frequency distributions for both the complete dataset and the annual maxima were overlapping, the parameters of the probability distribution of the annual maximum soil loss and the quantiles of a given return period were estimated using all available data. The empirical frequency distribution of x was described by a two-component distribution (Figure 11), with the two components discriminated by a return period of 25 years. The soil loss of a given return period was estimated by multiplying the mean soil loss by a frequency factor determined by fitting Gumbel's distribution to the two components of the frequency distribution.

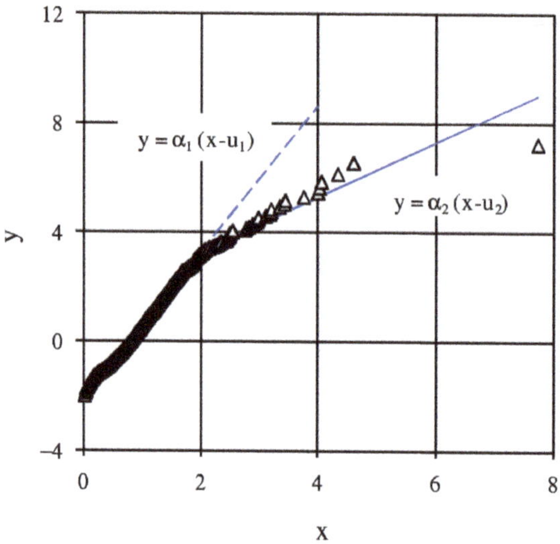

Figure 11. Gumbel's plot for the sequence of annual maximum soil losses ($x = A_e/\mu(A_e)$, $y = -\ln \ln(1/F(x))$ is the normalized Gumbel's variable, α_i and u_i are the two parameters of Gumbel's distribution, modified from [44]).

The subsequent study by Bagarello et al. [45] was conducted using soil loss measurements from individual plots in the period from November 1999 to January 2009. First, this study confirmed the results by Bagarello et al. [44] that the distribution of the normalized soil loss does not vary with λ, and the parameters of the probability distribution of the annual maximum soil loss and the quantiles of a given return period can be estimated using all available data. Then, the authors deduced that, according to the USLE scheme applied at the event temporal scale, the soil loss of a given return period is obtained by amplifying the mean soil loss for a sequence of events by a frequency factor estimated with the frequency distribution of the rainfall erosivity index. The reliability of this deduction was verified by the available measurements of rainfall erosivity index and event soil loss for each considered plot length. Assuming that this result is also valid at the annual temporal scale, an estimating criterion of the annual soil loss of a given return period was developed. By this criterion, the probability distribution of the yearly rainfall erosivity factor, which is available for the Sicilian region, can be used to design soil conservation practices.

4. Measurement of Rill Erosion at Sparacia

4.1. Measurement Methods

Rill erosion measurements were firstly performed by the traditional direct method and, more recently, applying an indirect technique (image-based) [46–60]. In the former case [1,61–65], the rill channels were visually identified in the field and divided into segments bounded by two measurement cross-sections with a varying distance interval depending on the variability of rill depth, width, and appearance of tributaries along the rill length (Figure 12). On average, this distance interval was equal to 2 m. In order to measure the rill segment features, i.e., mean width w, maximum scour depth H, and cross-section area σ, a rillmeter [61] was used. The rill segment length $L_{r,s}$ was measured by a metric ruler. Specifically, the values of w, H, and σ were set equal to the mean of the values measured at the bounding sections, and the rill segment volume $V_{r,s}$ was equal to σ $L_{r,s}$. The plan-altimetric survey of the rill network was carried out using a total station, and the segment slope was determined along the rill thalweg. The widely used direct method is very simple, low-cost [66], and invasive as the operators tend to modify the surface of the rilled area. In addition, it was time-consuming due to the high number of experimental plots established at the Sparacia station.

(a) (b)

Figure 12. View of the plot (**a**) before and (**b**) after rill network formation. Rill channels are numbered from 1 to 10 (modified from [61]).

Advancements in automatic 3D-photo reconstruction techniques induced us to apply the Structure from Motion (SfM) technique (image-based) to measure rill erosion using low-altitude aerial [62,67] and terrestrial acquisition platforms [67]. Image-based modeling creates a three-dimensional Digital Terrain Model (3D-DTM) using a set of oblique photographs taken from the same rilled surface [68] with uncalibrated and non-metric cameras, coupled with photogrammetric software packages, such as Photoscan professional (Agisoft). Camera model parameters and scene geometry are simultaneously solved [69], using redundant information coming from oblique photographs. The generated 3D point cloud has to be scaled and georeferenced using some ground control points (GCPs) located within the monitored rilled area. Di Stefano et al. [67] deployed six GCPs on the plot perimeter (Figure 13) and collected their (x, y, z) coordinates with a total station. From the 3D point cloud, a 2.5D surface (i.e., Digital Elevation Model, DEM) and the orthophoto were also generated. Unlike DEM, which represents the surface of the rilled area from a vertical perspective, the 3D-DTM can also represent cross-section areas with undercut walls [68] where different z-elevations can be associated with the same couple of plane coordinates x–y. In this case, the DEM underestimates the area of the cross-section and, therefore, the eroded rill volume [70]. However, the DEM is often used for hydrological analyses, and it can be easily managed as it is a raster model.

In order to extract channel networks from DEM, Carollo et al. [62] applied both a manual method, which consisted of tracing rills from the visual interpretation of orthophotos, and an automatic procedure based on the theory of drop analysis by Broscoe [71]. In the last case, drainage paths were identified by following the maximum slope direction and by all of the cells of the DEM with a value of drainage area greater than an established threshold. The value of this threshold was obtained by the drop analysis. By using the hierarchical approach of Horton and Strahler, Broscoe [71] defined drop as the height difference between the two points defining a channel from the source/confluence to the confluence with another stream. According to Broscoe [71], the threshold value of drainage area is that for which the drop mean values, calculated for each stream order, are almost equal.

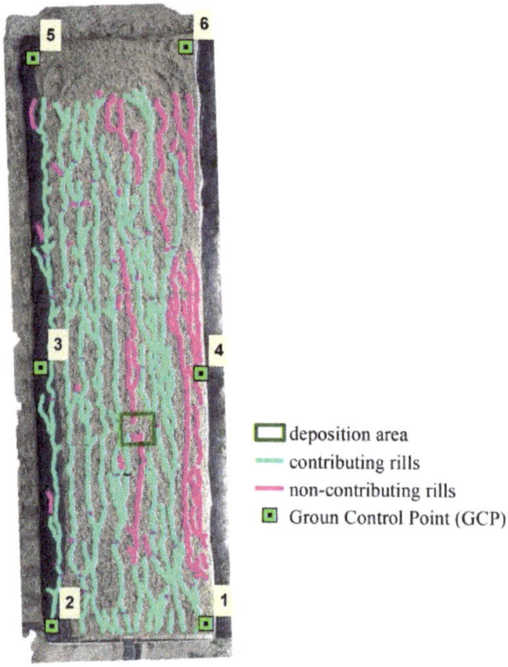

Figure 13. View of the P3 plot (22 m × 6 m, s = 26%) with Ground Control Points (GCPs) numbered from 1 to 6, contributing rills, non-contributing rills, and example of sediment deposition area (modified from [67]).

Di Stefano et al. [67] applied a method based on the combination of a drainage algorithm and the convergence index CI, which is defined by the following relationship:

$$CI = \left(\frac{1}{n-1} \sum_{i=1}^{n-1} \theta_i \right) - 90 \quad (3)$$

where n is the number of kernel cells, and for each external cell i, θ_i (°) is the angle between the aspect of the cell and the direction of the vector joining the center of the cell and the center of the kernel [72]. The convergence index ranges from −90° to 90°. Positive values relate to divergent areas, negative values relate to convergent areas, and null values represent areas without curvature as planar slopes. In order to determine cells that should correspond to the rill channel, an objective threshold, beyond which the CI values could be considered significantly different from a plane landform, was set equal to two times the standard deviation, std, of CI [73]. When considering that only negative CI values relate to convergent areas, significant flow convergence cells were characterized by $CI \leq -2\ std$. At the end of this step, the disconnected convergent areas were identified and considered as rill features. In order to achieve the connection, the flow accumulation algorithm was weighted with the threshold convergence grid following Thommeret et al. [72]. Finally, a grid representing the rill thalweg was obtained. Rills detected by Di Stefano et al. [67] were automatically separated into contributing rills, which were connected to the network reaching the plot outlet and contributed to total soil loss measured at the plot outlet and non-contributing rills, which were interrupted within the plot (Figure 13). Each contributing rill was divided into segments, each of them bounded by two 0.25 m long transects perpendicular to the rill thalweg and spaced 0.3 m apart, and $V_{r,s}$ was calculated as 0.3 σ. The use of the SfM technique allowed for obtaining accurate rill erosion measurements with rapid and not invasive field surveys but time-consuming image post-processing.

For both direct and indirect measurement methods, the total rill volume V was calculated by adding the volumes $V_{r,s}$, and converted into the weight of eroded soil by the soil bulk density. The latter was experimentally determined on different dates by collecting soil samples uniformly distributed on the erosion plots. Soil samples were taken by cylindric samplers [74], with a diameter of 8 cm and a height of 5 cm, and oven-dried at 105 °C for 48 h.

4.2. Experimental Results

In the monitoring period from September 2004 to February 2006, Bruno et al. [61] carried out measurements on rills formed on a 22% sloping plot in five erosion events. The authors verified the applicability of a power relationship between rill length L and rill volume V; thereby, rill length can be employed to estimate the severity of the rilling process. By applying the Π-theorem of the dimensional analysis and the incomplete self-similarity hypothesis [75,76], they also proposed a power dimensionless relationship among the rill morphological variables (length $L_{r,s}$, width w, depth H, and volume $V_{r,s}$). Di Stefano and Ferro [63] pointed out that the scale parameter of the L-V relationship varies with channel type (rills, Ephemeral Gullies (EGs), and gullies) while the exponent is constant. Moreover, the authors verified that the above power dimensionless relationship describes a morphological similarity between the channels of rills, EGs, and gullies.

The analysis of the rill cross-sections detected by Bruno et al. [61] along the channel highlighted changes from the narrow and deep upstream sections to enlarged downstream sections conserving the depth. In the terminal reach, the cross-section area diminished because of a decreasing flow transport capacity due to rill slope reduction.

Di Stefano et al. [1] extended the available database, including measurements collected in the year 2008 on all the plots established in the experimental area. The availability of measurements carried out on different slopes (14.9, 22, and 26%) allowed detection of the influence of the plot slope on rill width and depth. In particular, for a given flow discharge and increasing slope values, the rills deepened and narrowed. Di Stefano et al. [1] validated the L-V (Figure 14) and the power dimensionless (Figure 15) relationships previously calibrated by Di Stefano and Ferro [63].

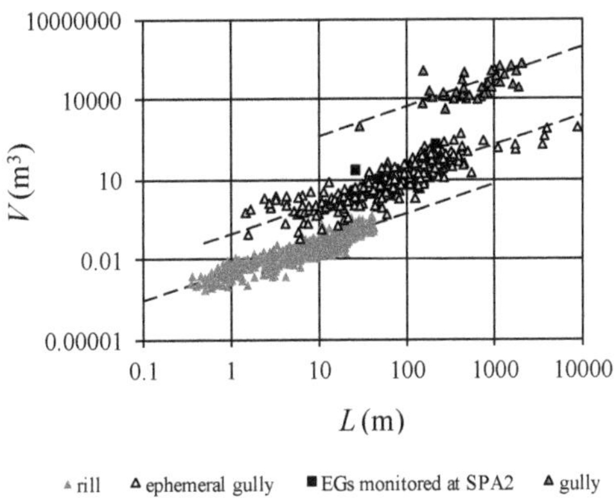

Figure 14. Relationship between the total length, L, and the total volume, V, for rills, EGs, and gullies (modified from [1]).

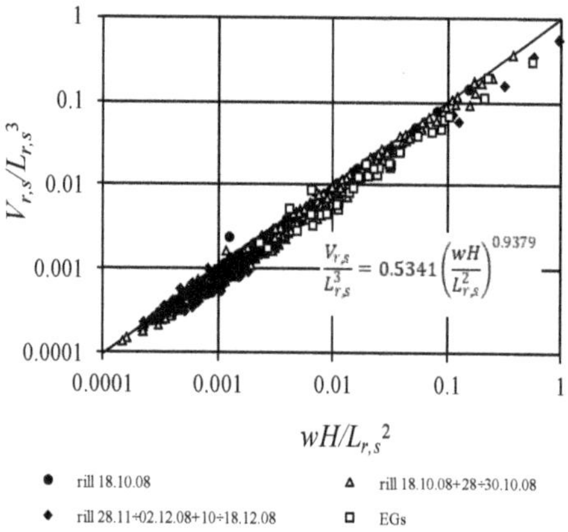

Figure 15. Relationship between the dimensionless groups $wH/L_{r,s}^2$ and $V_{r,s}/L_{r,s}^3$ (modified from [1]).

The reliability of these equations was further confirmed by Di Stefano et al. [64] using rill measurements performed in the event of October 2011. In this case, the indirect survey of the rill network was also conducted, for the first time at Sparacia, by an Unmanned Aerial Vehicle (UAV). Rills tracks were derived from the visual interpretation of orthophotos. For a single plot of 8 m × 44 m, Di Stefano et al. [64] compared the *L-V* pairs and the two groups of the power dimensionless relationship detected both with the direct and indirect measurement method and pointed out the consistency of the results obtained with the two methods.

Di Stefano et al. [64] proposed a simplified method to estimate the rill erodibility based on the simplifying assumptions concerning temporal variability of the bottom shear stress and wetted perimeter and neglecting the effect of sediment load on flow detachment. The application of the method requires only the knowledge of the geometric characteristics of the rills at the end of the erosion event and the duration of the event and gives a first approximation of the rill erodibility. Rill erodibility values were relatively low compared to those reported on a global dataset regarding field plot experiments on agricultural soils [77]. This result agrees with the relatively low value of the USLE erodibility factor for the Sparacia soil [78]. The rill erodibility variation between plots was negligible, and the plot erodibility was not stationary. The method was also validated by using the Water Erosion Prediction Project (WEPP) database for the condition 'rainfall plus flow addition at the head of each rill', which showed that a reliable value of rill erodibility could be obtained for 71 rills of 27 USA experimental sites [65].

For an 8 m × 44 m plot and the event of October 2011, Carollo et al. [62] compared the rill cross-sections obtained by both the direct method (rillmeter) and DEM jointly with the manual method to identify flow paths. By assuming the rillmeter measurements as a reference, the analysis showed that the indirect method provided lower values of depth and higher values of surface width, resulting in a substantial invariance of the cross-section area. The automatic extraction of the rill network by the drop analysis detected a higher number of rills and longer compared to those detected by the manual method, highlighting the difficulty of identifying rills in the downstream end of the plot from visual observation of the orthophoto. The measurements obtained with the automatic extraction of the rills for three 22, 33, and 44 m long plots confirmed the applicability of the *L-V* (Figure 16) and the power dimensionless (Figure 17) relationships, which were calibrated by Di Stefano and Ferro [63] by measurements performed by rillmeter and a metric ruler. In other words, the calibration coefficients of these equations were independent of the survey method.

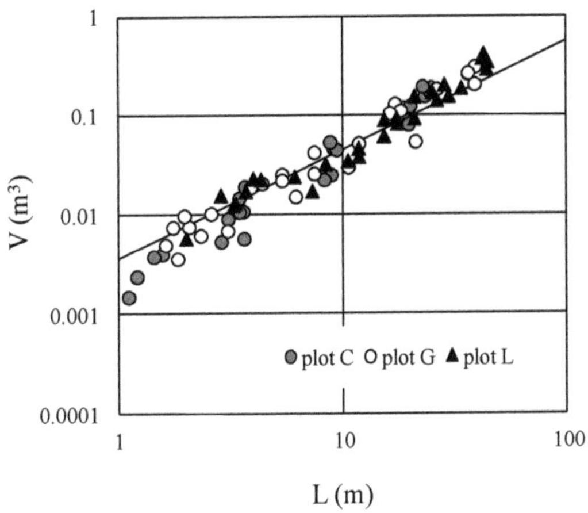

Figure 16. Comparison between the *L-V* power equation and the *L-V* pairs detected for the plots L (44 m × 8 m), G (33 m × 8 m), and C (22 m × 8 m) using the automatic extraction of the rills (modified from [62]).

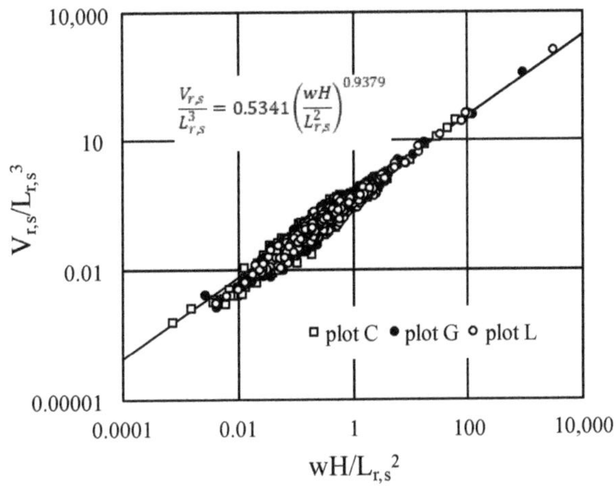

Figure 17. Comparison between the dimensionless power equation and the ($wH/L_{r,s}{}^2$, $V_{r,s}/L_{r,s}{}^3$) pairs detected for the plots L (44 m × 8 m), G (33 m × 8 m), and C (22 m × 8 m) using the automatic extraction of the rills (modified from [62]).

Di Stefano et al. [67] applied the SfM technique with a terrestrial (TS) and a UAV survey of rills formed on two 6 m × 22 m plots with the steepness of 22% and 26% during a rainfall event in October 2016. In order to increase the density of the UAV point cloud, for the 22% sloping plot, a further point cloud was built, merging all aerial photos and 40 terrestrial photos (UAV + TS survey). Rill network extraction was performed by the *CI* method, and the drainage density, i.e., the total length of rills per unit plot area, and the drainage frequency, i.e., the number of rill segments per unit plot area, were determined. For each of the four combinations plot—rill type (i.e., contributing and not contributing)—the highest values of drainage density (ranging from 0.74 to 1.45 m m^{-2}) and drainage frequency (ranging from 0.25 to 0.85 m^{-2}) were measured by TS, denoting its better capability of detecting rills, which was consistent with the highest density of the 3D point cloud. In order to investigate the plot slope, *s*, effect on the morphometric parameters, their mean values obtained by different surveys were used. The ratio between the contributing rill length and the total length was significantly higher for $s = 26\%$ (64%) than for $s = 22\%$ (45%). Moreover, for contributing rills, the available measurements and the literature data [79] indicated that the drainage density increases with plot steepness, whereas an opposite trend occurs for non-contributing rills (Figure 18). Both the results showed an increase in the connectivity degree and thus the efficiency of the sediment transport system, with plot steepness.

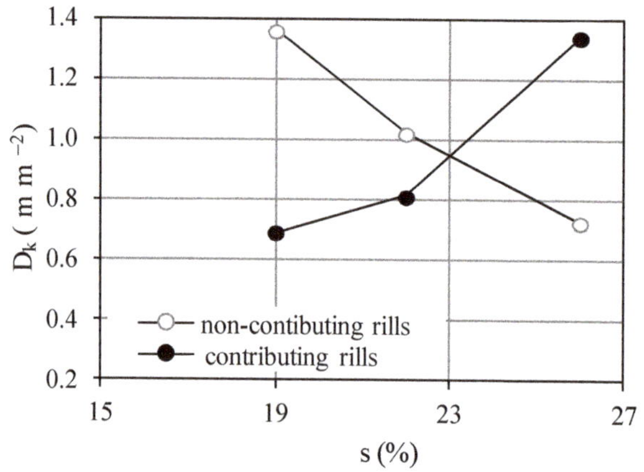

Figure 18. Effect of plot steepness s on drainage density D_k for contributing and non-contributing rills (modified from [67]).

A further analysis aimed to compare the amount of sediment stored in the tank with the rill component due to contributing rills. When considering the former as a reference value (DM, Figure 19), soil loss measurement error by the 3D model was equal to 6% (TS) and −6% (UAV) for the 26% sloping plot and equal to 13% (TS), −8% (UAV), and 10% (UAV + TS) for the 22% sloping plot (Figure 19). In any case, these errors were within the margin of error of the reference value for a probability level of 5% [23], which was equal to ±10.4% for the 26% sloping plot ($C_e = 390$ g L^{-1}), and equal to ±13.1% for the 22% sloping plot ($C_e = 282$ g L^{-1}). Finally, this analysis confirmed the reliability of the SfM measurement technique.

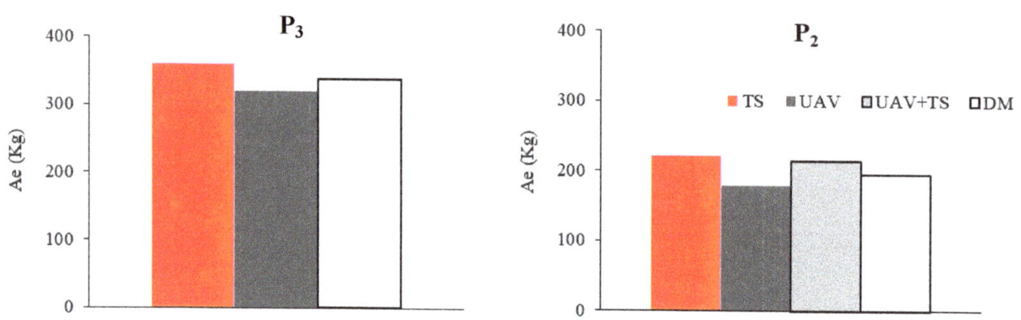

Figure 19. Comparison among the plot soil loss values measured by the different survey methods for the P3 (22 m × 6 m, $s = 26\%$) and P2 (22 m × 6 m, $s = 22\%$) plots (modified from [67]).

5. Comparing Interrill and Rill Erosion

A first attempt to determine the relative importance of interrill and rill erosion on total soil loss was carried out by Bagarello and Ferro [35] by using the data collected on the 8 m × 22 m plots and square microplots of different lengths (0.2 and 0.4 m) and areas.

The study examined three events in the period of April–December 2000, for which soil loss data were available for the three plot types. In the microplots, only interrill erosion likely occurred while rills developed in the USLE-plots. For a given event, a mean value of soil loss per unit area was determined by averaging the data collected in all operating plots of a given surface area. For each event, soil loss decreased in the passage from the smaller to the larger microplots. The straight line passing through the two data points representative of the microplots was extrapolated to the USLE-plot area to estimate the interrill soil erosion (Figure 20a). In this case, the rill component was obtained as the difference between total (rill + interrill) soil loss and the interrill component determined by extrapolation of the straight line. In all events, interrill soil erosion was a minor or negligible part of total soil erosion (0.1–7.1%). Methodological improvements, including the use of other microplot sizes (1 × 1, 1 × 2, 1 m × 5 m), were applied by Di Stefano et al. [64]. Their analysis was carried out using all soil loss measurements by microplots and USLE-plots established on the 14.9% slope simultaneously working for a given event in the period 2000–2013. The data points corresponding to 1 m × 5 m microplots aligned with those of other microplots in some events, whereas they were over the interrill line in other events (Figure 20b,c). Therefore, a plot length equal to 5 m was sufficient for the occurrence of rill erosion processes, at least in some events. The data points representative of the measurements carried out on USLE plots only aligned with those representatives of the microplots for four events, indicating that only interrill erosion occurred. For most events, rill erosion was dominant as it ranged from 70% to 99.9% of total soil erosion measured in the different plots. Moreover, the rill erosion rate was not dependent on the rainfall erosivity factor, as evaluated by the USLE approach, which varied from 42.7 MJ mm ha^{-1} h^{-1} to 686.3 MJ mm ha^{-1} h^{-1}.

The investigation by Bruno et al. [61] was conducted by measuring total soil loss and the direct survey of the rill channels in a plot 6 m × 22 m with s = 22% in the period September 2004–February 2006. For the five events for which a rill network developed, rill erosion was equal, on average, to 69.5% of total soil loss, and interrill erosion ranged from 193 to 4183 g m^{-2}. For the five events for which rills were not visually detectable, interrill erosion was assumed to be coincident with total soil loss and ranged from 10 to 96 g m^{-2}. The comparison between the interrill erosion rates for the two cases pointed out the interaction between rill and interrill processes. Indeed, the presence of a rill network promotes the transport efficiency of the sediments coming from the interrill areas. These sediments are transported by overland flow into the rill channels where the rill flow, which is characterized by a higher velocity than the overland flow, can effectively transport them and sediment particles detached from the rill wetted perimeter.

Di Stefano et al. [1] extended the investigation by Bruno et al. [61] by using the direct surveys of the rill channels performed in all the plots but two of the experimental station in the period 2004–2008 for a total of 41 plot-event combinations. The authors detected rill erosion higher than total soil loss for almost one-half of combinations and attributed this result to the occurrence of sediment delivery processes for the not contributing rills. Indeed, they calculated rill erosion from both contributing and not contributing rills, overestimating then the rill sediment that had reached the downstream tank. The occurrence of sediment delivery processes could not be excluded even if rill erosion was less than total soil loss. However, in this case, the rill erosion rates, calculated by neglecting possible sediment delivery processes, ranged between 24% and 94% of the total soil loss, with a mean value equal to 63%. As a general result, rill erosion was confirmed to be dominant relative to interrill erosion.

For testing the reliability of the aforementioned method based on the extrapolation of the interrill line, Di Stefano et al. [64] compared the rill erosion estimated by this method and that obtained by the direct survey of the rills. This comparison was possible for a single erosion event for which rill erosion was detected by the two methods. When assuming the direct measurement as a reference, the absolute error generally varied from 11% to 36%, indicating quite satisfactory reliability of the method based on the interrill line.

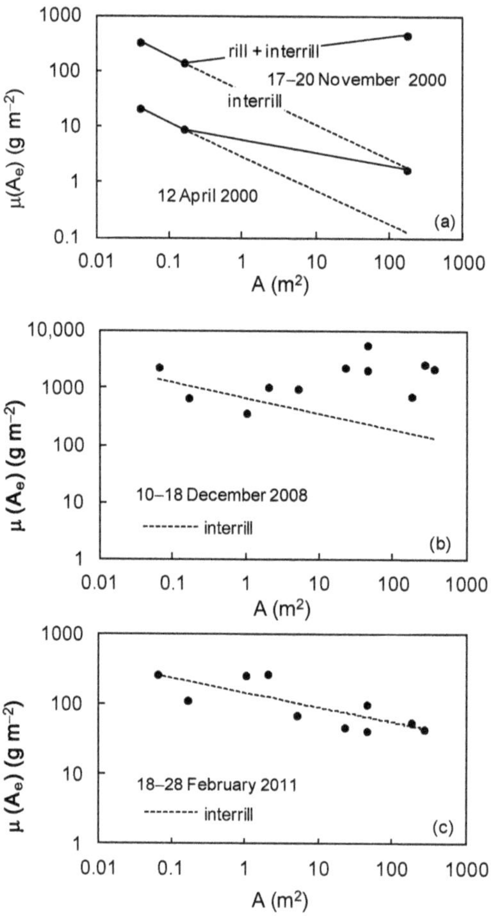

Figure 20. Relationship between mean soil loss, $\mu(A_e)$, and plot area, A, for some selected events (modified from (**a**) [35], and modified from (**b**,**c**) [64]).

6. Conclusions

Efficient management of a large experimental station for soil erosion measurement needs the use of methods and procedures that can guarantee a rapid and accurate field measurement. Currently, in the Sparacia plots, this is obtained by a quick sampling procedure using an efficient cylindrical sampler to measure total soil loss and by the Structure from Motion technique to measure the rill component.

The main results that were obtained during the extended monitoring period can be summarized by the following points:

- The long-term average soil loss per unit area generally did not vary appreciably with the plot length because the decrease in runoff coefficient was offset by an increase in sediment concentration. Moreover, the analysis at the event scale suggested that soil loss per unit area most frequently decreased in longer plots. In any case, these results were not consistent with the assumption made in the USLE/RUSLE model. On the contrary, the long-term steepness effect was well reproduced by the predictive relationships of the USLE/RUSLE slope steepness factor;
- The relative variability of soil loss, runoff volume, and sediment concentration decreased as the mean measured value increased, and soil loss measurements were

generally more variable than runoff volume and sediment concentration. In order to account for the natural variability of the soil loss measurements in the evaluation of the predictive capability of the erosion models, a new applicative criterion of the physical model concept was developed using measurements performed at the Sparacia and Masse experimental stations;

- In the sampled site, for a given plot length, the soil loss of a given return period can be estimated by multiplying the mean soil loss by a frequency factor. The latter is determined by fitting Gumbel's distribution to the two components of the frequency distribution of the normalized event soil loss;
- The measurements of the rill volumes and lengths allowed us to verify the reliability of a power relationship between the two variables. The measurements also confirmed the morphological similarity between the channels of the rills and EGs described by a power dimensionless relationship;
- The rill erodibility of the clay soil of Sparacia was estimated by a simplified method, which was also validated by using the WEPP database. Rill erodibility varied over time, maintaining relatively low values;
- The reliability of the SfM technique to measure rill erosion was positively tested;
- For the contributing rills, the drainage density increased, indicating a more efficient sediment transport system as plot steepness increased;
- As a general result, rill erosion was dominant relative to interrill erosion.

Further research could be developed to increase the reliability of the new applicative criterion of the physical model concept, e.g., by using data from the American database, while further soil loss measurements in the Sparacia station could be used to update the frequency analysis.

Moreover, the dynamics of a rill network during a rainfall season could be investigated. Rill channels are usually obliterated after each erosion event to re-establish plot standard conditions, while for this purpose, rills would not be deleted. Multi-temporal surveys, using the SfM technique, could allow for assessing the hydrological and erosive responses of plots with pre-existing rills when rainfall event occurs.

Finally, future research activities could benefit from the upgrade of the experimental site with an optical disdrometer to measure both the size and falling velocity of the raindrops. These measurements allow for determining the rainfall kinetic energy. Considering that, for given plot characteristics (soil, steepness, size), variations in soil loss exclusively depend on variations in rainfall characteristics, that is, according to the USLE-type models, on variations in rainfall erosivity, the soil loss measurement is proportional to the measurement of rainfall erosivity. Therefore, the comparison between soil loss measurements with intensity and kinetic energy of the rainfall can permit verifying if the rainfall erosivity is adequately represented by the current erosion models or if they need to be improved.

Author Contributions: Conceptualization, V.P. (Vincenzo Pampalone), F.G.C., A.N., V.P. (Vincenzo Palmeri), C.D.S., V.B. and V.F.; methodology, V.P. (Vincenzo Pampalone), F.G.C., A.N., V.P. (Vincenzo Palmeri), C.D.S., V.B. and V.F.; formal analysis, V.P. (Vincenzo Pampalone), F.G.C., A.N., V.P. (Vincenzo Palmeri), C.D.S., V.B. and V.F.; writing—review and editing, V.P. (Vincenzo Pampalone), F.G.C., A.N., V.P. (Vincenzo Palmeri), C.D.S., V.B. and V.F. All authors have read and agreed to the published version of the manuscript.

Funding: This research received no external funding.

Institutional Review Board Statement: Not applicable.

Informed Consent Statement: Not applicable.

Data Availability Statement: No new data were created or analyzed in this study. Data sharing is not applicable to this article.

Conflicts of Interest: The authors declare no conflict of interest.

References

1. Di Stefano, C.; Ferro, V.; Pampalone, V.; Sanzone, F. Field investigation of rill and ephemeral gully erosion in the Sparacia experimental area, South Italy. *Catena* **2013**, *101*, 226–234. [CrossRef]
2. Bagarello, V.; Ferro, V.; Pampalone, V. A comprehensive check of USLE-based soil loss prediction models at the Sparacia (south Italy) site. In *Innovative Biosystems Engineering for Sustainable Agriculture, Forestry and Food Production, MID-TERM AIIA 2019*; Coppola, A., Di Renzo, G., Altieri, G., D'Antonio, P., Eds.; Lecture Notes in Civil Engineering; Springer: Cham, Switzerland, 2020; Volume 67, pp. 3–11.
3. Bagarello, V.; Ferro, V.; Pampalone, V. A comprehensive analysis of USLE-based models at the Sparacia experimental area. *Hydrol. Process.* **2020**, *34*, 1545–1557. [CrossRef]
4. Bagarello, V.; Ferro, V.; Flanagan, D.C. Predicting plot soil loss by empirical and process-oriented approaches. A review. *J. Agric. Eng.* **2018**, *49*, 1–18. [CrossRef]
5. Bagarello, V.; Di Stefano, C.; Ferro, V.; Pampalone, V. Comparing theoretically supported rainfall-runoff erosivity factors at the Sparacia (South Italy) experimental site. *Hydrol. Process.* **2018**, *32*, 507–515. [CrossRef]
6. Bagarello, V.; Di Piazza, G.V.; Ferro, V. Manual sampling and tank size effects on the calibration curve of plot sediment storage tanks. *Trans. ASAE* **2004**, *47*, 1105–1112. [CrossRef]
7. Chaplot, V.; Le Bissonnais, Y. Field measurement of interrill erosion under different slopes and plot sizes. *Earth Surf. Process. Landf.* **2000**, *25*, 145–153. [CrossRef]
8. Brakensiek, D.L.; Osborn, H.B.; Rawls, W.J. *Field Manual for Research in Agricultural Hydrology. USDA Agricultural Handbook*; USDA: Washington, DC, USA, 1979.
9. Boix-Fayos, C.; Martínez-Mena, M.; Calvo-Cases, A.; Arnau-Rosalén, E.; Albaladejo, J.; Castillo, V. Causes and underlying processes of measurement variability in field erosion plots in Mediterranean conditions. *Earth Surf. Process. Landf.* **2007**, *32*, 85–101. [CrossRef]
10. Cerdà, A.; Brazier, R.; Nearing, M.; de Vente, J. Scales and Erosion. *Catena* **2013**, *102*, 1–2. [CrossRef]
11. Parsons, A.J.; Brazier, R.E.; Wainwright, J.; Powell, D.M. Scale relationships in hillslope runoff and erosion. *Earth Surf. Process. Landf.* **2006**, *31*, 1384–1393. [CrossRef]
12. Boix-Fayos, C.; Martínez-Mena, M.; Arnau-Rosalén, E.; Calvo-Cases, A.; Castillo, V.; Albaladejo, J. Measuring soil erosion by field plots: Understanding the sources of variation. *Earth Sci. Rev.* **2006**, *78*, 267–285. [CrossRef]
13. Sadeghi, S.H.R.; Bashari Seghaleh, M.; Rangavar, A.S. Plot sizes dependency of runoff and sediment yield estimates from a small watershed. *Catena* **2013**, *102*, 55–61. [CrossRef]
14. Bagarello, V.; Ferro, V. Measuring soil loss and subsequent nutrient and organic matter loss on farmland. In *Oxford Research Encyclopedias, Environmental Science*; Oxford University Press: Oxford, UK, 2017; p. 47.
15. Carter, C.E.; Parsons, D.A. Field test on the Coshocton-type wheel runoff sampler. *Trans. ASAE* **1967**, *10*, 133–135.
16. Bazzoffi, P. Fagna-type hydrological unit for runoff measurement and sampling in experimental plot trials. *Soil Technol.* **1993**, *6*, 251–259. [CrossRef]
17. Bonilla, C.A.; Kroll, D.G.; Norman, J.M.; Yoder, D.C.; Molling, C.C.; Miller, P.S.; Panuska, J.C.; Topel, J.B.; Wakeman, P.L.; Karthikeyan, K.G. Instrumentation for Measuring Runoff, Sediment and Chemical Losses. *J. Environ. Qual.* **2006**, *35*, 216–223. [CrossRef] [PubMed]
18. Pierson, F.B., Jr.; Van Vactor, S.S.; Blacbum, W.H.; Wood, J.C. Incorporating small scale spatial variability into predictions of hydrologic response on Sagebrush Rangelands. In *Variability of Rangeland Water Erosion Processes*; Soil Science Society of America Special Publication: Madison, WI, USA, 1994; Volume 38.
19. Bagarello, V.; Ferro, V. Calibrating storage tanks for soil erosion measurement from plots. *Earth Surf. Process. Landf.* **1998**, *23*, 1151–1170. [CrossRef]
20. Lang, R.D. Accuracy of two sampling methods used to estimate sediment concentrations in runoff from soil-loss plots. *Earth Surf. Process. Landf.* **1992**, *17*, 841–844. [CrossRef]
21. Zobisch, M.A.; Klingspor, P.; Oduor, A.R. The accuracy of manual runoff and sediment sampling from erosion plots. *J. Soil Water Conserv.* **1996**, *51*, 231–233.
22. Ciesiolka, C.A.A.; Yu, B.; Rose, C.W.; Ghadiri, H.; Lang, D.; Rosewell, C. Improvement in soil loss estimation in USLE type experiments. *J. Soil Water Conserv.* **2006**, *61*, 223–229.
23. Nikkami, D. Investigating sampling accuracy to estimate sediment concentrations in erosion plot tanks. *Turk. J. Agric. For.* **2012**, *36*, 583–590.
24. Carollo, F.G.; Di Stefano, C.; Ferro, V.; Pampalone, V.; Sanzone, F. Testing a new sampler for measuring plot soil loss. *Earth Surf. Process. Landf.* **2016**, *41*, 867–874. [CrossRef]
25. Todisco, F.; Vergni, L.; Mannocchi, F.; Bomba, C. Calibration of the soil loss measurement method at the Masse experimental station. *Catena* **2012**, *91*, 4–9. [CrossRef]
26. Wischmeier, W.H.; Smith, D.D. *Predicting Rainfall Erosion Losses. A Guide to Conservation Planning. USDA Agriculture Handbook No. 537*; USDA: Hyattsville, MD, USA, 1978.
27. Cerdà, A.; Bodì, M.B.; Burguet, M.; Segura, M.; Jovani, J. The plot size effect on soil erosion on rainfed agriculture land under different land uses in eastern Spain. In *Geophysical Research Abstract*; EGU2009-185-1; EGU General Assembly 2009; Copernicus Publications: Göttingen, Germany, 2009; Volume 11.

28. Bagarello, V.; Ferro, V. Analysis of soil loss data from plots of differing length for the Sparacia experimental area, Sicily, Italy. *Biosyst. Eng.* **2010**, *105*, 411–422. [CrossRef]
29. Bagarello, V.; Ferro, V.; Giordano, G.; Mannocchi, F.; Pampalone, V.; Todisco, F.; Vergni, L. Effect of plot size on measured soil loss for two italian experimental sites. *Biosyst. Eng.* **2011**, *108*, 18–27. [CrossRef]
30. Bagarello, V.; Ferro, V.; Pampalone, V. Testing assumptions and procedures to empirically predict bare plot soil loss in a Mediterranean environment. *Hydrol. Process.* **2015**, *29*, 2414–2424. [CrossRef]
31. Yair, A.; Raz-Yassif, N. Hydrological processes in a small arid catchment: Scale effects of rainfall and slope length. *Geomorphology* **2004**, *61*, 155–169. [CrossRef]
32. Bagarello, V.; Ferro, V. Scale effects on plot runoff and soil erosion in a mediterranean environment. *Vadose Zone J.* **2017**, *16*, 1–14. [CrossRef]
33. Chen, L.; Sela, S.; Svoray, T.; Assouline, S. Scale dependence of Hortonian rainfall–runoff processes in a semiarid environment. *Water Resour. Res.* **2016**, *52*, 5149–5166. [CrossRef]
34. Nearing, M.A.; Govers, G.; Norton, L.D. Variability in soil erosion data from replicated plots. *Soil Sci. Soc. Am. J.* **1999**, *63*, 1829–1835. [CrossRef]
35. Bagarello, V.; Ferro, V. Plot-scale measurement of soil erosion at the experimental area of Sparacia (southern Italy). *Hydrol. Process.* **2004**, *18*, 141–157. [CrossRef]
36. Bagarello, V.; Di Piazza, G.V.; Ferro, V.; Giordano, G. Predicting unit plot soil loss in Sicily, South Italy. *Hydrol. Process.* **2008**, *22*, 586–595. [CrossRef]
37. Warrick, A.W. Appendix 1: Spatial variability. In *Environmental Soil Physics*; Hillel, D., Ed.; Academic Press: San Diego, CA, USA, 1998; pp. 655–675.
38. Nearing, M.A. Why soil erosion models over-predict small soil losses and underpredict large soil losses. *Catena* **1998**, *32*, 15–22. [CrossRef]
39. Nearing, M.A. Evaluating soil erosion models using measured plot data: Accounting for variability on the data. *Earth Surf. Process. Landf.* **2000**, *25*, 1035–1043. [CrossRef]
40. Bagarello, V.; Ferro, V. Testing the "physical model concept" by soil loss data measured in Sicily. *Catena* **2012**, *95*, 1–5. [CrossRef]
41. Bagarello, V.; Ferro, V.; Giordano, G.; Mannocchi, F.; Pampalone, V.; Todisco, F. A modified applicative criterion of the physical model concept for evaluating plot soil erosion predictions. *Catena* **2015**, *126*, 53–58. [CrossRef]
42. Bagarello, V.; Ferro, V.; Pampalone, V. Comparing two applicative criteria of the soil erosion physical model concept. *Vadose Zone J.* **2017**, *16*, 1–10. [CrossRef]
43. Larson, W.E.; Lindstrom, M.J.; Schumacher, T.E. The role of severe storms in soil erosion: A problem needing consideration. *J. Soil Water Conserv.* **1997**, *52*, 90–95.
44. Bagarello, V.; Di Stefano, C.; Ferro, V.; Pampalone, V. Statistical distribution of soil loss and sediment yield at Sparacia experimental area, Sicily. *Catena* **2010**, *82*, 45–52. [CrossRef]
45. Bagarello, V.; Di Stefano, C.; Ferro, V.; Pampalone, V. Using plot soil loss distribution for soil conservation design. *Catena* **2011**, *86*, 172–177. [CrossRef]
46. Di Stefano, C.; Ferro, V.; Palmeri, V.; Pampalone, V. Flow resistance equation for rills. *Hydrol. Process.* **2017**, *31*, 2793–2801. [CrossRef]
47. Di Stefano, C.; Ferro, V.; Palmeri, V.; Pampalone, V. Flow resistance in step-pool rills. *Vadose Zone J.* **2017**, *16*, 1–10. [CrossRef]
48. Di Stefano, C.; Ferro, V.; Palmeri, V.; Pampalone, V. Testing slope effect on flow resistance equation for mobile bed rills. *Hydrol. Process.* **2018**, *32*, 664–671. [CrossRef]
49. Palmeri, V.; Pampalone, V.; Di Stefano, C.; Nicosia, A.; Ferro, V. Experiments for testing soil texture effects on flow resistance in mobile bed rills. *Catena* **2018**, *171*, 176–184. [CrossRef]
50. Nicosia, A.; Di Stefano, C.; Pampalone, V.; Palmeri, V.; Ferro, V.; Nearing, M.A. Testing a new rill flow resistance approach using the Water Erosion Prediction Project experimental database. *Hydrol. Process.* **2019**, *33*, 616–626. [CrossRef]
51. Di Stefano, C.; Nicosia, A.; Pampalone, V.; Palmeri, V.; Ferro, V. Rill flow resistance law under equilibrium bed-load transport conditions. *Hydrol. Process.* **2019**, *33*, 1317–1323. [CrossRef]
52. Di Stefano, C.; Nicosia, A.; Palmeri, V.; Pampalone, V.; Ferro, V. Comparing flow resistance law for fixed and mobile bed rills. *Hydrol. Process.* **2019**, *33*, 3330–3348. [CrossRef]
53. Carollo, F.G.; Di Stefano, C.; Nicosia, A.; Palmeri, V.; Pampalone, V.; Ferro, V. Flow resistance in mobile bed rills shaped in soils with different texture. *Eur. J. Soil Sci.* **2021**, *72*, 2062–2075. [CrossRef]
54. Di Stefano, C.; Nicosia, A.; Palmeri, V.; Pampalone, V.; Ferro, V. Estimating flow resistance in steep slope rills. *Hydrol. Process.* **2021**, *35*, e14296. [CrossRef]
55. Palmeri, V.; Di Stefano, C.; Ferro, V. A Maximizing Hydraulic Radius (MHR) method for defining cross-section limits in rills and ephemeral gullies. *Catena* **2021**, *303*, 105347. [CrossRef]
56. Pampalone, V.; Di Stefano, C.; Nicosia, A.; Palmeri, V.; Ferro, V. Analysis of rill step–pool morphology and its comparison with stream case. *Earth Surf. Process. Landf.* **2021**, *46*, 775–790. [CrossRef]
57. Nicosia, A.; Guida, G.; Di Stefano, C.; Pampalone, V.; Ferro, V. Slope threshold for overland flow resistance on sandy soils. *Eur. J. Soil Sci.* **2022**, *73*, e13182. [CrossRef]

58. Di Stefano, C.; Nicosia, A.; Palmeri, V.; Pampalone, V.; Ferro, V. Rill flow resistance law under sediment transport. *J. Soils Sediments* **2022**, *22*, 334–347. [CrossRef]
59. Nicosia, A.; Palmeri, V.; Pampalone, V.; Di Stefano, C.; Ferro, V. Slope threshold in rill flow resistance. *Catena* **2022**, *208*, 105789. [CrossRef]
60. Nicosia, A.; Di Stefano, C.; Palmeri, V.; Pampalone, V.; Ferro, V. Evaluating the effects of the rill longitudinal profile on flow resistance law. *Water* **2022**, *14*, 326. [CrossRef]
61. Bruno, C.; Di Stefano, C.; Ferro, V. Field investigation on rilling in the experimental Sparacia area, South Italy. *Earth Surf. Process. Landf.* **2008**, *33*, 263–279. [CrossRef]
62. Carollo, F.G.; Di Stefano, C.; Ferro, V.; Pampalone, V. Measuring rill erosion at plot scale by a drone-based technology. *Hydrol. Process.* **2015**, *29*, 3802–3811. [CrossRef]
63. Di Stefano, C.; Ferro, V. Measurements of rill and gully erosion in Sicily. *Hydrol. Process.* **2011**, *25*, 2221–2227. [CrossRef]
64. Di Stefano, C.; Ferro, V.; Pampalone, V. Modelling rill erosion at the Sparacia experimental area. *J. Hydrol. Eng.* **2015**, *20*, C5014001. [CrossRef]
65. Di Stefano, C.; Ferro, V.; Pampalone, V. Measuring field rill erodibility by a simplified method. *Land Degrad. Dev.* **2016**, *27*, 239–247. [CrossRef]
66. Casalí, J.; Loizu, J.; Campo, M.A.; De Santisteban, L.M.; Àlvarez-Mozos, J. Accuracy of methods for field assessment of rill and ephemeral gully erosion. *Catena* **2006**, *67*, 128–138. [CrossRef]
67. Di Stefano, C.; Palmeri, V.; Pampalone, V. An automatic approach for rill network extraction to measure rill erosion by terrestrial and low-cost UAV photogrammetry. *Hydrol. Process.* **2019**, *33*, 1883–1895.
68. Frankl, A.; Stal, C.; Abraha, A.; Nyssen, J.; Rieke-Zapp, D.; De Wulf, A.; Poesen, J. Detailed recording of gully morphology in 3D through image-based modelling. *Catena* **2015**, *127*, 92–101. [CrossRef]
69. James, M.R.; Robson, S. Straightforward reconstruction of 3D surfaces and topography with a camera: Accuracy and geosciences applications. *J. Geophys. Res.* **2012**, *117*, 1–17. [CrossRef]
70. Di Stefano, C.; Ferro, V.; Palmeri, V.; Pampalone, V.; Agnello, F. Testing the use of an image-based technique to measure gully erosion at Sparacia experimental area. *Hydrol. Process.* **2017**, *31*, 573–585. [CrossRef]
71. Broscoe, A.J. *Quantitative Analysis of Longitudinal Stream Profiles of Small Watersheds*; Technical Report 389-042; Department of Geology, Columbia University: New York, NY, USA, 1959.
72. Thommeret, N.; Bailly, J.S.; Puech, C. Extraction of thalweg networks from DTMs: Application to badlands. *Hydrol. Earth Syst. Sci.* **2010**, *14*, 1527–1536. [CrossRef]
73. Pirotti, F.; Tarolli, P. Suitability of LiDAR point density and derived landform curvature maps for channel network extraction. *Hydrol. Process.* **2010**, *24*, 1187–1197. [CrossRef]
74. Bryan, R.B.; Hawke, R.M.; Rockwell, D.L. The influence of subsurface moisture on rill system evolution. *Earth Surf. Process. Landf.* **1998**, *23*, 773–789. [CrossRef]
75. Barenblatt, G.I. *Similarity, Self-Similarity and Intermediate Asymptotics*; Consultants Bureau: New York, NY, USA, 1979.
76. Barenblatt, G.I. *Dimensional Analysis*; Gordon and Breach: Amsterdam, The Netherlands, 1987.
77. Knapen, A.; Poesen, J.; Govers, G.; Gyssels, G.; Nachtergaele, J. Resistance of soils to concentrated flow erosion: A review. *Earth-Sci. Rev.* **2007**, *80*, 75–109. [CrossRef]
78. Bagarello, V.; Di Stefano, C.; Ferro, V.; Giordano, G.; Iovino, M.; Pampalone, V. Estimating the USLE soil erodibility factor in Sicily, South Italy. *Appl. Eng. Agric.* **2012**, *28*, 199–206. [CrossRef]
79. Gessesse, G.; Fuchs, H.; Mansberger, R.; Klik, A.; Rieke-Zapp, D.H. Assessment of erosion, deposition and rill development on irregular soil surfaces using close range digital photogrammetry. *Photogramm. Rec.* **2010**, *25*, 299–318. [CrossRef]

water

MDPI

Article

Microbiological Properties in Cropping Systems and Their Relationship with Water Erosion in the Brazilian Cerrado

Marina Neves Merlo [1], Junior Cesar Avanzi [2,*], Lucas de Castro Moreira da Silva [3], Osnar Obede da Silva Aragão [4], Emerson Borghi [5], Fatima Maria de Souza Moreira [2], Michael Silveira Thebaldi [1], Álvaro Vilela de Resende [5], Marx Leandro Naves Silva [2] and Bruno Montoani Silva [2,*]

1 Department of Water Resources, Federal University of Lavras, Lavras 37200-900, Brazil; marinanevesmerlo@gmail.com (M.N.M.); michael.thebaldi@ufla.br (M.S.T.)
2 Department of Soil Science, Federal University of Lavras, Lavras 37200-900, Brazil; fmoreira@ufla.br (F.M.d.S.M.); marx@ufla.br (M.L.N.S.)
3 Department of Agricultural Engineering, Federal University of Viçosa, Viçosa 36570-900, Brazil; lucascmoreira91@gmail.com
4 Federal Institute of Pará, Campus Breves, Breves 68800-000, Brazil; osnar.silva@gmail.com
5 Brazilian Agricultural Research Corporation (Embrapa), Sete Lagoas 35701-970, Brazil; emerson.borghi@embrapa.br (E.B.); alvaro.resende@embrapa.br (Á.V.d.R.)
* Correspondence: junior.avanzi@ufla.br (J.C.A.); brunom.silva@ufla.br (B.M.S.)

Abstract: Many researchers have reported relationships of physical and chemical properties with water erosion; however, little is known about microbiological properties in this context. Thus, the aim of this study was to evaluate soil properties in relation to erosion in areas with different cropping system practices under no-tillage in the Brazilian Cerrado. The experiment has been carried out since 2014 in a Typic Haplustox under soybean (S) and maize (M) monoculture, maize/soybean annual rotation (MS), maize/brachiaria/soybean/brachiaria rotation (MBSB), two of those treatments with high input of fertilizer (MBSB-HI and MS-HI), and bare soil (BS). Soil losses were quantified in erosion plots. The design was completely randomized. The greater vegetation cover crop, provided by intercropping/succession with brachiaria, increased microbial biomass carbon. The lack of vegetation cover affected the basal soil respiration and metabolic quotient. Basal soil respiration proved to be inversely related to soil and water losses. Vegetation cover was a key factor regulating water erosion. Penetration resistance and aggregate stability correlated with soil and water losses. Thus, not only physical and chemical, but also biological properties are deeply affected by erosion, aiding in early monitoring of water erosion. Soil quality improvement in ecologically supported management contributes to mitigating erosion.

Keywords: soil erosion; microbial biomass; basal soil respiration; cover crop; no-tillage

check for updates

Citation: Merlo, M.N.; Avanzi, J.C.; Silva, L.d.C.M.d.; Aragão, O.O.d.S.; Borghi, E.; Moreira, F.M.d.S.; Thebaldi, M.S.; Resende, Á.V.d.; Silva, M.L.N.; Silva, B.M. Microbiological Properties in Cropping Systems and Their Relationship with Water Erosion in the Brazilian Cerrado. *Water* **2022**, *14*, 614. https://doi.org/10.3390/w14040614

Academic Editors: Vito Ferro and Alessio Nicosia

Received: 30 December 2021
Accepted: 15 February 2022
Published: 17 February 2022

1. Introduction

The sustainable use of environmental resources and their conservation is a current issue that has attracted worldwide attention. Soils play a central role in this context, as they are an important part of the water cycle and the basis for healthy food production. Nonetheless, soil functions are threatened by land use changes combined with the intensification of agricultural systems, which have proved to lead to soil degradation, reducing its production capacity. High soil erosion rates and a decline in soil physical and chemical quality are examples of harmful impacts that directly affect the food production capacity necessary for a growing population [1,2]. However, there is a lack of information on the relationship between soil microbiological properties and soil erosion processes.

It is evident, therefore, that there is a need to quantify and assess the environmental impacts resulting from land use changes to determine the real capacity of soil use, exploiting production potential while mitigating degradation of natural resources. In this respect soil conservation practices can be important tools for farmers to intensify land use while avoiding acceleration of erosion processes. Otherwise, inadequate soil management can promote erosion rates that deplete soil quality and, consequently, the production capacity of the land [2–5].

The no-tillage system is a set of conservation practices already widely used in the Brazilian Cerrado and with notable benefits in terms of soil and water conservation. Maintaining crop stover on the soil surface reduces the impact of raindrops, preventing surface sealing, favoring infiltration, and decreasing surface runoff, and consequently controlling the removal of the soil surface layer by water erosion. Considering that the topsoil of the soil layer is the richest in nutrients, organic carbon, and biodiversity, the no-tillage system, when well-managed, acts as a provider of soil quality. An example of improvement of soil physical, chemical, and biological quality is the increase in carbon stocks, which has great importance for soil aggregation and porosity, as observed by Mota et al. [6], for the release of nutrients, and for the maintenance and survival of microbiota.

Microbial properties can be considered good and quick-response indicators of soil quality [7], although they are still little assessed in studies evaluating soil conservation techniques. The metabolic activity of the microbiota can be detected before some significant evidence of change in the soil physical and chemical properties. For that reason, the activity of soil microorganisms is a sensitive indicator of soil quality [8–11] and may be related to soil susceptibility to erosion processes. The relationship between soil erosion and microbial properties is not yet well understood, although some efforts toward understanding have been made recently, such as the study of Qui et al. [12], which reported negative impacts related to soil erosion on soil microbial communities.

In this context, we hypothesized that microbiological properties can aid in early monitoring of the effects of water erosion, since they can be more sensitive and indicate changes in soil quality more quickly. Therefore, the aim of this study was to relate microbiological properties with soil losses due to water erosion in areas under different cropping practices in no-tillage systems in the Brazilian Cerrado. Different crop rotation systems were investigated, assessing overall soil quality but focusing on the relationship between biological quality and susceptibility to water erosion.

2. Materials and Methods

The experiment was conducted on the Brazilian Agricultural Research Corporation (Embrapa) experimental farm in the municipality of Sete Lagoas in the central region of the state of Minas Gerais, Brazil, at 19°28′30″ S and 44°15′08″ W and average elevation of 732 m. The soil in the experimental site was classified as Typic Haplustox [13], corresponding to a *Latossolo Vermelho distrófico típico* [14], found in a landscape with an average slope of 8%, characterized as gently sloping. The predominant climate in the region is humid subtropical (Cwa) according to the Köppen classification, with an average annual temperature of 20 °C and mean annual rainfall of 1416 mm [15], mainly concentrated from October to April (Figure 1). This region has a characteristic dry season with frequent occurrence of dry spells during the rainy season [16].

The treatments were set up in the 2014/2015 agricultural year in a total area of 4.4 ha (Figure 2). Prior to the experiment setting up, the area had been used for maize and soybean production under conventional soil tillage for more than two decades. The experiment was prepared first by chisel plowing the soil to a depth of 25 cm to mitigate compaction. Dolomitic limestone and agricultural gypsum were applied and incorporated into the soil profile with a moldboard plow and a disk plow. The experimental site was then separated into stripes with terraces between them, creating a single large experimental block, with each stripe corresponding to a treatment (Figure 2).

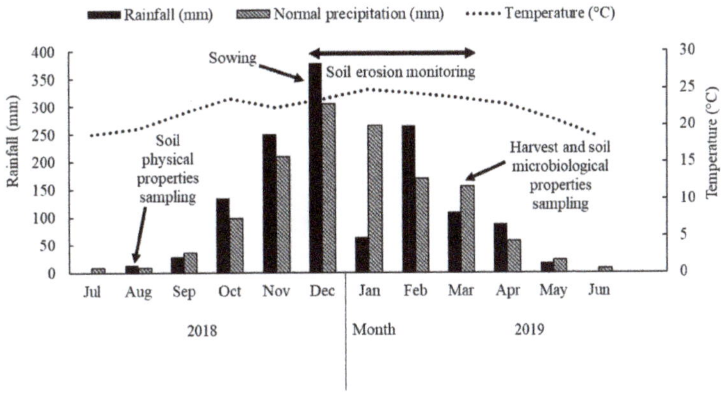

Figure 1. Daily average temperature and monthly precipitation data during the monitoring period (August 2018 to July 2019). The normal precipitation corresponds to the period from 1927 to 2013. Adapted from Silva et al. [16].

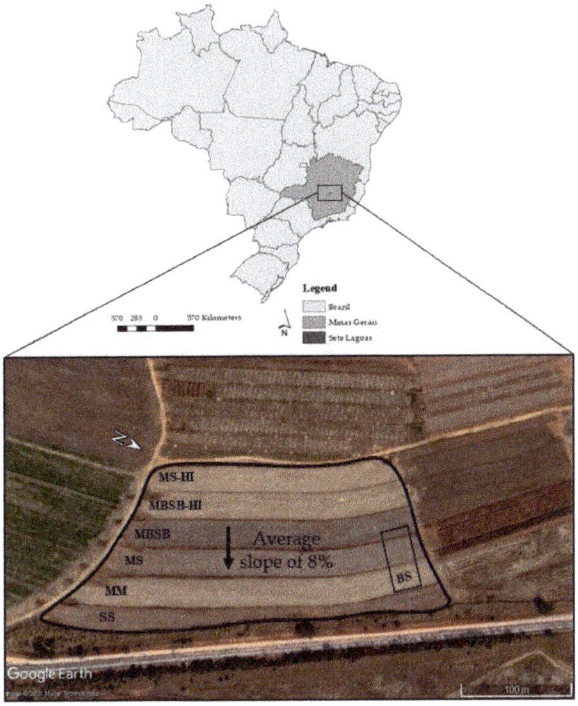

Figure 2. Aerial image of the experimental site in Sete Lagoas, central region of the state of Minas Gerais, Brazil. SS = soybean monoculture, MM = maize monoculture, MS = maize/soybean annual rotation, MBSB = maize/brachiaria/soybean/brachiaria rotation, BS = bare soil, and HI = high input of fertilizers.

The treatments differed by soil management practices such as different crop rotation systems and different levels of fertilizer input. Monocultures of soybean (SS) and maize (MM) were compared to rotation systems of maize-soybean (MS) and maize-*Brachiaria*-soybean-*Brachiaria* (MBSB) and two additional treatments with a high level of fertilizer input, designated MS-HI and MBSB-HI, while the other treatments were maintained with a medium fertilizer input. The amounts of fertilizers applied were 148, 250, 250, 250, 415 and 415 kg ha^{-1} of NPK 08-28-16 + 0.3% boron and 2.1% sulfur for the treatments SS, MM, MS, MBSB, MBSB-HI, and MS-HI, respectively. For the 2018/2019 crop season, maize was grown in MS and MBSB treatments, and soybeans in MSSB-HI and MS-HI.

In the 2018/2019 crop season, the seeds of brachiaria (*Urochloa brizantha* syn. *Brachiaria brizantha*) cv. BRS Piatã were sown at different times according to the crop used in intercropping, for treatments MBSB and MBSB-HI. *Brachiaria* seeds were mixed with fertilizers and sown in maize rows, simultaneously with the maize sowing. However, when *Brachiaria* was intercropped with soybean, the grass seeds were sown when the soybean plants reached the R5 growth stage, corresponding to the phenological stage of grain filling.

2.1. Erosion Assessment and Soil Sampling

In December 2018, three unitary erosion plots of 1 m^2 (0.5 m width × 2 m length) were implemented for each treatment, arranged with their length in the direction of the slope. Erosion plots were delimited by galvanized zinc sheet metal inserted 0.20 m into the soil and kept 0.20 m above the soil surface. At the lower end of each plot, there was a 20 L container for storing water and sediments from runoff and the transported soil particles.

From December 2018 to March 2019, samples were collected after rainfall events. The runoff volume was fully measured. Meanwhile, for evaluation of soil loss, after homogenizing the solution, we sampled 250 mL in each collecting container positioned at the plots, to later quantify the sediments in the laboratory by the gravimetric method. The slope of each plot was measured individually, ranging from 7% to 10%. For standardization, the values of water and soil losses were corrected considering a slope of 9%, according to Wishmeier and Smitsh [17].

Soil and plant residues were sampled within each erosion plot. The plant residues on the surface were manually collected to quantify the soil cover rate. Soil samples were collected in crop rows and in interrows at a 0–10 cm depth, for a total of 39 disturbed and 21 undisturbed soil samples. A separation between crop row and interrow was not made for the bare soil plots, as this treatment remained without plants throughout the experiment. Samples were taken when maize and soybean were at the grain maturation phase.

2.2. Soil Physical Quality Indicators

Soil penetration resistance was assessed under field conditions using a Stolf impact penetrometer [18] within the erosion plots. Simultaneously, soil cores were collected at 0–20 cm, 20–40 cm, and 40–60 cm deep, to quantify the soil water content in a laboratory by the gravimetric method.

The undisturbed soil samples were saturated for determination of total porosity. Microporosity was determined using Buchner funnel suction units for a matric potential of −6 kPa, and macroporosity was calculated by the difference between total porosity and microporosity [19,20]. Soil bulk density was determined after oven-drying the undisturbed samples at 105 °C [21].

For aggregate stability analysis, aggregates with diameters between 8 and 4.75 mm were subjected to the wet-sieving method with Yoder-type equipment [22]. Particle size analysis was performed by the pipette method [21].

2.3. Soil Chemical Quality Indicators

Particles of 2 mm were separated for soil fertility and organic matter analysis. The following chemical attributes were evaluated: pH in water; phosphorus and potassium, obtained by the Mehlich-1 method; calcium and magnesium, determined using the potassium chloride (KCl) extractor; and organic matter content, estimated by dry combustion. All analyses were performed using the protocols described in Teixeira et al. [23].

2.4. Soil Biological Quality Indicators

For analysis of microbial biomass carbon and soil basal respiration, particles smaller than 1 mm were separated and kept under refrigeration. Basal soil respiration (BSR) was estimated by incubation [24], using sodium hydroxide to capture the carbon dioxide released by the microorganisms. After 3 days, 0.5 M barium chloride was added to stop the reaction with carbon dioxide from air. The reaction indicator used was phenolphthalein and the titrant was 0.05 M hydrochloric acid.

Microbial biomass carbon (MBC) was quantified by the fumigation-extraction method [25], dividing the samples into control and fumigated samples and incubating them for 24 h with the addition of potassium sulfate and stirring them at 120 rpm for 30 min. The samples were then filtered to obtain the carbon extract, which was boiled for 5 min with potassium dichromate, sulfuric acid, and phosphoric acid. The indicator used for titration was diphenylamine with ammoniacal ferrous sulfate. The metabolic quotient was determined according to Anderson and Domsch [26] as the ratio between BSR and MBC.

The hydrolysis of fluorescein diacetate (FDA) was determined by the method proposed by Dick, Breakwell, and Turco [27]. Sodium phosphate buffer solution was added to the soil samples at pH 7.0 with fluorescein diacetate and incubated under rotation at 3840 rpm for 24 h at 35 °C. The reaction was stopped with acetone and samples were stirred for 5 min. Then, the samples were filtered and read in a spectrophotometer at 490 nm. For the control samples, the same procedures were performed, except for addition of fluorescein diacetate.

2.5. Data Analysis

Considering the study area formed by a large experimental block, a completely randomized design was adopted, as recommended by Ferreira et al. [28] and successfully used in previous studies of long-term field experiments [29,30] and also in the same experiment in anterior studies [6,16,31]. Thus, for soil losses, water losses, penetration resistance, total porosity, macroporosity, microporosity, and soil bulk density, the different cropping systems (seven treatments) were evaluated. As for the other variables (i.e., chemical and biological indicators), the same treatments were evaluated considering two different positions from the crop rows and interrows. Soil losses by erosion and water losses by runoff were log-transformed due to heteroscedasticity. After checking the assumptions, analysis of variance (ANOVA) and Tukey's test at 5% probability were performed. Pearson's correlation matrix and a principal component analysis (PCA) were performed as an aid for understanding data variance and the relationships between the variables evaluated.

3. Results

3.1. Soil Losses, Water Losses, and Soil Physical Indicators

Table 1 shows the soil losses (SL), water losses (WL), cover rate (CR), total porosity (TP), macroporosity (Ma), microporosity (Mi), and bulk density (Bd) for the treatments evaluated during the field experiment. The highest values of SL and WL were found for the soybean monoculture (SS) and bare soil (BS), both showing similar losses. No differences were found for the soil physical properties shown in Table 1.

The in-field penetration resistance (PR) and the soil moisture at this evaluation moment are shown in Figure 3. Considering the top soil layer (0 to 10 cm), the most prone to water erosion, it is observed that the highest PR were in the SS and BS treatments, with averages of 3.88 and 4.26 MPa, respectively. In contrast, the lowest PR was observed in the MBSB-HI treatment (2.02 MPa), where the smallest losses of sediment and water were also obtained. Soil moisture showed significant differences at the surface (0–20 cm depth) and subsurface (20–40 cm depth) soil layers. These differences may restrict conclusions about the mechanical impedance of PR itself, as PR is strongly affected by soil moisture; the two are negatively correlated. However, maximum variation in the soil moisture range among treatments is only 27.2–32.7% at the 0–20 cm depth. Although soil moisture affects the PR, the BS showed greater PR at the 0–20 cm depth. Furthermore, MBSB-HI showed lower PR than SS, even though it had higher soil moisture than SS.

Table 1. Soil losses, water losses, cover rate, total porosity, macroporosity, microporosity, and bulk density for each treatment evaluated.

Soil Management System	SL	WL	CR	TP	Ma	Mi	Bd
	Mg ha^{-1}	mm	Mg ha^{-1}	m^3 m^{-3}	m^3 m^{-3}	m^3 m^{-3}	g cm^{-3}
SS	20.364 a	144.747 a	1.216 cd	0.612 ^{ns}	0.196 ^{ns}	0.416 ^{ns}	0.958 ^{ns}
MM	1.025 b	13.290 b	1.75 bcd	0.593	0.18	0.414	1.018
MS	0.630 c	11.959 b	1.5 cd	0.589	0.171	0.418	1.013
MBSB	0.156 d	7.032 bc	3.55 ab	0.589	0.156	0.433	0.993
MBSB-HI	0.122 d	3.893 c	4.650 a	0.599	0.184	0.415	0.902
MS-HI	0.251 d	16.005 b	2.433 bc	0.581	0.15	0.432	1.038
BS	20.144 a	120.442 a	0.000 d	0.606	0.192	0.414	0.951

Means followed by different letters differ by Tukey's test ($p < 0.05$). SS = soybean monoculture, MM = maize monoculture, MS = maize/soybean annual rotation, MBSB = maize/brachiaria/soybean/brachiaria rotation, BS = bare soil, HI = high input of fertilizers, and ^{ns} = not significant. The samples used for TP, Ma, Mi, and Bd were collected randomly in each treatment, without distinction between crop rows or interrows.

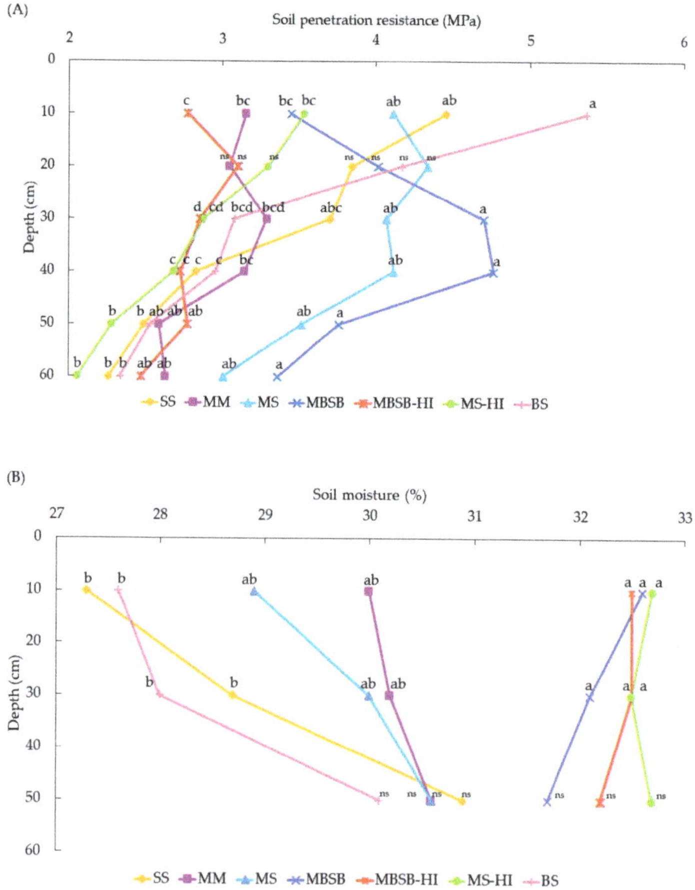

Figure 3. Soil penetration resistance (PR) for each treatment assessing 0–60 cm (**A**) and soil moisture (**B**) for three different depths (0–20 cm, 20–40 cm, and 40–60 cm) at the time of evaluation of soil penetration resistance. Means followed by different letters among treatments differ by Tukey's test ($p < 0.05$). SS = soybean monoculture, MM = maize monoculture, MS = maize/soybean annual rotation, MBSB = maize/brachiaria/soybean/brachiaria rotation, BS = bare soil, HI = high input of fertilizers, and ^{ns} = not significant.

3.2. Physical, Chemical, and Microbiological Indicators in Crop Rows

The soil physical, chemical, and microbiological indicators for the crop rows are shown in Table 2. The treatments MBSB-HI and MBSB had higher MBC and FDA than the other treatments. A high variance among replicates was found for BSR, which restricted the evidence of differences between treatments. In general, regarding BSR, the MM treatment had a higher basal respiration rate than the SS and BS treatments. MS-HI also showed higher basal respiration than SS did. The MM and MS-HI treatments exhibited high qCO_2, which is an ecosystem stress indicator. In addition, MM had lower MBC than BS did and this was lower than all the treatments that used crop rotation. The FDA in MM had the lowest value, but it did not differ from SS, MS-HI, and BS. The lowest concentration of phosphorus was observed in the MBSB treatment, differing from MS. We did not find differences among the treatments for organic carbon, calcium, GMD, and MWD.

Table 2. Physical, chemical, and microbiological indicators evaluated in crop rows for each treatment.

Soil Management System	MBC	BSR	qCO_2	FDA	P	Sb	m	K	Mg	Ca	OC	GMD	MWD
SS	272 cd	194 c	0.72 bc	110 c	21.6 ab	6.41 ab	1.27 ab	148 ab	1.36 ab	4.68 ^{ns}	1.63 ^{ns}	4.58 ^{ns}	4.75 ^{ns}
MM	196 d	424 a	2.19 a	107 c	45.1 ab	5.58 b	1.49 a	153 ab	0.89 b	3.98	1.56	4.39	4.77
MS	311 bc	288 abc	0.93 bc	127 b	68.8 a	6.58 ab	0.71 cd	154.6 ab	1.49 a	4.70	1.43	4.76	4.91
MBSB	361 ab	292 abc	0.81 bc	137 ab	20.6 b	5.57 b	1.18 abc	66.6 b	1.28 ab	4.12	1.42	4.49	4.86
MBSB-HI	423 a	272 abc	0.65 c	147 a	27.6 ab	6.73 ab	0.57 d	124.7 ab	1.38 ab	5.03	1.47	4.60	4.85
MS-HI	316 bc	384 ab	1.24 ab	111 c	22.1 ab	7.60 ab	0.70 d	199.7 a	1.39 ab	5.71	1.75	4.68	4.90
BS	311 bc	251 bc	0.82 bc	113 c	29.1 ab	8.58 a	0.85 bcd	70.3 b	1.61 a	4.83	1.63	4.64	4.85

Means followed by different letters differ by Tukey's test ($p < 0.05$). SS = soybean monoculture, MM = maize monoculture, MS = maize/soybean annual rotation, MBSB = maize/brachiaria/soybean/brachiaria rotation, BS = bare soil, and HI = high input of fertilizers. MBC = microbial biomass carbon (mg C kg dry soil^{-1}), BSR = basal soil respiration (mg C-CO_2 kg of dry soil^{-1} d^{-1}); qCO_2 = metabolic quotient (mg C-CO_2 g MBC^{-1} d^{-1}), FDA = fluorescein diacetate hydrolysis (mg F kg^{-1} of dry soil d^{-1}), P = phosphorus (mg dm^{-3}), Sb = sum of bases (cmol$_c$ dm^{-3}), m = aluminum saturation of soil (%), K = potassium (mg dm^{-3}), Mg = magnesium (cmol$_c$ dm^{-3}), Ca = calcium (cmol$_c$ dm^{-3}), OC = organic carbon (dag kg^{-1}), GMD = geometric mean diameter (mm), and MWD = mean weight diameter (mm), ^{ns} = not significant.

The correlation matrix between the variables observed in crop rows is shown in Figure 4 and the PCA is presented in Figure 5. According to Vieira (2018), the linear correlation coefficient can have an inverse correlation (negative values) or direct correlation (positive values), and correlations can be classified as small ($0 \leq r < 0.25$), weak ($0.25 \leq r < 0.50$), moderate ($0.50 \leq r < 0.75$), strong ($0.75 \leq r < 1.00$), or perfect ($r = 1.00$).

SL correlated directly and strongly with WL and moderately with PR_{10}, and it moderately and inversely correlated with CR, BSR, and SM, and weakly correlated with FDA. The same correlations were observed for WL, indicating that these soil physical and microbiological properties are associated with soil and water losses. The soil vegetation cover rate (CR), however, was positively correlated with MBC, FDA, and SM, and negatively correlated with PR_{10}. In both cases, the correlation was moderate.

MBC was strongly correlated with FDA; moderately correlated with qCO_2, Sb, and m, and weakly correlated with Mg and MWD. qCO_2 was negatively correlated with FDA (moderate) and Mg (weak), although the FDA had a moderate correlation coefficient with SM and Bd.

The soil chemical indicators, i.e., organic carbon, sum of bases, aluminum saturation of soil, potassium, phosphorus, calcium, magnesium; other soil physical indicators, i.e., total porosity, macroporosity, microporosity, GMD, MWD, and bulk density; and some microbiological indicators, MBC and qCO_2, do not show correlations with soil and water losses, indicating that they do not provide evidence of association with water erosion under the conditions of our study.

The PCA reinforces, in general, that the BS and SS treatments were strong components of the WL, SL, and PR_{10} variations, and inversely related to the soil cover rate. Among the variables, the basal respiration rate was negatively correlated with soil and water losses and was positively correlated with soil cover rate, providing useful insights regarding the relationship between soil erosion and microbial activity.

SL																					
0.90	WL																				
-0.65	**-0.62**	CR																			
-0.23	-0.24	**0.54**	MBC																		
-0.68	**-0.62**	0.27	-0.01	BSR																	
-0.36	-0.32	-0.12	**-0.73**	**0.64**	qCO_2																
-0.46	**-0.45**	**0.72**	**0.79**	-0.06	**-0.54**	FDA															
0.10	0.24	-0.21	0.03	-0.05	-0.12	-0.15	OC														
0.36	0.31	-0.33	0.18	0.14	-0.16	-0.15	**0.44**	Sb													
0.17	0.22	-0.20	**-0.66**	-0.21	0.41	-0.38	-0.05	-0.41	m												
-0.24	-0.17	0.14	-0.33	0.37	**0.44**	-0.27	0.18	0.17	-0.06	K											
-0.26	-0.30	-0.21	-0.43	0.04	0.38	-0.17	-0.21	-0.28	0.02	0.23	P										
0.03	-0.01	0.00	0.23	0.33	-0.07	0.02	0.43	**0.78**	**-0.46**	0.43	-0.24	Ca									
0.30	0.24	-0.21	**0.45**	-0.06	**-0.49**	0.20	0.15	**0.70**	**-0.59**	-0.09	-0.32	**0.63**	Mg								
0.07	0.05	-0.10	0.38	-0.04	-0.37	0.11	0.16	0.43	**-0.49**	0.12	-0.09	0.27	**0.51**	GMD							
-0.26	-0.26	0.10	**0.49**	0.12	-0.32	0.30	0.20	0.20	**-0.57**	0.00	0.00	0.07	0.26	**0.69**	MWD						
0.62	**0.58**	**-0.75**	-0.28	-0.22	-0.06	-0.38	0.15	**0.54**	0.01	-0.15	0.05	0.11	0.40	0.34	0.14	PR_{10}					
-0.65	**-0.67**	**0.71**	0.43	0.42	0.056	**0.48**	-0.10	-0.30	-0.16	0.04	-0.28	-0.09	-0.31	0.01	0.43	**-0.52**	SM				
0.20	**0.45**	-0.25	0.03	-0.32	-0.23	0.01	0.35	0.01	0.09	-0.18	-0.14	-0.06	0.17	-0.14	-0.29	0.02	**-0.43**	TP			
0.21	**0.43**	-0.30	0.00	-0.37	-0.22	0.01	0.36	0.02	0.10	-0.15	-0.07	-0.18	0.02	-0.10	-0.15	0.11	-0.33	**0.89**	Ma		
-0.15	-0.28	0.27	0.05	0.31	0.14	0.00	-0.27	-0.02	-0.09	0.07	-0.04	0.27	0.17	0.02	-0.051	-0.19	0.11	**-0.49**	**-0.84**	Mi	
-0.13	-0.31	-0.12	-0.38	0.38	**0.47**	**-0.46**	-0.43	-0.23	0.06	0.14	0.31	-0.18	-0.27	0.06	0.072	-0.01	0.10	**-0.65**	**-0.65**	**0.46**	Bd

Figure 4. Pearson correlation matrix for variables sampled in crop rows. SL = soil losses, WL = water losses, CR = cover rate, MBC = microbial biomass carbon, BSR = basal soil respiration, qCO_2 = metabolic quotient, FDA = fluorescein diacetate hydrolysis, OC = soil organic carbon, Sb = sum of bases, m = aluminum saturation of soil, K = potassium, P = phosphorus, Ca = calcium, Mg = magnesium, MWD = mean weight diameter, GMD = geometric mean diameter, PR_{10} = soil resistance to root penetration at 10 cm depth, SM = soil moisture in situ, TP = total porosity, Ma = macroporosity, Mi = microporosity, and Bd = bulk density. Values in bold print: significant at 5%.

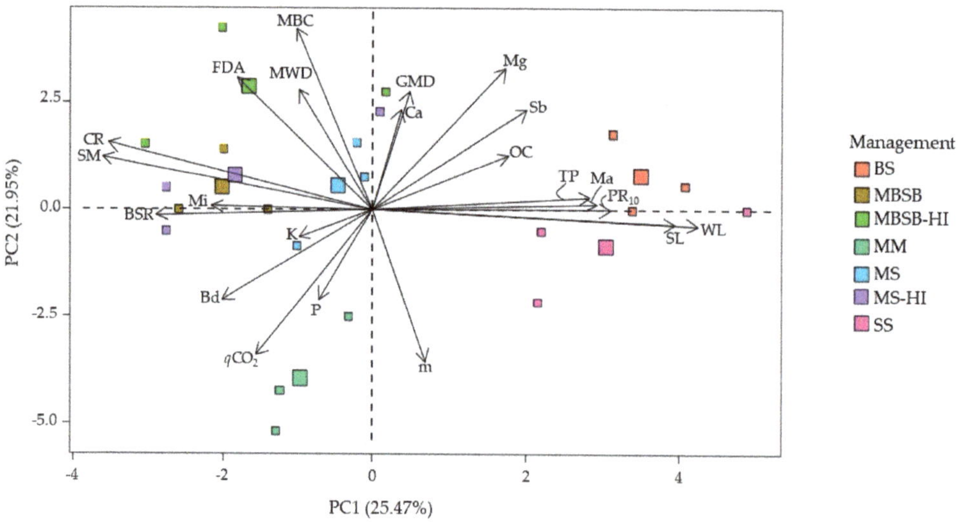

Figure 5. Principal component analysis for variables sampled in the crop rows. SL = soil losses, WL = water losses, CR = cover rate, MBC = microbial biomass carbon, BSR = basal soil respiration, qCO_2 = metabolic quotient, FDA = fluorescein diacetate hydrolysis, OC = soil organic carbon, Sb = sum of bases, m = aluminum saturation of soil, K = potassium, P = phosphorus, Ca = calcium, Mg = magnesium, MWD = mean weight diameter, GMD = geometric mean diameter, PR_{10} = soil resistance to root penetration at 10 cm depth, SM = soil moisture in situ, TP = total porosity, Ma = macroporosity, Mi = microporosity, Bd = bulk density, SS = soybean monoculture, MM = maize monoculture, MS = maize/soybean annual rotation, MBSB = maize/brachiaria/soybean/brachiaria rotation, BS = bare soil, and HI = high input of fertilizers.

3.3. Physical, Chemical, and Microbiological Indicators in Crop Interrows

The soil physical, chemical, and microbiological indicators observed at the crop interrows are shown in Table 3. The MBSB-HI treatment, as per what occurred for crop rows, had the highest MBC and the lowest qCO_2 among the treatments. The FDA of MBSB-HI had the lowest value among the treatments, being statistically equal to SS, MS, and BS, differently to what was observed in the crop rows. The lowest MBC value, in this case, was observed in the MS treatment, which also had high BSR and the highest qCO_2, indicating possible agroecosystem stress. For the crop rows, however, this response was observed for the MM treatment. In general, the values obtained for MBC, BSR, and qCO_2 for the interrow were higher than the values for the crop row (Table 2), showing that vegetation cover is necessary to the agroecosystem's balance. The MS-HI treatment did not show significant differences regarding K and Ca in relation to the other treatments, except for MM (for both elements) and MBSB (only for K). For Mg, however, only MBSB-HI and MS differed from SS and MM. Organic carbon, GMD, and MWD did not show differences among the treatments.

Table 3. Physical, chemical, and microbiological indicators evaluated in crop interrows for each treatment.

Soil Management System	MBC	BSR	qCO_2	FDA	P	Sb	m	K	Mg	Ca	OC	GMD	MWD
SS	395 b	352 c	0.90 cd	112.1 c	21.5 ab	6.71 abc	0.70 bc	123.3 ab	1.37 b	4.81 ab	1.48 ^{ns}	4.39 ^{ns}	4.78 ^{ns}
MM	380 b	480 b	1.27 bc	142.6 a	26.3 ab	6.49 abc	1.95 a	87.8 b	1.22 b	4.05 b	1.80	4.66	4.90
MS	178 d	483 ab	2.7 a	118.5 bc	51.1 a	5.50 c	0.74 bc	122.6 ab	1.95 a	5.32 ab	1.73	4.58	4.84
MBSB	256 cd	459 b	1.83 b	126.1 b	18.8 b	7.59 abc	1.01 b	79.0 b	1.50 ab	4.56 ab	1.53	4.71	4.89
MBSB-HI	498 a	424 b	0.85 d	117.1 c	18.6 b	6.26 bc	0.49 c	135.1 ab	1.95 a	5.98 ab	1.80	4.76	4.92
MS-HI	389 b	532 a	1.37 b	99.2 d	22.9 ab	8.27 ab	0.58 bc	199.1 a	1.60 ab	6.47 a	1.46	4.76	4.93
BS	311 bc	251 c	0.82 d	112.7 c	29.1 ab	8.58 a	0.85 bc	105.6 ab	1.61 ab	4.83 ab	1.63	4.64	4.85

Means followed by different letters differ by Tukey's test ($p < 0.05$). SS = soybean monoculture, MM = maize monoculture, MS = maize/soybean annual rotation, MBSB = maize/brachiaria/soybean/brachiaria rotation, BS = bare soil, and HI = high input of fertilizers. MBC = microbial biomass carbon (mg C kg dry soil^{-1}), BSR = basal soil respiration (mg C-CO_2 kg of dry soil^{-1} d^{-1}); qCO_2 = metabolic quotient (mg C-CO_2 g MBC^{-1} d^{-1}), FDA = fluorescein diacetate hydrolysis (mg F kg^{-1} of dry soil d^{-1}), P = phosphorus (mg dm^{-3}), Sb = sum of bases (cmol$_c$ dm^{-3}), m = aluminum saturation of soil (%), K = potassium (mg dm^{-3}), Mg = magnesium (cmol$_c$ dm^{-3}), Ca = calcium (cmol$_c$ dm^{-3}), OC = organic carbon (dag kg^{-1}), GMD = geometric mean diameter (mm), and MWD = mean weight diameter (mm), ^{ns} = not significant.

The Pearson correlation matrix showing the correlation between the soil erosion indicators and the soil physical, chemical, and microbiological properties evaluated in the interrows is shown in Figure 6, and its PCA in Figure 7. In the interrows SL correlated strongly and positively with WL, just as in the row, and these losses correlated moderately and negatively with CR, resulting in the same linear correlation coefficient values for both. In addition, soil and water losses were positively and moderately correlated with PR_{10}.

Both SL and WL showed a strong inverse correlation with BSR (Figure 6), stronger than that observed for these indicators for the crop row data (Figure 4). Furthermore, SL and WL were significantly inversely correlated with CR, qCO_2, GMD, MWD, and SM. CR was not correlated with MBC, but had a strong positive correlation with BSR. BSR was moderately correlated with PR_{10}, and SM; and FDA was inversely correlated with K and Ca.

The PCA (Figure 7) shows that soil and water losses were correlated with soil physical properties, positively correlated with PR_{10} and negatively correlated with GMD and MWD. In addition, the BSR microbiological property was negatively correlated with SL and WL, just as observed for data from the crop rows (Figure 5). In general, treatments with cover crops or crop rotation led to improved soil microbiological and chemical properties, whereas the BS and SS monoculture systems led to a loss of soil physical quality (mainly PR_{10}).

SL																					
0.90	WL																				
-0.65	**-0.62**	CR																			
0.06	0.04	0.35	MBC																		
-0.87	**-0.80**	**0.52**	0.03	BSR																	
-0.50	**-0.47**	0.03	**-0.80**	**0.48**	qCO_2																
-0.24	-0.31	0.07	-0.10	0.11	0.12	FDA															
-0.17	-0.26	0.11	0.05	0.08	0.07	0.30	OC														
0.30	**0.22**	-0.20	0.15	-0.32	-0.42	-0.32	-0.30	Sb													
-0.13	-0.14	-0.18	-0.10	0.10	0.01	**0.82**	0.10	-0.10	m												
-0.23	-0.15	0.22	0.33	**0.44**	-0.07	**-0.64**	-0.22	0.20	-0.42	K											
-0.06	-0.10	-0.31	**-0.57**	0.04	**0.63**	0.04	0.27	-0.15	-0.09	-0.06	P										
-0.17	-0.27	0.25	0.24	0.30	0.00	**-0.61**	0.27	0.11	**-0.67**	**0.62**	0.12	Ca									
-0.22	-0.36	0.28	-0.10	0.13	0.31	-0.34	0.40	-0.29	**-0.64**	0.09	0.36	**0.65**	Mg								
-0.46	**-0.53**	0.41	0.19	0.28	-0.05	0.01	0.04	0.25	0.06	0.14	-0.38	0.12	0.05	GMD							
-0.47	**-0.56**	0.41	0.27	0.37	-0.07	0.06	0.05	0.18	0.13	0.18	-0.40	0.10	-0.01	**0.97**	MWD						
0.62	**0.58**	**-0.75**	**-0.45**	**-0.60**	-0.02	-0.20	0.00	0.26	-0.03	-0.27	0.27	-0.11	-0.04	-0.18	-0.27	PR_{10}					
-0.65	**-0.67**	**0.71**	0.27	**0.58**	0.02	0.00	-0.15	0.10	-0.13	0.29	-0.36	0.21	0.11	**0.59**	**0.61**	**-0.52**	SM				
0.20	**-0.45**	-0.25	-0.09	-0.33	-0.21	-0.08	-0.04	0.02	-0.07	-0.12	-0.04	-0.18	-0.26	-0.20	-0.30	0.02	**-0.43**	TP			
0.21	**0.43**	-0.30	0.12	-0.31	-0.22	-0.01	-0.03	-0.06	0.01	-0.21	-0.14	-0.22	-0.28	-0.04	-0.13	0.11	-0.33	**0.89**	Ma		
-0.15	-0.28	0.27	-0.11	0.20	0.17	-0.07	0.00	0.14	-0.09	0.26	0.21	0.21	0.23	-0.17	-0.11	-0.19	0.11	**-0.49**	**-0.84**	Mi	
-0.13	-0.31	-0.12	-0.23	0.36	0.33	0.13	-0.26	0.11	0.29	0.12	0.22	-0.07	-0.11	-0.02	0.13	-0.01	0.10	**-0.65**	**-0.65**	**0.46**	Bd

Figure 6. Pearson correlation matrix for variables sampled in crop interrows. SL = soil losses, WL = water losses, CR = cover rate, MBC = microbial biomass carbon, BSR = basal soil respiration, qCO2 = metabolic quotient, FDA = fluorescein diacetate hydrolysis, Sb = sum of bases, m = aluminum saturation of soil, Mg = magnesium, MWD = mean weight diameter, PR_{10} = soil resistance to root penetration at 10 cm depth, SM = soil moisture in situ, and Bd = bulk density. Values in bold print: significant at 5%.

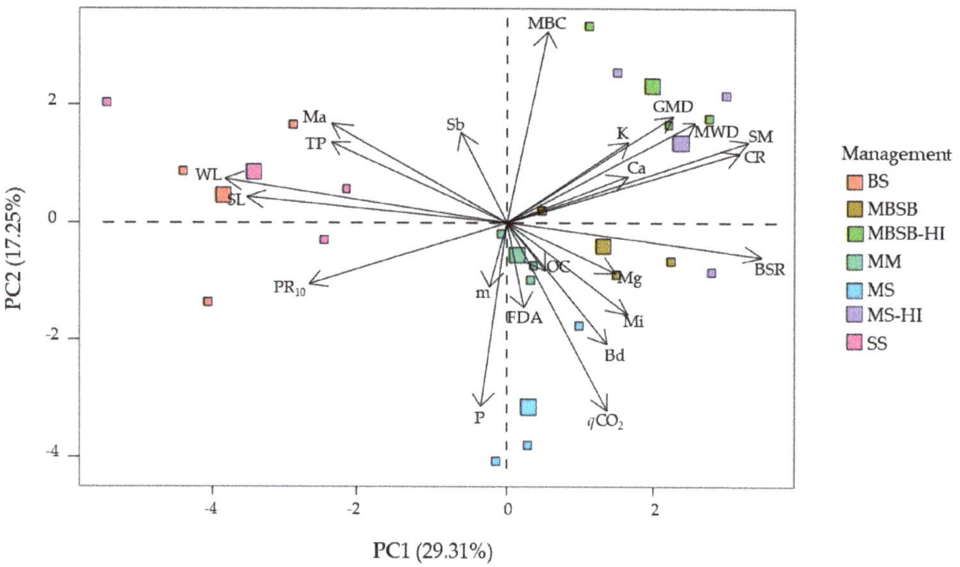

Figure 7. Principal component analysis for variables sampled in crop interrows. SL = soil losses, WL = water losses, CR = cover rate, MBC = microbial biomass carbon, BSR = basal soil respiration, qCO_2 = metabolic quotient, FDA = fluorescein diacetate hydrolysis, OC = soil organic carbon, Sb = sum of bases, m = aluminum saturation of soil, K = potassium, P = phosphorus, Ca = calcium, Mg = magnesium, MWD = mean weight diameter, GMD = geometric mean diameter, PR_{10} = soil resistance to root penetration at 10 cm depth, SM = soil moisture in situ, TP = total porosity, Ma = macroporosity, Mi = microporosity, Bd = bulk density, SS = soybean monoculture, MM = maize monoculture, MS = maize/soybean rotation, MBSB = maize/brachiaria/soybean/brachiaria rotation, BS = bare soil, and HI = high input of fertilizers.

4. Discussion

The high soil and water losses for the SS treatment indicate that soybean monoculture is a system of low sustainability in the Cerrado and high susceptible to water erosion, and it similar to a bare soil surface (like the BS treatment). Those conditions are explained by the low production of biomass, thus not promoting enough covering to protect the soil surface against the impact of raindrops. Soybean stover not only has low biomass production compared to grasses (e.g., maize and brachiaria), it also has high nitrogen content, i.e., a low C/N ratio, therefore being rapidly decomposed without providing permanent protection to the soil surface [32]. Furthermore, the surface sealing due to the direct impact of raindrop causes a reduction in soil water infiltration rates, increasing direct runoff and, consequently, water and soil losses [33]. Conversely, treatments with more stover production, e.g., due to inclusion of *Urochloa* sp. or due to maize biomass production [34], reduce water and soil losses.

No differences were found for the soil physical properties shown in Table 1. Therefore, the duration of our study was not sufficient to provide strong evidence of differences in soil physical quality evaluated by porosity and static indicators, even though it showed strong differences when evaluating dynamic attributes, such as soil and water losses, as discussed above. In the same experimental area, Mota et al. [6] observed that porosity-based indicators did not correlate with grain yield, but the pore size distribution indicator and organic carbon were effective in distinguishing the cropping systems. Silva et al. [16], in the same experimental area, observed that soil and water losses were negatively correlated with crop yield and vegetation cover rate. Therefore, these findings and the absence of differences in porosity-based indicators further reinforces the role of vegetation cover as a key factor in controlling the erosive process. The soil aggregation indicators GMD and MWD, the first representing the size of aggregates and the second the ratio of large size aggregates, can provide evidence regarding soil resistance to water erosion [35–38], since they can indicate the physical condition of the soil surface, providing insights regarding soil structural quality and infiltration capacity.

The soil physical indicator PR generated different insights. The lower PR observed in the MBSB-HI treatment can be explained by the abundance of grass roots in the upper soil layers and the time that had passed since the experiment was first set up (since the 2014/2015 crop year). This created so-called biopores through periodic renewal of the aggressive root system of brachiaria grass [39,40]. These biopores contribute to soil structure alleviation and improve soil porosity [41,42], which is crucial for water percolation in the soil profile and reduction of direct runoff. However, it is noteworthy that PR is strongly affected by soil moisture of the moment of its assessment, which showed significant differences, though in a relatively small range. Thus, it is difficult to draw conclusions based only on the results of PR, since treatments with lower soil moisture will have greater PR and vice versa, even if there are no differences in the mechanical impediment itself [18]. Meanwhile, Moura et al. [31], in the same experimental area, found that the least limiting water range indicator, which considers PR, distinguished the monoculture treatments (SS and MM) from the diversified crop systems already early in 2015/2016. PR could also contribute to these insights or predictions, since it may also encompass changes in soil structure due to biopores in diversified cropping systems, as already discussed.

Considering that the water supply is homogeneous throughout the area (only rainfed), the difference in in-field soil moisture can also provide useful insights about the differences among the cropping systems. The reduced soil moisture in the SS and BS treatments may be related to little or no stover over the soil surface (i.e., low cover rate), which causes an increase in soil temperature and evaporation. Furthermore, the absence of pores with diameters greater than that of the soil, coming from the stover, generates continuity of upward movement of water from the soil to the atmosphere, resulting in higher water loss by evaporation.

Soil and water losses also showed a negative correlation with in-field soil moisture. Several studies have reported the effect of antecedent soil moisture content on soil losses by erosion [43–46]. Soil water infiltration decreases as the soil moisture content increases. Thus, infiltration rate reduction leads to greater runoff and consequently greater soil loss.

The positive correlation between SL and WL highlights the direct impact of raindrops on the soil, which break down aggregates and carry away soil particles, organic material, and nutrients [47], and lead to surface sealing, reducing infiltration capacity and increasing direct runoff. The more elevated the water losses by runoff and soil losses by erosion transport, the greater the potential for decline in agricultural yield, and pollution of aquatic ecosystems [48].

The strong inverse relationship of SL and WL with BSR is a result of reduction in the richest organic layer of the soil. Organic compounds are a source of energy for microbial biomass, and reducing their quantity will inevitably reduce decomposition and, consequently, respiration [8,49]. The carryover of microorganisms previously attached to soil particles may also be a factor contributing to reduction in BSR. The increased WL due to surface sealing also contributes to BSR reduction, since there is less water infiltration and, consequently, soil water storage [50]. These findings were not reported for tropical soils, and could be relevant information for early diagnosis and monitoring of erosion processes. Future studies can investigate this microbiological indicator and calibrate reference values or relative indexes for on-farm management tools.

The higher MBC found in treatments with inclusion of brachiaria (MBSB-HI and MBSB) may be associated with the role of the grass in thermal and water insulation [34], which affects moisture, temperature, and plant residues on the soil surface. The higher values of MBC in more diversified cropping systems, considering the use of intercropped *Urochloa* sp., have already been reported by Araujo et al. [51], who found similar results, especially in the soil surface layer (0–10 cm), corroborating our results.

The study of Hoffmann et al. [8] states that the soil organic matter, mainly by quantifying the MBC, can reveal important data on the stock of soil organic components, which are closely linked to soil quality. Cândido et al. [52] also reported the importance of soil organic carbon for increasing soil quality and reducing soil erosion. This confirms the important role played by organic carbon in the sustainability of agricultural systems, affecting physical, chemical, and biological properties, with reflections on the stability of yield on production systems [52]. Thus, the maintenance of conservation management systems (e.g., no-tillage with intercropped brachiaria grass) is linked to the sustainability of production systems and a better plant development, leading to improvements in the agroecosystems and the environment itself.

Araujo et al. [51], working with maize in monoculture, also reported high qCO_2, which indicates stress conditions of the soil microbial community, since a small microbial mass has a high energy consumption to maintain metabolic activity. In our study, the MBSB-HI and MBSB treatments had higher MBC and FDA compared to the other treatments. Thus, the sensitivity of MBC to different land uses indicates that treatments including brachiaria were more stable in terms of soil moisture and availability of organic matter and nutrients, resulting in an increase in the microbial population and, consequently, in metabolic activity. Therefore, qCO_2 combined with high BSR may indicate environmental change due to higher energy consumption for soil organic matter decomposition.

Considering high BSR values, Odum [53] reported that respiration of the microbial community can be the first sign of imbalance, since energy is directed towards cell maintenance instead of growth and reproduction, leading to loss of C in the form of CO_2. Furthermore, Hoffmann et al. [8] highlighted that cropping systems with greater plant diversity and less soil disturbance favor the MBC increase, possibly due to the greater offer of soil organic matter, thus providing greater soil quality.

5. Conclusions

Monoculture grain cropping systems of soybeans and maize contrasted with diverse cropping systems with soybean/maize rotation, intercropped brachiaria grass, and higher fertilization, regarding soil quality and water erosion. The higher soil losses by erosion

(SL) and water losses by runoff (WL) occurred for soybean monoculture (SS) and bare soil treatment (BS), confirming the importance of maintaining crop stover over the soil surface over time.

Soil penetration resistance (PR) was important for distinguishing cropping systems, allowing recognition of better soil quality in cropping systems with intercropped grass. PR and aggregate stability indicators (GMD and MWD) were correlated with SL and WL, and this contributes to identifying erosion issues.

Cropping systems with intercropped brachiaria grass showed higher microbial biomass carbon (MBC) and basal soil respiration (BSR). BSR was negatively correlated with SL and WL, due to the removal of surface organic matter, which acts as source of energy for the microbiota. Soil without surface organic matter can lead to surface sealing, which reduces water infiltration and soil moisture, directly affecting microbiological communities and the erosion process. Therefore, higher BSR indicates lower soil and water losses and, consequently, better soil quality. These results provide useful insights regarding the close relationship between soil microbiological quality and intensity of the soil erosion process in cropping systems. Thus, not only physical and chemical, but also biological properties are deeply affected by erosion, aiding in early monitoring of water erosion. Soil quality improvement in ecologically supported management contributes to mitigating erosion.

Author Contributions: Conceptualization, J.C.A., M.L.N.S. and B.M.S.; methodology, M.N.M., J.C.A., L.d.C.M.d.S., O.O.d.S.A., E.B., F.M.d.S.M., M.S.T., Á.V.d.R., M.L.N.S. and B.M.S.; validation, M.N.M. and L.d.C.M.d.S.; formal analysis, M.N.M. and L.d.C.M.d.S.; investigation, M.N.M., J.C.A., L.d.C.M.d.S., O.O.d.S.A., E.B., F.M.d.S.M., M.S.T., Á.V.d.R., M.L.N.S. and B.M.S.; resources, J.C.A., E.B., F.M.d.S.M., Á.V.d.R. and B.M.S.; data curation, M.N.M., J.C.A., L.d.C.M.d.S. and B.M.S.; writing—original draft preparation, M.N.M. and L.d.C.M.d.S.; writing—review and editing, M.N.M., J.C.A., L.d.C.M.d.S., O.O.d.S.A., E.B., F.M.d.S.M., M.S.T., Á.V.d.R., M.L.N.S. and B.M.S.; supervision, J.C.A. and B.M.S.; project administration, J.C.A., E.B., Á.V.d.R. and B.M.S.; funding acquisition, J.C.A., E.B., Á.V.d.R. and B.M.S. All authors have read and agreed to the published version of the manuscript.

Funding: This study was funded by the Minas Gerais State Research Funding Foundation (FAPEMIG), processes CAG-APQ 01053-15, CAG—APQ-02075-16 and PPM-00774-18; the Brazilian National Council of Technological and Scientific Development (CNPq), processes 422538/2016-7, 408204/2016-8 and 306511-2017-7; the Agrisus Foundation, process 1491/15; and the Coordination for the Improvement of Higher Education Personnel (CAPES), process 001.

Acknowledgments: We acknowledge the financial support of FAPEMIG, CAPES, Agrisus, and CNPq. We are also grateful to the Department of Soil Science at the Federal University of Lavras, and the Brazilian Agricultural Research Corporation (Embrapa) for collaboration in setting up and conducting the field experiment.

Conflicts of Interest: The authors declare no conflict of interest.

References

1. Khanchoul, K.; Boubehziz, S. Spatial variability of soil erodibility at el hammam catchment, northeast of algeria. *Environ. Ecosyst. Sci.* **2019**, *3*, 17–25. [CrossRef]
2. Mello, G.; Bueno, C.R.P.; Pereira, G.T. Variabilidade espacial de perdas de solo, do potencial natural e risco de erosão em áreas intensamente cultivadas. *Rev. Bras. Eng. Agríc. Ambient.* **2006**, *10*, 315–322. [CrossRef]
3. Blake, W.H.; Rabinovich, A.; Wynants, M.; Kelly, C.; Nasseri, M.; Ngondya, I.; Patrick, A.; Mtei, K.; Munishi, L.; Boeckx, P.; et al. Soil erosion in East Africa: An interdisciplinary approach to realising pastoral land management change. *Environ. Res. Lett.* **2018**, *13*, 124014. [CrossRef]
4. Borrelli, P.; Märker, M.; Panagos, P.; Schütt, B. Modeling soil erosion and river sediment yield for an intermountain drainage basin of the Central Apennines, Italy. *CATENA* **2014**, *114*, 45–58. [CrossRef]
5. Oliveira, F.G.; Seraphim, O.J.; Borja, M.E.L. Estimativa de perdas de solo e do potencial natural de Erosão da bacia de contribuição da microcentral Hidrelétrica do Lageado, Botucatu—SP. *Eng. Agríc.* **2015**, *30*, 302–309. [CrossRef]
6. Mota, P.K.; Silva, B.M.; Borghi, E.; Viana, J.H.M.; Resende, Á.V.; Moura, M.S. Soil physical quality in response to intensification of grain production systems. *Rev. Bras. Eng. Agríc. Ambient.* **2020**, *24*, 647–655. [CrossRef]
7. Raghavendra, M.; Sharma, M.P.; Ramesh, A.; Richa, A.; Billore, S.D.; Verma, R.K. Soil Health Indicators: Methods and Applications. In *Soil Analysis: Recent Trends and Applications*; Springer: Singapore, 2020; pp. 221–253.

8. Hoffmann, R.B.; Moreira, E.E.A.; Hoffmann, G.S.S.; Araújo, N.S.F. Efeito do manejo do solo no carbono da biomassa microbiana. *Braz. J. Anim. Environ. Res.* **2018**, *1*, 168–178.
9. Gonçalves, V.A.; Melo, C.A.D.; Assis, I.R.; Ferreira, L.R.; Saraiva, D.T. Biomassa e atividade microbiana de solo sob diferentes sistemas de plantio e sucessões de culturas. *Rev. Ciênc. Agrár.* **2019**, *62*. [CrossRef]
10. Souza, R.A.; Telles, T.S.; Machado, W.; Hungria, M.; Filho, J.T.; Guimarães, M.F. Effects of sugarcane harvesting with burning on the chemical and microbiological properties of the soil. *Agric. Ecosyst. Environ.* **2012**, *155*, 1–6. [CrossRef]
11. Venzke Filho, S.P.; Feigl, B.J.; Piccolo, M.C.; Siqueira Neto, M.; Cerri, C.C. Biomassa microbiana do solo em sistema de plantio direto na região de Campos Gerais—Tibagi, PR. *Rev. Bras. Ciênc. Solo* **2008**, *32*, 599–610. [CrossRef]
12. Qiu, L.; Zhang, Q.; Zhu, H.; Reich, P.B.; Banerjee, S.; van der Heijden, M.G.A.; Sadowsky, M.J.; Ishii, S.; Jia, X.; Shao, M.; et al. Erosion reduces soil microbial diversity, network complexity and multifunctionality. *ISME J.* **2021**, *15*, 2474–2489. [CrossRef] [PubMed]
13. Soil Survey Staff. *Keys to Soil Taxonomy*, 12th ed.; USDA—Natural Resources Conservation Service: Washington, DC, USA, 2014.
14. Santos, H.G.; Jacomine, P.K.T.; Anjos, L.H.C.; Oliveira, V.Á.; Lumbreras, J.F.; Coelho, M.R.; Almeida, J.A.; Araújo Filho, J.C.; Oliveira, J.B.; Cunha, T.J.F. *Sistema Brasileiro de Classificação de Solos*, 5th ed.; Embrapa Solos: Brasília, Brazil, 2018; ISBN 978-85-7035-800-4.
15. Alvares, C.A.; Stape, J.L.; Sentelhas, P.C.; Gonçalves, J.L.M.; Sparovek, G. Köppen's climate classification map for Brazil. *Meteorol. Z.* **2013**, *22*, 711–728. [CrossRef]
16. Silva, L.C.M.; Avanzi, J.C.; Peixoto, D.S.; Merlo, M.N.; Borghi, E.; Resende, Á.V.; Acuña-Guzman, S.F.; Silva, B.M. Ecological intensification of cropping systems enhances soil functions, mitigates soil erosion, and promotes crop resilience to dry spells in the Brazilian Cerrado. *Int. Soil Water Conserv. Res.* **2021**, *9*, 591–604. [CrossRef]
17. Wischmeier, W.H.; Smith, D.D. *Predicting Rainfall Erosion Losses: A Guide to Conservation Planning*; Department of Agriculture: Washington, DC, USA, 1978.
18. Stolf, R.; Fernandes, J.; Furlani Neto, V.L. Recomendação Para o uso do Penetrômetro de Impacto—Modelo IAA/Planalsucar—Stolf; Piracicaba, Brazil. 1983. Available online: https://www.servidores.ufscar.br/hprubismar/arquivos10.htm (accessed on 29 December 2021).
19. Grohmann, F. Distribuição e tamanho de poros em três tipos de solos do estado de São Paulo. *Bragantia* **1960**, *19*, 319–328. [CrossRef]
20. Oliveira, L.B. Determinação da macro e microporosidade pela mesa de tensão em amostras de solo com estrutura indeformada. *Pesqui. Agropecu. Bras.* **1968**, *3*, 197–200.
21. Flint, L.E.; Flint, A.A. Porosity. In *Methods of Soil Analysis: Part 4, Physical Methods*; Dane, J.H., Topp, G.C., Eds.; Soil Science Society of America Book Series; Soil Science Society of America: Madison, WI, USA, 2002; pp. 241–254.
22. Kemper, W.D.; Rosenau, R.C. Aggregate Stability and Size Distribution. In *Methods of soil analysis. Part 1, Physical and Mineralogical Methods*, 2nd ed.; Klute, A., Ed.; Soil Science Society of America Book Series; American Society of Agronomy: Madison, WI, USA, 1986; pp. 425–442.
23. Teixeira, P.C.; Donagemma, G.K.; Fontana, A.; Teixeira, W.G. *Manual de Métodos de Análise de Solo*, 3rd ed.; Embrapa Solos: Brasília, Brazil, 2017.
24. Jenkinson, D.S.; Powlson, D.S. The effects of biocidal treatments on metabolism in soil—V. *Soil Biol. Biochem.* **1976**, *8*, 209–213. [CrossRef]
25. Vance, E.D.; Brookes, P.C.; Jenkinson, D.S. An extraction method for measuring soil microbial biomass C. *Soil Biol. Biochem.* **1987**, *19*, 703–707. [CrossRef]
26. Anderson, T.-H.; Domsch, K.H. Application of eco-physiological quotients (qCO2 and qD) on microbial biomasses from soils of different cropping histories. *Soil Biol. Biochem.* **1990**, *22*, 251–255. [CrossRef]
27. Dick, R.P.; Breakwell, D.P.; Turco, R.F. Soil Enzyme Activities and Biodiversity Measurements as Integrative Microbiological Indicators. *Methods Assess. Soil Qual.* **1997**, *49*, 247–271.
28. Ferreira, D.F.; Cargnelutti Filho, A.; Lúcio, A.D. Procedimentos estatísticos em planejamentos experimentais com restrição na casualização. In *Boletim Informativo Sociedade Brasileira de Ciência do Solo*; Sociedade Brasileira de Ciência do Solo: Campinas, Brazil, 2012; pp. 16–19.
29. Cecagno, D.; Costa, S.E.V.G.A.; Anghinoni, I.; Kunrath, T.R.; Martins, A.P.; Reichert, J.M.; Gubiani, P.I.; Balerini, F.; Fink, J.R.; Carvalho, P.C.F. Least limiting water range and soybean yield in a long-term, no-till, integrated crop-livestock system under different grazing intensities. *Soil Tillage Res.* **2016**, *156*, 54–62. [CrossRef]
30. Peixoto, D.S.; Silva, B.M.; Oliveira, G.C.; Moreira, S.G.; Silva, F.; Curi, N. A soil compaction diagnosis method for occasional tillage recommendation under continuous no tillage system in Brazil. *Soil Tillage Res.* **2019**, *194*, 104307. [CrossRef]
31. Moura, M.S.; Silva, B.M.; Mota, P.K.; Borghi, E.; Resende, A.V.; Acuña-Guzman, S.F.; Araújo, G.S.S.; Silva, L.C.M.; Oliveira, G.C.; Curi, N. Soil management and diverse crop rotation can mitigate early-stage no-till compaction and improve least limiting water range in a Ferralsol. *Agric. Water Manag.* **2021**, *243*, 106523. [CrossRef]
32. Andrade, A.T.; Torres, J.L.R.; Torres, J.L.R.; Paes, J.M.V.; Teixeira, C.M.; Condé, A.B.T. Desafios do Sistema Plantio Direto no Cerrado. *Inf. Agropecu.* **2018**, *39*, 18–26.

33. Debiasi, H.; Franchini, C.F.; Gonçalves, S.L. *Manejo da Compactação do Solo em Sistema de Produção de Soja Sob Semeadura Direta*; Embrapa: Londrina, Brazil, 2008; Available online: https://www.infoteca.cnptia.embrapa.br/handle/doc/470946 (accessed on 29 December 2021).
34. Wu, G.-L.; Yang, Z.; Cui, Z.; Liu, Y.; Fang, N.-F.; Shi, Z.-H. Mixed artificial grasslands with more roots improved mine soil infiltration capacity. *J. Hydrol.* **2016**, *535*, 54–60. [CrossRef]
35. Le Bissonnais, Y.; Arrouays, D. Aggregate stability and assessment of soil crustability and erodibility: II. Application to humic loamy soils with various organic carbon contents. *Eur. J. Soil Sci.* **1997**, *48*, 39–48. [CrossRef]
36. Wilson, H.A.; Browning, G.M. Soil Aggregation, Yields, Runoff, and Erosion as Affected by Cropping Systems. *Soil Sci. Soc. Am. J.* **1946**, *10*, 51–57. [CrossRef]
37. Wischmeier, W.H.; Mannering, J.V. Relation of Soil Properties to its Erodibility. *Soil Sci. Soc. Am. J.* **1969**, *33*, 131–137. [CrossRef]
38. Yoder, R.E. A Direct Method of Aggregate Analysis of Soils and a Study of the Physical Nature of Erosion Losses 1. *Agron. J.* **1936**, *28*, 337–351. [CrossRef]
39. Crusciol, C.A.C.; Nascente, A.S.; Mateus, G.P.; Pariz, C.M.; Martins, P.O.; Borghi, E. Intercropping soybean and palisade grass for enhanced land use efficiency and revenue in a no till system. *Eur. J. Agron.* **2014**, *58*, 53–62. [CrossRef]
40. Silva, R.F.; Severiano, E.C.; Oliveira, G.C.; Barbosa, S.M.; Peixoto, D.S.; Tassinari, D.; Silva, B.M.; Silva, S.H.G.; Dias Júnior, M.S.; Figueiredo, T.A.F.R. Changes in soil profile hydraulic properties and porosity as affected by deep tillage soil preparation and Brachiaria grass intercropping in a recent coffee plantation on a naturally dense Inceptisol. *Soil Tillage Res.* **2021**, *213*, 105127. [CrossRef]
41. Williams, S.M.; Weil, R.R. Crop Cover Root Channels May Alleviate Soil Compaction Effects on Soybean Crop. *Soil Sci. Soc. Am. J.* **2004**, *68*, 1403–1409. [CrossRef]
42. Ajayi, A.E.; Horn, R.; Rostek, J.; Uteau, D.; Peth, S. Evaluation of temporal changes in hydrostructural properties of regenerating permanent grassland soils based on shrinkage properties and μCT analysis. *Soil Tillage Res.* **2019**, *185*, 102–112. [CrossRef]
43. Vaezi, A.; Haghani, Z. Effect of Soil Water Content on Runoff Production and Soil Loss from Rills in Field Experimental Plots in Different Slopes. *Water Soil Sci.* **2021**, *31*, 31–43.
44. Nishigaki, T.; Sugihara, S.; Kilasara, M.; Funakawa, S. Surface Runoff Generation and Soil Loss Under Different Soil and Rainfall Properties in The Uluguru Mountains, Tanzania. *Land Degrad. Dev.* **2017**, *28*, 283–293. [CrossRef]
45. Wei, L.; Zhang, B.; Wang, M. Effects of antecedent soil moisture on runoff and soil erosion in alley cropping systems. *Agric. Water Manag.* **2007**, *94*, 54–62. [CrossRef]
46. Luk, S.-H.; Hamilton, H. Experimental effects of antecedent moisture and soil strength on rainwash erosion of two luvisols, Ontario. *Geoderma* **1986**, *37*, 29–43. [CrossRef]
47. Bertoni, J.; Lombardi Neto, F. *Conservação do Solo*, 9th ed.; Ícone: São Paulo, Brazil, 2014.
48. Baldassarini, J.S.; Nunes, J.O.R. Estimação da perda de solo por processos erosivos em parcelas de monitoramento utilizando pinos de erosão em propriedades rurais do interior do Estado de São Paulo. *Confins* **2018**. Available online: https://journals.openedition.org/confins/16084 (accessed on 29 December 2021). [CrossRef]
49. Cattelan, A.J.; Vidor, C. Flutuações na biomassa, atividade e população microbiana do solo, em função de variações ambientais. *Rev. Bras. Ciênc. Solo* **1990**, *14*, 133–142.
50. Silva, E.E.; Azevedo, P.H.S.; De-Polli, H. *Determinação de Respiração Basal (RBS) e Quociente Metabólico do solo (qCO2)*; Comunidado Técnico 99—EMBRAPA Agrobiol: Seropédica, Brazil, 2007.
51. Araujo, T.S.; Gallo, A.S.; Araujo, F.S.; Santos, L.C.; Guimarães, N.F.; Silva, R.F. Biomassa e atividade microbiana em solo cultivado com milho consorciado com leguminosas de cobertura. *Rev. Ciênc. Agrár.* **2019**, *42*, 347–357.
52. Cândido, B.M.; Silva, M.L.N.; Curi, N.; Freitas, D.A.F.; Mincato, R.L.; Ferreira, M.M. Métodos de indexação de indicadores na avaliação da qualidade do solo em relação à erosão hídrica. *Rev. Bras. Ciênc. Solo* **2015**, *39*, 589–597. [CrossRef]
53. Odum, E.P. Trends Expected in Stressed Ecosystems. *Bioscience* **1985**, *35*, 419–422. [CrossRef]

water

Article

Evaluating the Effects of the Rill Longitudinal Profile on Flow Resistance Law

Alessio Nicosia *, Costanza Di Stefano, Vincenzo Palmeri, Vincenzo Pampalone and Vito Ferro

Department of Agricultural, Food and Forestry Sciences, University of Palermo, Viale delle Scienze, Building 4, 90128 Palermo, Italy; costanza.distefano@unipa.it (C.D.S.); agrovincenzo.palmeri@gmail.com (V.P.); vincenzo.pampalone@unipa.it (V.P.); vito.ferro@unipa.it (V.F.)
* Correspondence: alessio.nicosia@unipa.it

Abstract: In this paper, for the first time, the effect of the longitudinal profile shape of the rill (uniform, concave, and convex) on flow resistance law was studied. The first part of the paper is based on a theoretical equation to estimate the Darcy–Weisbach friction factor *f*, deduced from the power velocity distribution and rill measurements performed on a plot. At first, the equation to estimate the Γ parameter of the velocity profile was calibrated using all available measurements. Then an analysis of the hydraulic characteristics at reach scale, for comparable values of discharge, was carried out, comparing the different profile shapes. To assess the influence of the rill profile shape on flow resistance law, this calibration was also carried out using the data categorized by profile. In the second part of the paper, an analysis of the scour depth and eroded rill volume was developed for four rills of each configuration. The results showed that an accurate estimate of *f* can be obtained by calibrating the flow resistance equation for each profile shape. The component of the Darcy–Weisbach friction factor that is due to the profile shape varied from 0.68 to 14.6% of the overall friction factor for the concave profile, and from 3.4 to 26.9% for the convex profile. The analysis also showed that, for the convex profile, the scour was concentrated downstream of the slope change, while for the uniform and concave profiles, it was uniformly distributed. Furthermore, the scour depth measured in correspondence of the rill thalweg had a generally increasing trend, with discharge for all the investigated profile shapes. Total eroded rill volume of the concave profile was lower than those detected for the uniform and convex profiles and was characterized by a reduction of 57.9%, as compared to the uniform profile.

Keywords: slope profile; concave profile; convex profile; rill erosion; rill hydraulics

Citation: Nicosia, A.; Di Stefano, C.; Palmeri, V.; Pampalone, V.; Ferro, V. Evaluating the Effects of the Rill Longitudinal Profile on Flow Resistance Law. *Water* **2022**, *14*, 326. https://doi.org/10.3390/w14030326

Academic Editor: Manuel Seeger

Received: 20 December 2021
Accepted: 21 January 2022
Published: 23 January 2022

1. Introduction

Many researchers have studied the effects of length and slope gradient on runoff and soil loss using uniform plots, i.e., having a constant slope, while a small number of studies have focused on the effect of profile shape (concave, convex) [1,2].

Young and Mutchler [3,4] investigated experimentally the effect of slope shape (convex, concave, uniform) on runoff and soil loss at plot scale and showed that concave hillslopes tends to have reduced total sediment loss compared to a uniform profile slope.

Rieke-Zapp and Nearing [2] carried out laboratory experiments to determine the relationship between slope shapes and soil erosion using five slope shape treatments (convex-linear, concave-linear, nose slope, head slope and uniform). These experiments demonstrated that rill patterns, sediment yield and runoff, all changed with slope shape. In particular, these authors reported a decrease of 75% of soil loss for concave slopes, as compared to uniform ones with similar surface area.

Sensoy and Kara [5] carried out an experimental investigation using nine experimental field plots with different profile shapes, established on a 30% hillslope, subjected to natural rainfall. This field experiment indicated that slope shape (uniform, concave, convex)

affected both runoff and soil loss. In particular, the uniform slope was characterized by the highest values of runoff and soil loss, while the concave slope produced the lowest values. The experimental results also indicated that the soil eroded from the uniform plots was mainly composed of fine particles (<2 mm), while both the concave and convex slopes were characterized by a large size variability of the eroded particles.

Jeldes et al. [6] compared concave and uniform hillslopes having the same degree of mechanical stability and concluded that concave slopes produced 15–40% less sediment than uniform slopes. This value was lower than that reported in the literature, which varied from 50% [7] to 80% [8].

Mombini et al. [9] studied the effect of surface roughness on soil loss using different complex hillslopes in terms of plan shapes (convergent, parallel, and divergent) and profiles (convex, concave and uniform). This research, carried out in laboratory conditions using three different soil surface roughness values and a single simulated rainfall intensity, indicated that the highest soil loss due to soil roughness was observed on uniform parallel hillslopes.

Despite the current availability of techniques able to produce a more accurate and detailed representation of the hillslope surface [10–13], the rill erosion processes on non-linear profiles (concave, convex) have been barely studied by the scientific community.

Rill erosion results from soil particle detachment and transport by channelized flows. When overland flow tends to convert into a concentrated flow, the formation of rills occurs. Rills are eroded channels, characterized by rapid morphological evolution, that represent a relevant sediment source and a transport path for eroded particles from hillslopes [14,15]. The increased flow shear stress and velocity within the rill increase sediment yield promoting the transport of soil particles that are detached by rill flow and delivered from the interrill areas to the rill channels [16–23]. Rill erosion has both on-site (reduction of soil fertility and productivity) and off-site (river over-sedimentation, reservoir sedimentation and degradation of water quality) impacts [24] and is dominant as compared to interrill erosion [21].

In previous papers, a theoretical approach based on the integration of a power-velocity profile, was proposed to deduce the Darcy–Weisbach friction factor [25,26], and its applicability was tested for rill flows [27–35] on uniform plots with different slope gradients.

In this paper, for the first time, the theoretical approach by Ferro [25,26] is applied to rill measurements carried out on three plots having different profiles (concave, convex and uniform).

The specific aims of this study are to: (a) test the reliability of the abovementioned theoretical approach; (b) estimate the component of Darcy–Weisbach friction factor corresponding to the profile shape; (c) investigate the evolution of the rill thalweg for the three examined profile shapes and (d) compare the volumes eroded by flow with similar discharges for the different profiles.

2. Materials and Methods

2.1. The Experimental Plots and Measurements

A plot, 2 m wide and 7 m long (Figure 1), located in the experimental area of the Department of Agriculture, Food and Forest Sciences of the University of Palermo, was used to carry out the experiments.

Figure 1. View of the experimental plot with uniform (**a**), convex (**b**), and concave (**c**) profiles.

The plot was prepared with three different profiles (concave, convex and uniform) (Figure 1) using a mean plot slope s_p of 18%. In particular, for the concave and convex profiles, the plot was divided into two 3.27 m long parts, and two slope values (12 and 24%) were set up to obtain a mean value of 18% (Figure 2). The soil was a clay loam characterized by 32.7%, 30.9% and 36.4% of clay, silt, and sand, respectively.

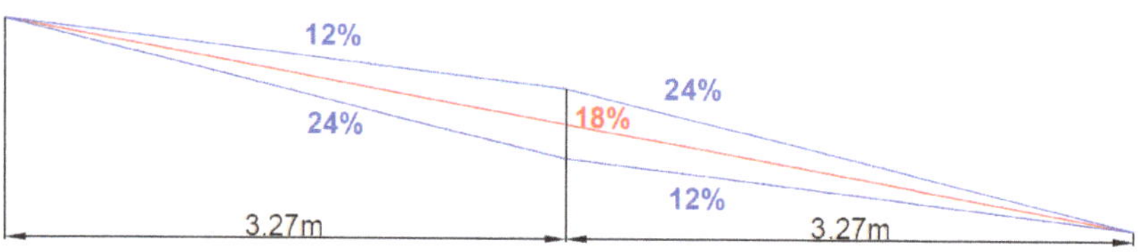

Figure 2. Scheme of the investigated profiles.

For all the profile shapes, soil was manually hoed and raked and then compacted by a woody bar. The bulk density of the soil was determined by sampling the soil, using a metal cylinder of known volume, and determining its weight after drying. A uniform soil moisture condition for all the experiments was achieved by wetting the soil until infiltration stopped. Then, rills were manually incised along the maximum slope direction of the plot and were shaped using a clear flow discharge of 0.1 L s^{-1}, applied for 3 min. The rills had the same longitudinal profile as the plot. For each rill, nine longitudinal segments, bounded by two cross-sections spaced 0.624 m apart (Figure 3), were determined.

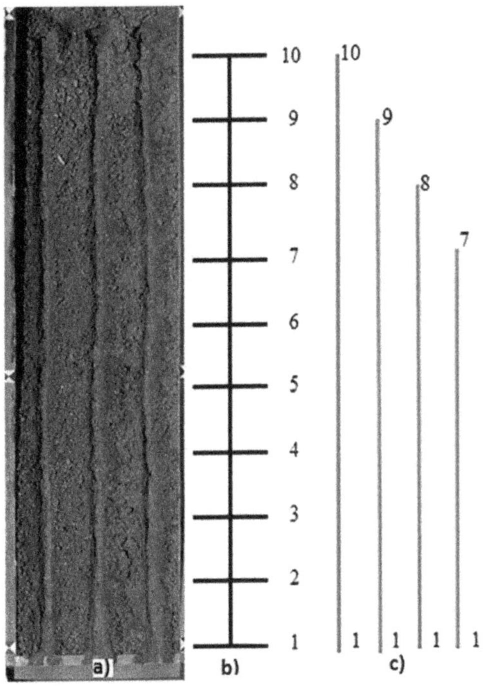

Figure 3. View of the plot area with some tested rills (**a**), scheme of cross-sections (**b**), and examples of rill reaches (**c**).

A constant clear inflow discharge, ranging from 0.18 to 0.63 L s^{-1}, was used for the experiments.

Measurements of water depth h, cross-section area σ, wetted perimeter C, bed slope s, and flow velocity V were carried out for rill reaches (i.e., the channel between a given measured cross-section and the rill end) on uniform (99 reaches), concave (71 reaches), and convex (72 reaches) rills. The indirect measurements of Reynolds number $Re = V\,h/\nu_k$, in which ν_k is the kinematic viscosity, Froude number $F = V/\sqrt{(g\,h)}$, in which g is the acceleration due to gravity, and Darcy–Weisbach friction factor $f = 8\,gRs\,/V^2$, where R is the hydraulic radius, were also obtained.

For each experimental run, a 3D-DTM of the plot area was built from a set of 70 photographs using a 3D-photo reconstruction technique with the image-processing software Agisoft Photoscan Professional (Version 1.1.6, Agisoft, St. Petersburg, Russia). The slope gradient of each rill segment was determined using the rill thalweg imaged by the 3D-DTM. The slope of each rill reach was calculated by averaging the values measured in the considered reach.

The mean water depth h and the mean hydraulic radius R in the reach were determined using the method developed by Di Stefano et al. [36]. This method couples an accurate ground survey of the rill channel, obtained by close-range photogrammetry, with the survey of the water tracks inside the channel marked by a dye solution. The measured water depths and geometric cross-sections obtained by the 3D-DTM, at a distance interval of 6.2 cm within the reach, were used to calculate the values of the hydraulic cross-section area σ and wetted perimeter C in the reach and the values of h and R using the following equations [36]:

$$h = \frac{\sum_{i=2}^{N-1}\left(\frac{\sigma_1}{2} + \sigma_i + \frac{\sigma_N}{2}\right)}{\sum_{i=2}^{N-1}\left(\frac{w_1}{2} + w_i + \frac{w_N}{2}\right)} \quad (1)$$

$$R = \frac{\sum_{i=2}^{N-1}\left(\frac{\sigma_1}{2} + \sigma_i + \frac{\sigma_N}{2}\right)}{\sum_{i=2}^{N-1}\left(\frac{C_1}{2} + C_i + \frac{C_N}{2}\right)} \quad (2)$$

in which w is the surface width and N is the number of the cross-sections in the reach.

A methylene blue solution was applied as a dye tracer to measure the flow velocity [28,37–40]. The measured surface velocity was converted to the mean flow velocity V by a correction factor of 0.8 [41–43].

For four rills of each profile shape, Digital Elevation Models (DEMs) at the ends of the shaping phase (*D1*) and the experimental runs (*D2*) were obtained from the 3D-DTM. The DEM of the difference (DoD) between *D2* and *D1* was used to measure the scour depth values inside the whole rill channel and calculate the total eroded rill volume *RV*. Considering that the minimum level of detection of the used DEMs was ±0.005 m, variations of bed elevation in the range ±0.005 m were not detectable.

Finally, for the same rills, the scour depth values *SD* were measured, at 0.005 m intervals along the rill thalweg, as the difference in height between the longitudinal profiles at the end of the shaping phase (*D1*) and at the end of the experimental runs (*D2*) (an example for one rill with a convex profile shape is plotted in Figure 4). For each rill, the mean scour depth SD_m was also calculated as the mean of the *SD* values.

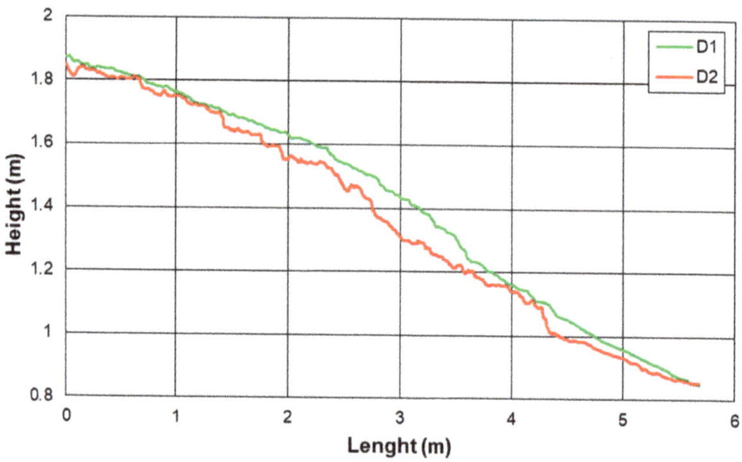

Figure 4. Comparison, using a rill with a convex profile, between the rill thalweg at the end of the shaping phase (D1) and at the end of the experimental run (D2).

2.2. The Rill Flow Resistance Equation

For an open channel flow, the local flow velocity profile $v(y)$ along a given vertical is represented by the following functional relationship [44–46]:

$$\varphi\left(\frac{dv}{dy}, y, h, d, u_*, s, \rho, \mu, g\right) = 0 \quad (3)$$

where φ is a functional symbol, v is the local velocity, y is the distance from the bottom, d is a characteristic bed particle diameter, $u_* = \sqrt{(g\,R\,s)}$ is the shear velocity, ρ is the water density and μ is the dynamic water viscosity.

Applying the Π-Theorem of the dimensional analysis and using as dimensional independent variables y, u_* and μ, Equation (3) can be expressed by the following dimensionless functional relationship [30]:

$$\frac{y}{u_*}\frac{dv}{dy} = \varphi_1\left(\frac{h}{d}, s, \frac{u_* y}{\nu_k}, Re, F\right) \quad (4)$$

in which φ_1 is a functional symbol.

Hypothesizing the Incomplete Self-Similarity in $u_* y/\nu_k$ [25,47–50], neglecting the influence of Re (turbulent flow regime) and considering that the flow Froude number accounts also for the ratio h/d [26], the integration of Equation (4) yields to the following velocity profile:

$$\frac{v}{u_*} = \Gamma(s, F)\left(\frac{u_*\ y}{\nu_k}\right)^{\delta} \quad (5)$$

where $\Gamma(s, F)$ is a function to be defined by velocity measurements and the exponent δ is calculated by the following theoretical equation [51,52]

$$\delta = \frac{1.5}{\ln Re} \quad (6)$$

The following expression of the Darcy–Weisbach friction factor f is deduced [25,48,53] by integrating Equation (5):

$$f = 8\left[\frac{2^{1-\delta}\ \Gamma\ Re^{\delta}}{(\delta+1)\ (\delta+2)}\right]^{-2/(1+\delta)} \quad (7)$$

From the velocity profile (Equation (5)) the following estimate Γ_v of Γ function [25,53] is obtained by setting equal to $\alpha\, h$ the distance y at which the local velocity is equal to the cross-section average velocity V:

$$\Gamma_v = \frac{V}{u_*\left(\frac{u_* \alpha h}{\nu_k}\right)^{\delta}} \quad (8)$$

The coefficient α, which is less than 1, takes into account that both V is located below the water surface and the mean velocity profile in the cross-section is considered. The coefficient α has to be calculated by the following theoretical relationship deduced by Ferro [25]:

$$\alpha = \left[\frac{2^{1-\delta}}{(\delta+1)\ (\delta+2)}\right]^{1/\delta} \quad (9)$$

For this investigation α varied in a narrow range and the mean value of 0.124 was assumed. Considering that, according to Equation (5), Γ theoretically depends only on s and F [26,54], Γ_v can be estimated using the following power equation:

$$\Gamma_v = a\,\frac{F^b}{s^c} \quad (10)$$

where a, b and c are coefficients to be determined from experimental measurements.

3. Results

3.1. Rill Flow Resistance for Different Profile Shapes

The experimental runs were carried out for Reynolds numbers corresponding to turbulent flow ($2764 \le Re \le 11015$ for the uniform profile, $2282 \le Re \le 14346$ for the concave one, and $2486 \le Re \le 9487$ for the convex one) and for Froude number values corresponding to subcritical and supercritical flow conditions ($0.54 \le F \le 1.70$ for the uniform profile, $0.64 \le F \le 1.63$ for the concave one, and $0.44 \le F \le 1.71$ for the convex one) (Figure 5). In other words, the investigated ranges of *Re* and *F* are very similar for the three different profiles.

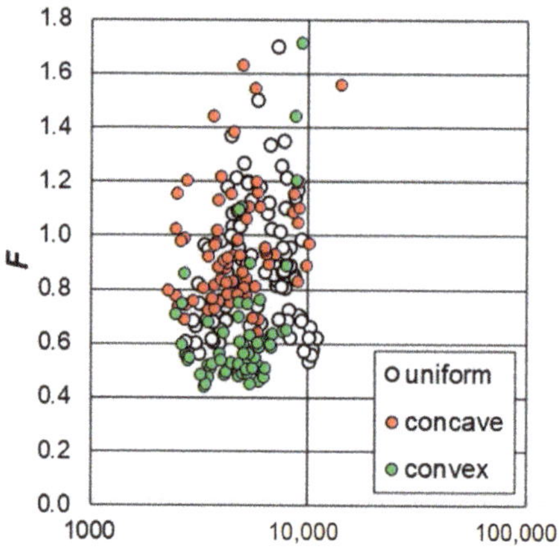

Figure 5. Experimental values of Froude and Reynolds number for each profile shape.

Using all the 242 measurements, i.e., neglecting the influence of the profile shape, Equation (10) was calibrated obtaining the following relationship:

$$\Gamma_v = 0.4322 \frac{F^{1.0857}}{s^{0.5235}} \quad (11)$$

characterized by a coefficient of determination equal to 0.9894. The comparison between the 242 measured Γ_v values, obtained by Equation (8) and those calculated applying Equation (11) is plotted in Figure 6a.

Introducing Equation (11) into Equation (7), the following flow resistance law was obtained:

$$f = 8\left[\frac{(\delta+1)\ (\delta+2)\ s^{0.5235}}{2^{1-\delta}\ Re^{\delta} 0.4322\ F^{1.0857}}\right]^{2/(1+\delta)} \quad (12)$$

A good agreement, characterized by a root mean square error *RMSE* equal to 0.1743, between the measured Darcy–Weisbach friction factor values and those calculated by Equation (12) was detected (Figure 6b), even if Equation (12) tended to slightly overestimate the *f* values for the uniform profile. The friction factor values calculated by Equation (12) were characterized by errors less than or equal to ±20% for 96.3% of cases and less than or equal to ±10% for 78.5% of cases. Figure 6 highlights that the *f* measurements obtained for the convex profile covered a different range than those obtained from the uniform and concave profiles. Different characteristics of the convex profile are also revealed by the frequency distributions of the hydraulic variables *F*, *f*, *s* and *V* plotted in Figure 7.

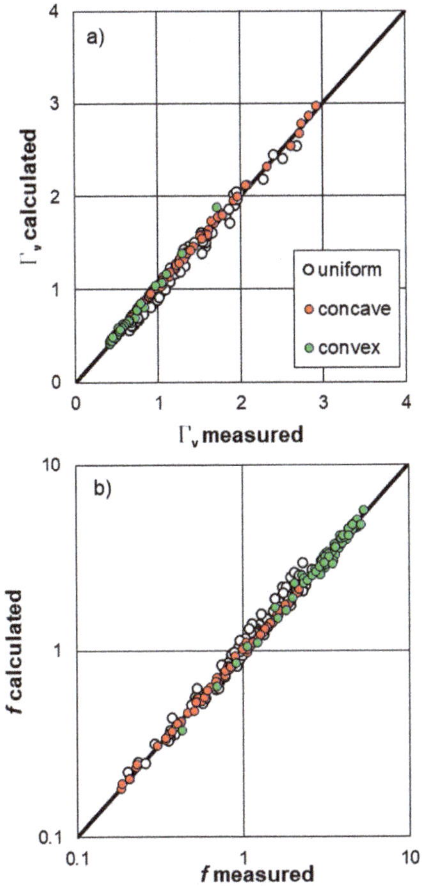

Figure 6. Comparison between the measured Γ_v values obtained by Equation (8) and those calculated applying Equation (11) (**a**) and measured Darcy–Weisbach friction factor values and those calculated by Equation (12) (**b**).

As this analysis showed that profile shape influences flow characteristics, the theoretical flow resistance law was calibrated for each investigated condition. Table 1 lists the a, b and c coefficients of Equation (10) obtained for the uniform, concave, and convex profiles.

Table 1. Values of a, b and c coefficients of Equation (10) obtained for the uniform, concave, and convex profiles.

Profile Shape	a	b	c
Uniform	0.4419	1.0127	0.5224
Concave	0.4025	1.1084	0.5469
Convex	0.5187	1.0437	0.3795

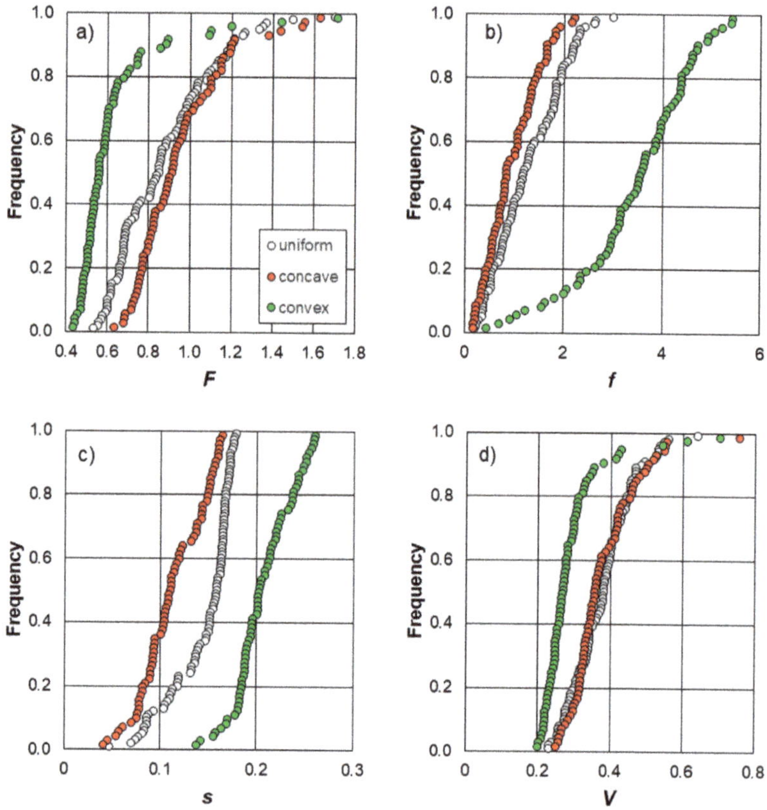

Figure 7. Frequency distributions of Froude number (**a**), friction factor (**b**), reach slope (**c**) and flow velocity (**d**).

Figure 8a shows the comparison between the measured Γ_v values obtained by Equation (8) and those calculated applying Equation (10), with the specific *a*, *b* and *c* coefficients (Table 1) for the different profiles. The coefficient of determination of the obtained relationships is equal to 0.9812, 0.9979 and 0.9882 for the uniform, concave, and convex profiles, respectively.

Figure 8b shows good agreement, characterized by a RMSE equal to 0.1234, between the measured values of the Darcy–Weisbach friction factor and those calculated by Equations (7) and (10), with *a*, *b* and *c* coefficients obtained for each profile shape. In this case, the calculated friction factor values are characterized by errors always less than or equal to ±20% and less than or equal to ±10% for 86.8% of cases.

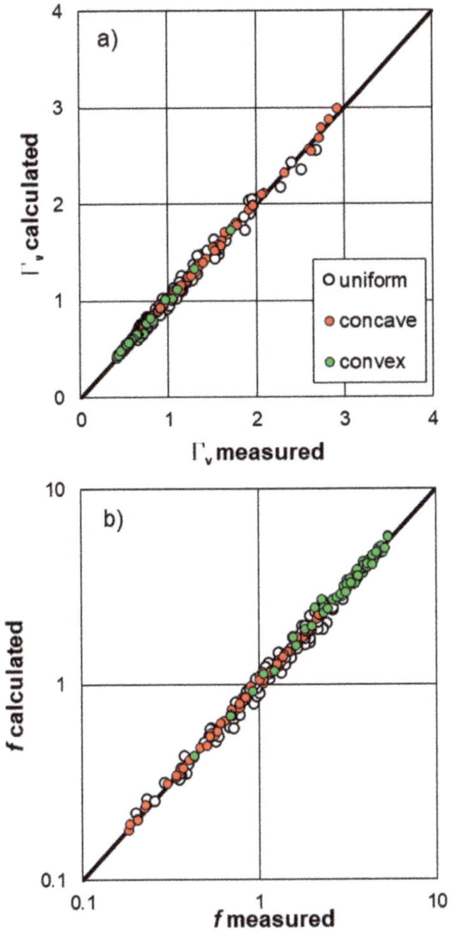

Figure 8. Comparison between the measured Γ_v values obtained by Equation (8) and those calculated by Equation (10) with the shape profile-specific a, b and c coefficients (**a**) and measured Darcy–Weisbach friction factor values and those calculated by Equations (7) and (10) with the shape profile-specific a, b and c coefficients (**b**).

Moreover, to study the influence of profile shape on rill flow resistance, the additive method of shear stresses [55], already used to investigate the effects of sediment transport for rills [29,56], and of riparian vegetation for a vegetated channel [57], was applied. In particular, the Darcy–Weisbach friction factor, f_{mor}, corresponding to the effect of the profile on flow resistance, was calculated as the difference f-f_u between the measured f value of the concave or convex profile and that of, f_u, calculated for the uniform profile in the same hydraulic conditions (Equations (7) and (10), with a, b and c coefficients obtained for the uniform profile). This f_{mor} value is almost equal to zero for 9.9% (concave) and 4.2% (convex) of the measured values. The ratio f_{mor}/f varies from 0.68 to 14.6% for concave profile and from 3.4 to 26.9% for convex profile (Figure 9).

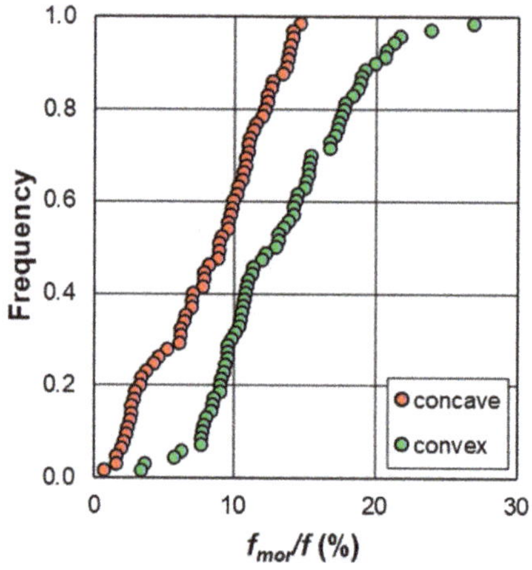

Figure 9. Frequency distribution of the ratio f_{mor}/f.

3.2. Analysis of Rill Scour Depth and Eroded Volume for Different Profile Shapes

Figure 10 shows the DoD obtained for the uniform (a), concave (b) and convex (c) profiles were characterized by comparable ranges of flow discharge (0.26–0.58 L s^{-1} for uniform, 0.18–0.46 L s^{-1} for concave, and 0.21–0.60 L s^{-1} for convex profiles). This figure clearly shows that the concave profile is characterized by scour depth values (ranging from 0.005 to 0.08 m) lower than those of uniform (0.005–0.153 m) and convex (0.005–0.176 m) profiles.

For all three investigated profile shapes, Figure 10 also shows that rills are generally scoured by flow. The blue areas of this figure identify rill zones where the variation of bed elevation is lower than the minimum level of detection (±0.005 m), or lack variation. Moreover, for the convex profile, the scour is concentrated downstream of the slope change, while for uniform and concave profiles, it is uniformly distributed.

The knowledge of the scour depth values obtained by the DoDs allows calculation of the total eroded volume *RV*. Figure 11 shows the comparison between the applied discharge *Q* and *RV* using an example of four rills analysed for each profile. The evident trend, as expected, is that *RV* increases with *Q* for all the investigated profiles. Finally, for the uniform, concave, and convex profiles, the sum of the four *RV* values resulted in 0.0644, 0.0271 and 0.0654 m^3, respectively.

For the forementioned four experimental runs for each profile shape, the frequency distribution of the scour depth values *SD* was plotted in Figure 12. This figure highlights that scour depth, measured in corresponding rill thalwegs, has a generally increasing trend with discharge for all investigated profile shapes.

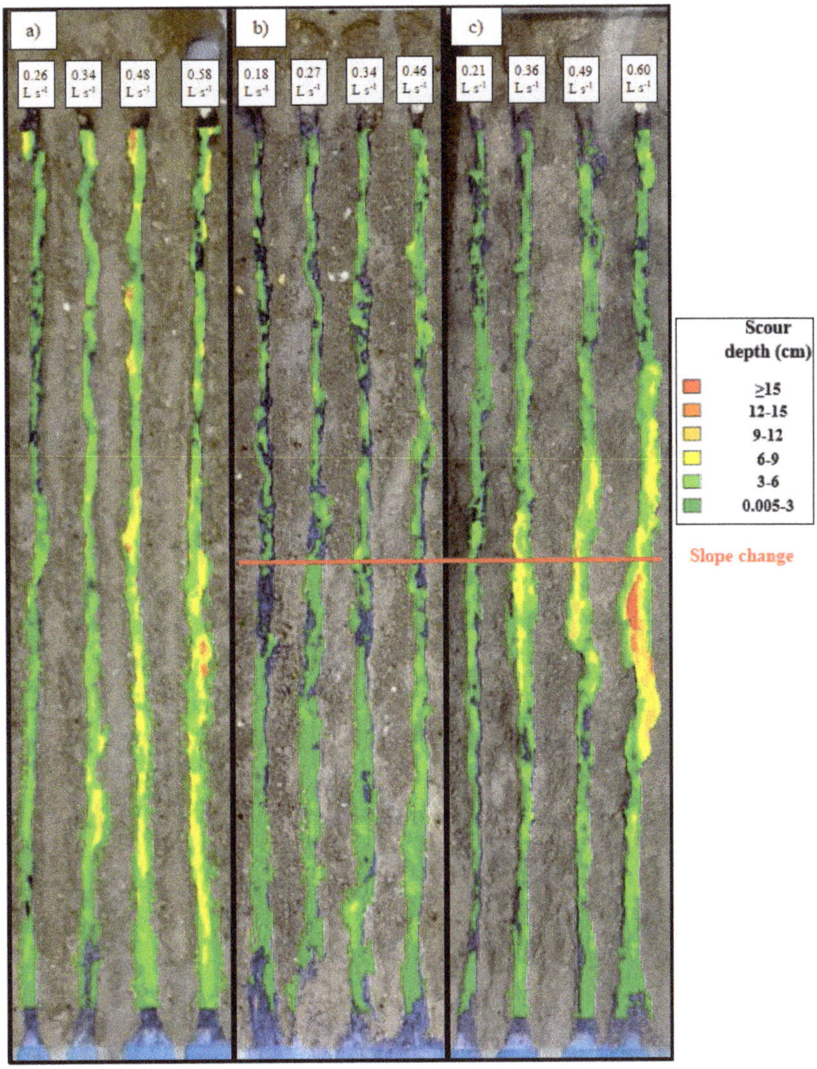

Figure 10. DoD obtained for the uniform (**a**), concave (**b**) and convex (**c**) profiles.

Figure 13 shows the frequency distribution of the SD/SD_m ratio for the investigated discharges. This figure demonstrates that the variable SD/SD_m can be considered independent of discharge.

Figure 14 shows the empirical frequency distribution of SD/SD_m for each profile shape, using an example for a single discharge value ($\approx$0.35 L s^{-1}). The overlapping of the three distributions suggested that SD/SD_m does not depend on profile shape.

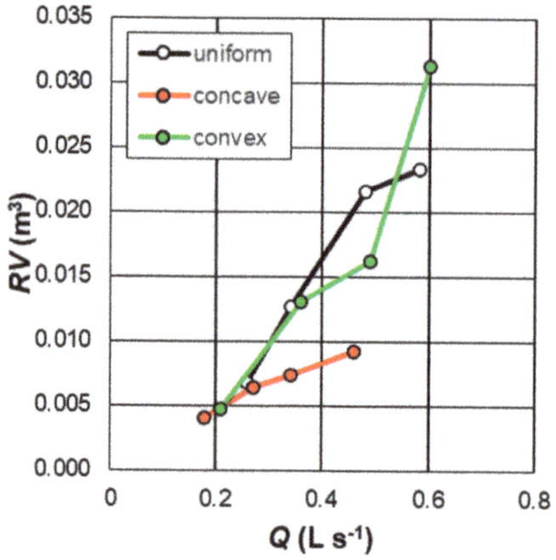

Figure 11. An example of the comparison between the applied discharge Q and the total eroded rill volume RV, for four rill channels for each profile shape.

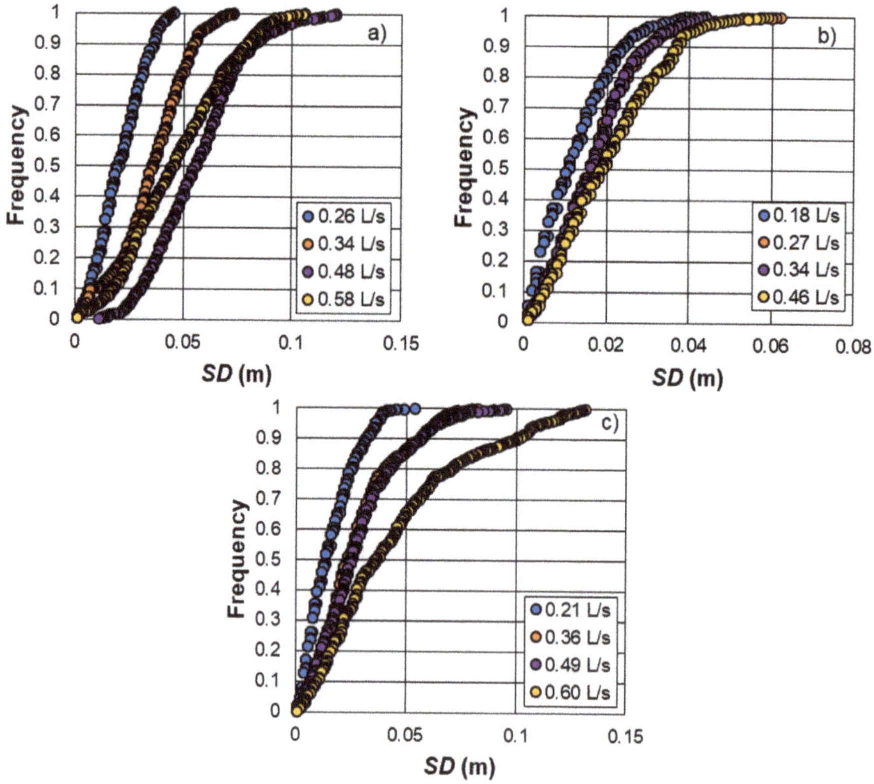

Figure 12. An example of the frequency distribution of the scour depth SD for four discharge values, for uniform (**a**), concave (**b**), and convex (**c**) profiles.

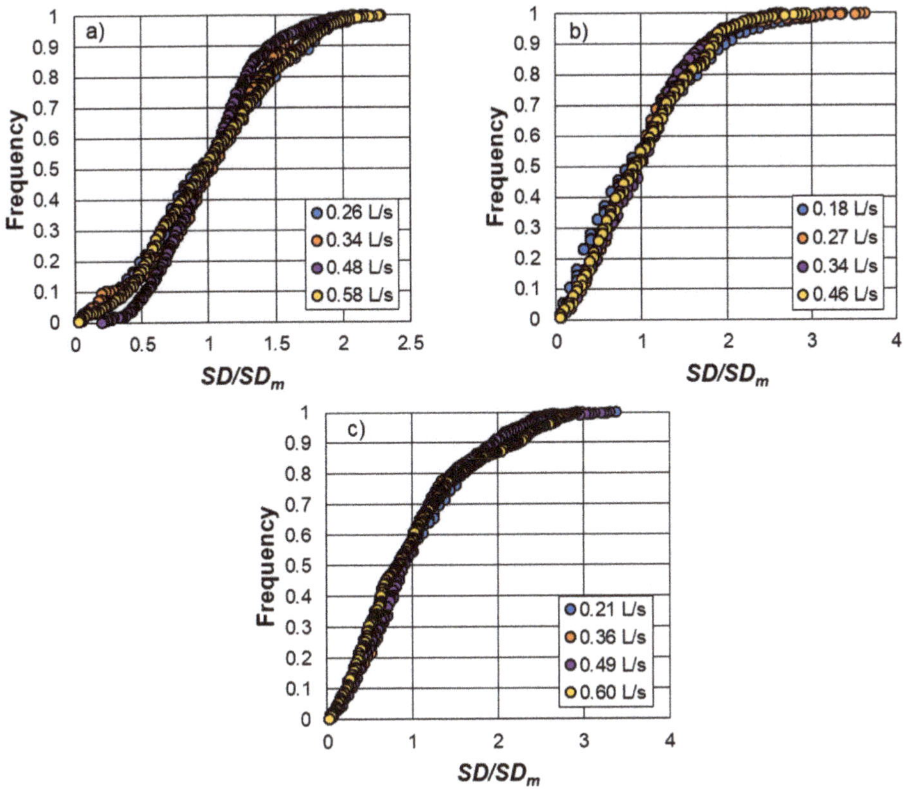

Figure 13. An example of the frequency distribution of the SD/SD_m ratio, for four discharge values, for uniform (**a**), concave (**b**), and convex (**c**) profiles.

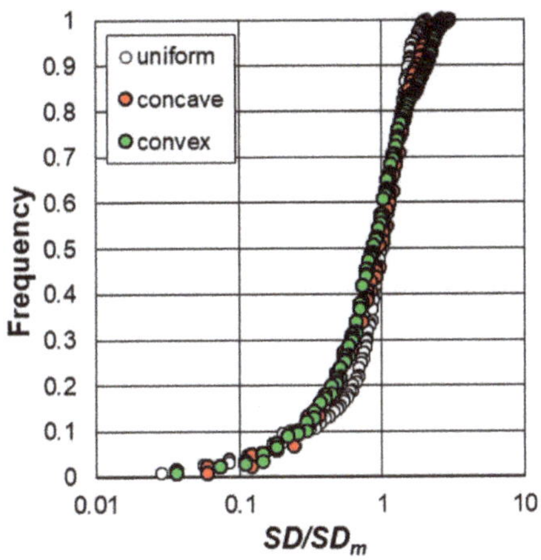

Figure 14. Frequency distribution of the SD/SD_m ratio for a fixed value of discharge ($\approx$0.35 L s^{-1}).

The relationship between SD_m and the discharge was determined for each profile shape (Figure 15). The mean scour depth increased with discharge according to a power relationship depending on the profile.

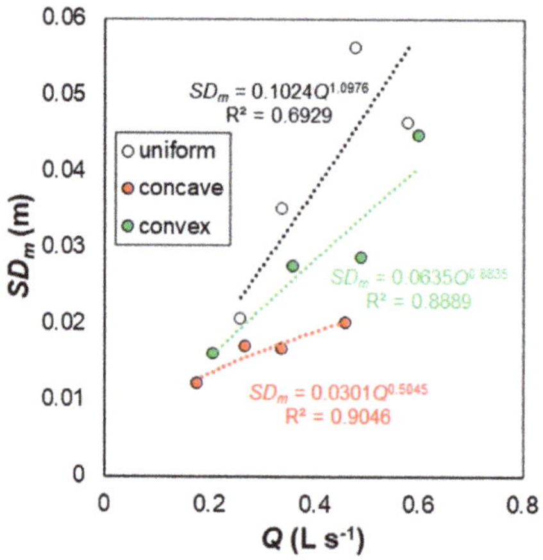

Figure 15. Relationship between mean scour depth SD_m and discharge for each profile shape.

4. Discussion

4.1. Rill Flow Resistance for Different Profile Shapes

Figure 5 shows that the rill flows for the three examined profiles were characterized by comparable Reynolds numbers, while for the convex profile, they were characterized, on average, by Froude numbers lower than those related to the uniform and concave profiles, and less than 1 for 93% of cases. For the latter, the pairs (Re, F) were generally overlapped. The detected trend for Froude number confirmed that the profile shape influences the kinematic flow characteristics (Figure 7a), while the comparable Reynolds number values (Figure 5) suggest that the profile does not affect the flow regime.

Figure 7 confirmed this overlapping trend for the measured flow velocity (Figure 7d) and Froude number (Figure 7a) on uniform and concave profiles. The slope values of the concave profile were less than those of the uniform profile, while the s values of the convex profile were greater than those measured on the uniform profile (Figure 7c). For the convex profile, the occurrence of the highest slope values and the lowest velocity values can be explained by the relevant erosion phenomena. Indeed, the highest RV values detected for the convex profile at the single rill and plot scale, demonstrate that for this profile, the erosion phenomena determine a major increase of grain roughness and sediment transport that lower flow velocity and increase the Darcy–Weisbach friction factor (Figure 7b). The overlapping values of flow velocity and friction factor f for uniform and concave profiles can be explained by the similar distribution of erosion phenomena occurring in the rill channels (Figure 10a,b).

The comparison between Figures 6b and 8b and the error statistics highlight that a single flow resistance equation (Equation (12)) is applicable to all investigated profiles, despite the differences highlighted by the analysis of hydraulic characteristics at reach scale. However, the f estimate is more accurate if specific calibration coefficients (a, b and c) are applied for each profile.

Figure 9 demonstrates that the ratio f_{mor}/f was greater than 10% for 39.1% and 69.6% of the investigated cases for the concave and convex profiles, respectively. Thus, for the examined conditions, the effect of profile shape on flow resistance can be considered relevant.

The convex profile was characterized by the highest values of the ratio f_{mor}/f (3.4–26.9%) (Figure 9), which demonstrates that the contribution of the component of total flow resistance due to the profile was more pronounced than for the concave profile. This result can be explained as higher eroded volumes were detected for the convex profile and f_{mor} accounts for sediment transport and roughness effects on the flow resistance.

4.2. Analysis of Rill Scour Depth and Eroded Volume for Different Profile Shapes

Previous studies [23,58] considered a slope value of 18% as a threshold to discriminate between the gentle (GS) and the steep slope (SS) condition. This distinction is explained as differences in hydraulic (flow depth, velocity, Reynolds number, Froude number) and sediment transport (flow transport capacity, actual sediment load) variables occur. For the SS condition, results by Nicosia et al. [58] supported the conclusion that the sediment transport capacity T_c was so high that it was not a limiting factor for the actual sediment transport, while for the GS case, the actual sediment transport could be limited by both T_c and the soil particle detachability and transportability.

For the convex profile, Figure 10c shows that the scour was concentrated, while it was evenly distributed for uniform (Figure 10a) and concave profiles (Figure 10b). In particular, the noticeable scour of the rills in convex profiles was located downstream of the slope change from a gentle (12%) to a steep (24%) slope condition, because in the upstream half of the rill, the actual sediment transport was limited by T_c, while in the downstream half, it was not. Notwithstanding, the spatial distribution of the scour was similar for uniform and concave profiles, the scour depth values of concave profiles were lower than those of uniform and convex profiles, which instead were similar (Figure 10). Consequently, also the eroded volume of the rill for the concave profile was lower than the eroded volumes for the uniform and convex profiles (Figure 11). In particular, the concave profile resulted in 57.9% less soil loss than the uniform profile. This result confirms the literature findings [2,6–8], which have reported 50 to 80% less soil loss from concave hillslopes than from uniform slopes.

Figure 11 also shows that increasing values of flow discharge determine a higher RV. This result agrees with that obtained by Di Stefano et al. [56] who found that when discharge increased, the contribution of sediment transport to flow resistance and the eroded rill volume also increased.

The results shown in Figures 13 and 14 suggest that SD/SD_m was independent of discharge and profile shape, respectively. In other words, the mean value SD_m accounted for the effects of both discharge and profile shape on scour depth.

Finally, Figure 15 demonstrates that profile shape determined a different increasing relationship between mean scour depth along the rill thalweg and discharge. The mean scour depth for the uniform profile was always higher than that obtained for the convex profile. Instead, the total eroded volume obtained for the uniform profile was slightly lower than that obtained for the convex profile. This result can be explained as the DoD information accounts for the spatial distribution of scour depth, while SD_m represents the mean value of measurements performed along a single linear flow path.

5. Conclusions

In this paper, the effect of the rill profile on flow resistance law was evaluated using measurements obtained from uniform, concave, and convex profiles.

At first, the equation to estimate the velocity profile parameter Γ was calibrated by all the available measurements and an accurate estimate of the Darcy–Weisbach friction factor (Equation (12)) was obtained. Then, to assess the influence of each profile on flow resistance law, the equation to estimate Γ was also calibrated when the data for each profile shape was distinguished. This choice determined an appreciable improvement of the friction factor estimate.

The Darcy–Weisbach friction factor corresponding to the effect of the profile on flow resistance, varied from 0.68 to 14.6% of the overall f factor for the concave profile, and from 3.4 to 26.9% of f for the convex profile.

The developed analysis also showed that for convex profiles, scour was concentrated downstream of the slope change, while for uniform and concave profiles, it was uniformly distributed. Furthermore, the scour depth measured in the corresponding rill thalwegs, had a generally increasing trend with discharge for all investigated profiles.

The total eroded volume for the concave profile was lower than those obtained for uniform and convex profiles and was characterized by being 57.9% less than for the uniform profile.

Finally, the analysis indicated that the mean value of scour depth in the rill thalweg accounted for the effects of both discharge and profile shape, and increases with increased discharge according to a relationship varying with the profile shape.

The main limitations of the study are the use of a single soil type and one mean plot slope. These can be overcome by other experiments using different soils and values of mean plot slope.

The present investigation highlights a markedly less soil loss from concave slopes than from either uniform or convex slopes. Moreover, the proposed approach allows estimating rill flow velocity, which is an essential component of process-based soil erosion models.

Author Contributions: Conceptualization, A.N., C.D.S., V.P. (Vincenzo Palmeri), V.P. (Vincenzo Pampalone) and V.F.; formal analysis, A.N., C.D.S., V.P. (Vincenzo Palmeri), V.P. (Vincenzo Pampalone) and V.F.; investigation, A.N., C.D.S., V.P. (Vincenzo Palmeri), V.P. (Vincenzo Pampalone) and V.F.; methodology, A.N., C.D.S., V.P. (Vincenzo Palmeri), V.P. (Vincenzo Pampalone) and V.F.; project administration, V.F.; software, A.N., C.D.S., V.P. (Vincenzo Palmeri), V.P. (Vincenzo Pampalone) and V.F.; supervision, V.F.; writing—original draft, A.N., C.D.S., V.P. (Vincenzo Palmeri), V.P. (Vincenzo Pampalone) and V.F.; writing—review and editing, A.N., C.D.S., V.P. (Vincenzo Palmeri), V.P. (Vincenzo Pampalone) and V.F. All authors have read and agreed to the published version of the manuscript.

Funding: This research received no external funding.

Data Availability Statement: Data is contained within the article.

Conflicts of Interest: The authors declare no conflict of interest.

References

1. Liu, B.Y.; Nearing, M.A.; Risse, L.M. Slope Gradient Effects on Soil Loss for Steep Slopes. *Trans. ASAE* **1994**, *37*, 1835–1840. [CrossRef]
2. Rieke-Zapp, D.; Nearing, M.A. Slope Shape Effects on Erosion: A Laboratory Study. *Soil Sci. Soc. Am. J.* **2005**, *69*, 1463–1471. [CrossRef]
3. Young, R.A.; Mutchler, C.K. Effect of slope shape on erosion and runoff. *Trans. Am. Soc. Agric. Eng.* **1969**, *12*, 231–233.
4. Young, R.A.; Mutchler, C.K. Soil movement on irregular slopes. *Water Resour. Res.* **1969**, *5*, 1084–1089. [CrossRef]
5. Sensoy, H.; Kara, O. Slope shape effect on runoff and soil erosion under natural rainfall conditions. *iForest* **2014**, *7*, 110–114. [CrossRef]
6. Jeldes, I.A.; Drumm, E.C.; Yoder, D.C. Design of Stable Concave Slopes for Reduced Sediment Delivery. *J. Geotech. Geoenviron. Eng.* **2015**, *141*, 04014093. [CrossRef]
7. Williams, R.D.; Nicks, A.D. Using CREAMS to simulate filter strip effectiveness in erosion control. *J. Soil Water Conserv.* **1988**, *43*, 108–112.
8. Hancock, G.R.; Loch, R.J.; Willgoose, G.R. The design of post-mining landscapes using geomorphic principles. *Earth Surf. Process. Landf.* **2003**, *28*, 1097–1110. [CrossRef]
9. Mombini, A.; Amanian, N.; Talebi, A.; Kiani-Harchegani, M.; Rodrigo-Comino, J. Surface roughness effects on soil loss rate in complex hillslopes under laboratory conditions. *Catena* **2021**, *206*, 105503. [CrossRef]
10. Frankl, A.; Stal, C.; Abraha, A.; Nyssen, J.; Rieke-Zapp, D.; De Wulf, A.; Poesen, J. Detailed recording of gully morphology in 3D through image-based modelling. *Catena* **2015**, *127*, 92–101. [CrossRef]
11. Javernick, L.; Brasington, J.; Caruso, B. Modeling the topography of shallow braided rivers using Structure-from-Motion photogrammetry. *Geomorphology* **2014**, *213*, 166–182. [CrossRef]
12. Seiz, S.M.; Curless, B.; Diebel, J.; Scharstein, D.; Szeliski, R. A comparison an evaluation of multi-view stereo reconstruction algorithms. In Proceedings of the 2006 IEEE Computer Society Conference on Computer Vision and Pattern Recognition (CVPR'06), New York, NY, USA, 17–22 June 2006.
13. Di Stefano, C.; Ferro, V.; Palmeri, V.; Pampalone, V. Measuring rill erosion using structure from motion: A plot experiment. *Catena* **2017**, *156*, 383–392. [CrossRef]
14. Mutchler, C.K.; Young, R.A. Soil detachment by raindrops. In *Present and Prospective Technology for Prediction Sediment Yields and Sources*; USDAARS Publication ARS-S-40: Beltsville, MD, USA, 1975; pp. 113–117.
15. Zhang, P.; Tang, H.; Yao, W.; Zhang, N.; Xizhi, L.V. Experimental investigation of morphological characteristics of rill evolution on loess slope. *Catena* **2016**, *137*, 536–544. [CrossRef]

16. Bagarello, V.; Ferro, V. Plot-scale measurements of soil erosion at the experimental area of Sparacia (southern Italy). *Hydrol. Process.* **2004**, *18*, 141–157. [CrossRef]
17. Bagarello, V.; Ferro, V. Analysis of soil loss data from plots of different length for the Sparacia experimental area, Sicily, Italy. *Byosyst. Eng.* **2010**, *105*, 411–422. [CrossRef]
18. Govers, G.; Giménez, R.; Van Oost, K. Rill erosion: Exploring the relationship between experiments, modeling and field observations. *Earth Sci. Rev.* **2007**, *8*, 87–102. [CrossRef]
19. Bruno, C.; Di Stefano, C.; Ferro, V. Field investigation on rilling in the experimental Sparacia area, South Italy. *Earth Surf. Process. Landf.* **2008**, *33*, 263–279. [CrossRef]
20. Bagarello, V.; Di Stefano, C.; Ferro, V.; Pampalone, V. Establishing a soil loss threshold for limiting rilling. *J. Hydrol. Eng.* **2015**, *20*, C6014001. [CrossRef]
21. Di Stefano, C.; Ferro, V.; Pampalone, V.; Sanzone, F. Field investigation of rill and ephemeral gully erosion in the Sparacia experimental area, South Italy. *Catena* **2013**, *101*, 226–234. [CrossRef]
22. Di Stefano, C.; Ferro, V.; Pampalone, V. Modeling rill erosion at the Sparacia experimental area. *J. Hydrol. Eng.* **2015**, *20*, C5014001. [CrossRef]
23. Peng, W.; Zhang, Z.; Zhang, K. Hydrodynamic characteristics of rill flow on steep slopes. *Hydrol. Process.* **2015**, *29*, 3677–3686. [CrossRef]
24. Iradukunda, P.; Bwambale, E. Reservoir sedimentation and its effect on storage capacity–A case study of Murera reservoir, Kenya. *Cogent Eng.* **2021**, *8*, 1. [CrossRef]
25. Ferro, V. New flow resistance law for steep mountain streams based on velocity profile. *J. Irrig. Drain. Eng.* **2017**, *143*, 1–6. [CrossRef]
26. Ferro, V. Assessing flow resistance in gravel bed channels by dimensional analysis and self-similarity. *Catena* **2018**, *169*, 119–127. [CrossRef]
27. Di Stefano, C.; Ferro, V.; Palmeri, V.; Pampalone, V. Flow resistance equation for rills. *Hydrol. Process.* **2017**, *31*, 2793–2801. [CrossRef]
28. Di Stefano, C.; Ferro, V.; Palmeri, V.; Pampalone, V. Testing slope effect on flow resistance equation for mobile bed rills. *Hydrol. Process.* **2018**, *32*, 664–671. [CrossRef]
29. Di Stefano, C.; Nicosia, A.; Palmeri, V.; Pampalone, V.; Ferro, V. Comparing flow resistance law for fixed and mobile bed rills. *Hydrol. Process.* **2019**, *33*, 3330–3348. [CrossRef]
30. Di Stefano, C.; Nicosia, A.; Palmeri, V.; Pampalone, V.; Ferro, V. Estimating flow resistance in steep slope rills. *Hydrol. Process.* **2021**, *35*, e14296. [CrossRef]
31. Di Stefano, C.; Nicosia, A.; Palmeri, V.; Pampalone, V.; Ferro, V. Rill flow resistance law under equilibrium bed-load transport conditions. *Hydrol. Process.* **2019**, *33*, 1317–1323. [CrossRef]
32. Ferro, V.; Nicosia, A. Comment on "Rill erosion processes on steep colluvial deposit slope under heavy rainfall in flume experiments with artificial rain by F. Jiang et al.". *Catena* **2020**, *185*, 103–793. [CrossRef]
33. Nicosia, A.; Di Stefano, C.; Pampalone, V.; Palmeri, V.; Ferro, V.; Nearing, M.A. Testing a new rill flow resistance approach using the water erosion prediction project experimental database. *Hydrol. Process.* **2019**, *33*, 616–626. [CrossRef]
34. Palmeri, V.; Pampalone, V.; Di Stefano, C.; Nicosia, A.; Ferro, V. Experiments for testing soil texture effects on flow resistance in mobile bed rills. *Catena* **2018**, *171*, 176–184. [CrossRef]
35. Carollo, F.G.; Di Stefano, C.; Nicosia, A.; Palmeri, V.; Pampalone, V.; Ferro, V. Flow resistance in mobile bed rills shaped in soils with different texture. *Eur. J. Soil Sci.* **2021**, *72*, 2062–2075. [CrossRef]
36. Di Stefano, C.; Nicosia, A.; Pampalone, V.; Palmeri, V.; Ferro, V. New technique for measuring water depth in rill channels. *Catena* **2019**, *181*, 104090. [CrossRef]
37. Line, D.E.; Meyer, L.D. Flow velocities of concentrated runoff along cropland furrows. *Trans. ASAE* **1988**, *31*, 1435–1439.
38. Govers, G. Relationship between discharge, velocity and flow area for rills eroding loose, non-layered materials. *Earth Surf. Process. Landf.* **1992**, *17*, 515–528. [CrossRef]
39. Abrahams, A.D.; Li, G.; Parsons, A.J. Rill hydraulics on a semiarid hillslope, southern Arizona. *Earth Surf. Process. Landf.* **1996**, *21*, 35–47. [CrossRef]
40. Di Stefano, C.; Nicosia, A.; Palmeri, V.; Pampalone, V.; Ferro, V. Dye-tracer technique for rill flows by velocity profile measurements. *Catena* **2020**, *185*, 104313. [CrossRef]
41. Luk, S.H.; Merz, W. Use of the salt tracing technique to determine the velocity of overland flow. *Soil Technol.* **1992**, *5*, 289–301.
42. Li, G.; Abrahams, A.D. Effect of salting sediment load on the determination of the mean velocity of overland flow. *Water Resour. Res.* **1997**, *33*, 341–347. [CrossRef]
43. Zhang, G.; Luo, R.; Cao, Y.; Shen, R.; Zhang, X.C. Correction factor to dye-measured flow velocity under varying water and sediment discharges. *J. Hydrol.* **2010**, *389*, 205–213. [CrossRef]
44. Barenblatt, G.I. *Dimensional Analysis*; Gordon &Breach, Science Publishers Inc.: Amsterdam, The Netherlands, 1987.
45. Barenblatt, G.I. Scaling laws for fully developed turbulent shear flows, part 1, Basic hypothesis and analysis. *J. Fluid Mech.* **1993**, *248*, 513–520. [CrossRef]
46. Ferro, V. Applying hypothesis of self-similarity for flow-resistance law of small-diameter plastic pipes. *J. Irrig. Drain. Eng.* **1997**, *123*, 175–179. [CrossRef]

47. Barenblatt, G.I.; Monin, A.S. Similarity laws for turbulent stratified flows. *Arch. Ration. Mech. Anal.* **1979**, *70*, 307–317. [CrossRef]
48. Barenblatt, G.I.; Prostokishin, V.M. Scaling laws for fully developed turbulent shear flows, part 2. Processing of experimental data. *J. Fluid Mech.* **1993**, *248*, 521–529. [CrossRef]
49. Butera, L.; Ridolfi, L.; Sordo, S. On the hypothesis of self-similarity for the velocity distribution in turbulent flows. *Excerpta* **1993**, *8*, 63–94.
50. Ferro, V.; Pecoraro, R. Incomplete self-similarity and flow velocity in gravel bed channels. *Water Resour. Res.* **2000**, *36*, 2761–2770. [CrossRef]
51. Castaing, B.; Gagne, Y.; Hopfinger, E.J. Velocity probability density functions of high Reynolds number turbulence. *Phys. D* **1990**, *46*, 177–200. [CrossRef]
52. Barenblatt, G.I. On the scaling laws (incomplete self-similarity with respect to Reynolds numbers) for the developed turbulent flows in tubes. *C. R. Acad. Sci. Ser.* **1991**, *313*, 307–312.
53. Ferro, V.; Porto, P. Applying hypothesis of self-similarity for flow resistance law in Calabrian gravel bed rivers (Fiumare). *J. Hydraul. Eng.* **2018**, *144*, 1–11. [CrossRef]
54. Ferro, V. Assessing flow resistance law in vegetated channels by dimensional analysis and self-similarity. *Flow Meas. Instrum.* **2019**, *69*, 101610. [CrossRef]
55. Yalin, M.S. *Mechanics of Sediment Transport*, 2nd ed.; Pergamon Press: Oxford, UK, 1977.
56. Di Stefano, C.; Nicosia, A.; Palmeri, V.; Pampalone, V.; Ferro, V. Rill flow resistance law under sediment transport. *J. Soils Sediments* **2022**, *22*, 334–347. [CrossRef]
57. Nicosia, A.; Bischetti, G.B.; Chiaradia, E.; Gandolfi, C.; Ferro, V. A Full-Scale study of Darcy-Weisbach friction factor for channels vegetated by riparian species. *Hydrol. Process.* **2021**, *35*, e14009. [CrossRef]
58. Nicosia, A.; Palmeri, V.; Pampalone, V.; Di Stefano, C.; Ferro, V. Slope threshold in rill flow resistance. *Catena* **2022**, *208*, 105789. [CrossRef]

water

MDPI

Article

An Enhanced Flume Testing Procedure for the Study of Rill Erosion

Vinícius Naves de Oliveira [1], Gilson de F. N. Gitirana, Jr. [1,*], Marcia Maria dos Anjos Mascarenha [1], Mauricio Martines Sales [1], Luiz Felipe Ramos Varrone [1] and Marta Pereira da Luz [2,3]

1 School of Civil and Environmental Engineering, Universidade Federal de Goiás, Av. Universitária, Setor Universitário, Goiânia CEP 74605-220, GO, Brazil; viniciusnaves@discente.ufg.br (V.N.d.O.); marciamascarenha@ufg.br (M.M.d.A.M.); mmartines@ufg.br (M.M.S.); luizfrvarrone@discente.ufg.br (L.F.R.V.)

2 Dam Safety and Technology Department, Furnas Centrais Elétricas S.A., Eletrobras Furnas, BR153, Km 510, Zona Rural, Aparecida de Goiânia, Goiânia CEP 74923-650, GO, Brazil; martaluz@furnas.com.br

3 Industrial and Systems Engineering Postgraduate Program (MEPROS), Pontifical Catholic University of Goiás (PUC Goiás), Av. Universitária 1.440, Setor Universitário, Goiânia CEP 74605-010, GO, Brazil

* Correspondence: gilsongitirana@ufg.br

Abstract: This paper presents the development and verification of an improved and cost-effective flume apparatus and corresponding testing methodology. A rigorous analysis of the flow conditions during testing was considered and an interpretation of test results was carried out following the premises of the Water Erosion Prediction Project (WEPP) rill erosion model. The apparatus and methodology were verified using statically compacted specimens of a latosol from the central region of Brazil. Tests were performed on samples with void ratios of 1.0 and 1.5 and under variable hydraulic conditions to verify the repeatability and ideal analysis time for the soil loss curves. The soil loss curves presented hyperbolic behavior, with a maximum value that appeared to be randomly behaved, which can be attributed to the complex nature of the erosion processes at later stages. The equipment and testing methodology produced erosion curves with repeatability that were superior with respect to their initial linear and transition portions. Recommendations are made regarding the adequate interpretation of the testing data and the selection of the ideal elapsed time for soil loss analysis.

Keywords: soil erosion; water erosion; flume test; experimental tests; WEPP model

check for updates

Citation: de Oliveira, V.N.; Gitirana, G.d.F.N., Jr.; dos Anjos Mascarenha, M.M.; Sales, M.M.; Varrone, L.F.R.; da Luz, M.P. An Enhanced Flume Testing Procedure for the Study of Rill Erosion. *Water* **2021**, *13*, 2956. https://doi.org/10.3390/w13212956

Academic Editors: Vito Ferro and Alessio Nicosia

Received: 25 August 2021
Accepted: 7 October 2021
Published: 20 October 2021

1. Introduction

Water erosion of soils and its quantification through rational models poses a significant challenge due to the complexity of the processes involved, which include numerous soil properties and hydraulic conditions [1–4]. The Water Erosion Process Project (WEPP) model, proposed by Foster [5], is considered to be one of the main basic frameworks that is available for the quantification of water erosion [6–10]. The WEPP model evaluates water erosion by incorporating physical concepts associated with the erosion mechanisms and by dividing the erosion process into (a) interstitial erosion caused mainly by splashing and by water flow in areas adjacent to streams and (b) rill erosion resulting from the effects of hydraulic energy of the flow concentration on the stream bed.

According to the WEPP modeling approach, the evaluation of rill erosion parameters should be based on flume tests performed under controlled hydraulic conditions. The interpretation of such tests assumes that erosion takes place due to the hydraulic shear stress imposed by the flow in the flume bed, as shown in Figure 1. Based on these flow conditions, the hydraulic shear stress τ_h (Pa) in the flume bed can be given by

$$\tau_h = \gamma_w h \sin \alpha \quad (1)$$

where γ_w is the unit weight of the water (kN m^{-3}); h is the piezometric head (m); and α is the slope of the channel.

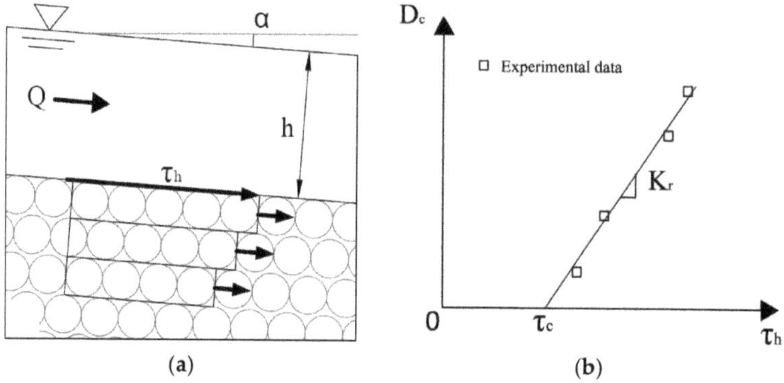

Figure 1. Erosion formation: (**a**) hydraulic conditions; (**b**) erosion function.

The general concepts shown in Figure 1 were first presented by Du Boys [11], who stated that particle detachment occurs when the hydraulic shear stress is higher than the shear strength generated by frictional forces due to the submerged unit weight of the soil. This shear strength limit is known as the critical shear stress, τ_c. The WEPP approach is in part based on the concepts introduced by Du Boys [11] and indicates that when is greater than τ_c, a progressive detachment of soil layers takes place along the stream bed (Figure 1a). The rate of mass loss, D_c (g cm^{-2} min^{-1}), is generally assumed to be proportional to the difference between the hydraulic shear stress acting on the flume bed and the critical hydraulic shear stress, which can be represented as follows:

$$D_c = K_r(\tau_h - \tau_c) \quad (2)$$

where K_r is the soil erosion coefficient (g cm^{-2} min^{-1} Pa^{-1}).

The proportionality between D_c and $\tau_h - \tau_c$ has been demonstrated by numerous authors, such as Partheniades [12], Foster [5], Nearing et al. [13], Zhang et al. [14], Bastos [15], Dey [8] and Kimiaghalam; Clark; and Ahmari [16]. There is also significant evidence in the literature showing that the relationship between the rate of mass loss and the shear stress may also be nonlinear for certain ranges of hydraulic shear stress [5,17–22]. However, according to Lane, Foster, and Nicks [23], assuming a linear relationship may be often acceptable for the approximate quantification of soil loss.

Lim and Khalili [24] state that the most used laboratory test for studying soil erodibility is the hydraulic flume. The soil erosion parameters τ_c and K_r may be determined based on flume tests where the value of τ_h is varied under strictly controlled conditions and where the corresponding values of D_c are measured. Such tests depend on the establishment of rigorous experimental protocols regarding sampling or remolding procedures and the control of the initial water content [25,26], the establishment of an adequate hydraulic regime [27], the monitoring of soil conditions and mass loss [28], and the interpretation methods of results, including the selection of adequate testing periods [29]. Several studies have presented variations in flume apparatus design and testing protocols, such as Lyle and Smerdon [25], Ghebreiyessus et al. [30], Kamphuis et al. [31], Zhu et al. [32], Cantalice et al. [33], Shepard et al. [34], and Ni et al. [35].

Some studies [15,33,36–43] have adopted flume apparatuses with relatively simpler construction, which leads to more cost-effective alternatives. The following common design characteristics are generally observed in these apparatuses: (a) the use of an open channel; (b) the piezometric head and corresponding hydraulic shear stress are varied by changing the flume tilt angle; and (c) the specimen is placed flush with the flume bed, and the specimen is not raised during testing. In this case, the soil loss may be measured by obtaining the dry weigh of the specimen or by collecting the water and sediment that flow over the specimen using a set of sieves.

Unfortunately, the aforementioned cost-effective design approach is often adopted with little regard for the determination of ideal flow regimes and frequently leads to erosion curves that are difficult to interpret. These flume designs are not standardized, and there are several methodologies that can be used to execute and interpret test results. The following aspects vary significantly between the different flume test methodologies that were previously presented: (a) geometry of the flume; (b) geometry of the specimen surface exposed to water flow; (c) position of the specimen with respect to the flume; (d) procedure for measuring soil loss; and (e) procedure for interpreting soil loss curves and obtaining erosion parameters. It is apparent that further standardization is required in order to promote testing methodologies that produce data that are consistent and comparable. In this context, this paper presents a new flume test apparatus and data interpretation procedure. Great focus is given to the establishment of highly controlled hydraulic conditions, the verification of testing repeatability, and to the establishment of sound procedures for the interpretation of test results.

2. Materials and Methods

2.1. General Apparatus Characteristics and Hydraulic Design

The general arrangement of the developed apparatus is presented in Figure 2. It consists of (a) a reservoir made of polymethyl methacrylate, which can be used to regulate the water flow; (b) a flume made of the same material; (c) systems for regulating the flow and leveling the flume; and (d) a sample port. The system has a ramp length of 2.0 m, a base width of 0.18 m, and sidewalls that are 0.20 m in height. These dimensions were defined according to a flow analysis that will be described later on. The specimen enclosure is cubic and has edges of 18.0 cm in length. Ports were included in the specimen enclosure and were placed along uniformly spaced positions in the direction orthogonal to the specimen surface, allowing the instrumentation of the specimen using tensiometers and/or water content probes. The flume was designed to operate with tilt angles varying between 5 and 60 degrees. The design flow rates vary between 1.0×10^{-4} m^3 s^{-1} and 2.4×10^{-3} m^3 s^{-1} when the flume is set with a tilt angle of 5 degrees and between 4.0×10^{-4} m^3 s^{-1} and 9.0×10^{-3} m^3 s^{-1} in the maximum slope of 60 degrees.

As a design requirement, it was established that the apparatus should ensure a uniform flow condition in a supercritical regime over the specimen surface. A uniform flow corresponds to a constant flow velocity at any instant and at every point along the flume, including the specimen area. This design condition has been generally overlooked in previous open channel flume designs. A supercritical regime would allow the reproduction of uniform and controlled hydraulic shear stresses along the specimen surface [44]. In addition, flow disturbance that may occur in the downstream portions of the specimen as the erosion process advances would not propagate to upstream regions, keeping the flow characteristics relatively uniform.

In previous studies, it was assumed that the critical flow depth was equal to the normal depth along the flume, corresponding to a constant uniform flow under a critical regime. However, feeding the canal in a typical flume causes a phenomenon known as hydraulic backwater, resulting in a non-uniform flow regime at the entrance point of the flume. According to Munson et al. [44], this flow regime tends to evolve rapidly towards a permanent and non-uniform flow in a critical regime, in which the normal flow height is affected, compromising the strict control of hydraulic shear stresses.

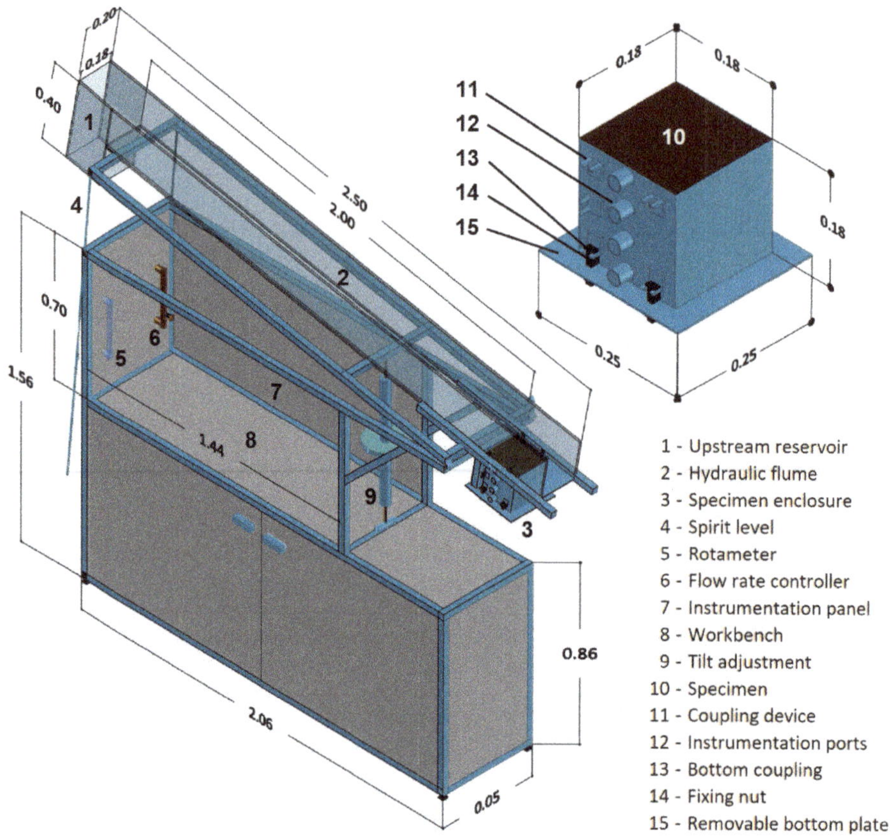

Figure 2. Flume apparatus; all dimensions are in meters.

The normal flow height value in the uniform flow region can be obtained using explicit solutions for trapezoidal channels, as presented in Vatankhah [45]. Taking the difference between the discharge in the uniform flow region and the feed discharge (Q), which depends on the flow height itself, the following equation is obtained:

$$f(y_0) = \frac{1}{n}\frac{(by_0)^{5/3}(\sin\alpha)^{1/2}}{(b+2y_0)^{2/3}} - Q \tag{3}$$

where $f(y_0)$ is the tolerance for discharge variation; n is Manning's roughness coefficient (assumed to be 0.085 for polymethyl methacrylate); b is the flume width (m); y_0 is the flow depth, taken normal to the flume bed (m); α is the tilt slope of the flume; and Q is the system discharge ($m^3\ s^{-1}$).

Equation (3) is a non-linear and indeterminate equation. The value of y_0 can be obtained iteratively, using Newton's method, as follows:

$$y_{0_{(n+1)}} = y_{0_{(n)}} - \frac{f(y_{0_{(n)}})}{f'(y_{0_{(n)}})} \tag{4}$$

where $y_{0(i)}$ is the ith value of y_0, which is evaluated during the iterative procedure, and $f'\left(y_{0(i)}\right)$ is the first-order derivative of $f(y_0)$. The derivative estimation can be completed using the complex step approach method, described in Martins et al. [46]:

$$f'(y_0) = \frac{\mathrm{Im}[f(y_0 + ip)]}{p} \quad (5)$$

where Im is the image of the complex function, and p is the finite difference interval.

To guarantee uniform flow in a supercritical regime along the entire flume, including its entry and exit sections, the flow rate and the slope of the flume must be constrained by certain limits so that the hydraulic backwater dissipates before the specimen insertion region. Therefore, for each selected discharge, the distance between the starting point of the flow and the point where the normal height reaches the value of y_0 was calculated. It was thus possible to confirm the occurrence of the specified flow regime. Such evaluation was conducted for each selected discharge using the direct step numerical integration method as presented by Munson et al. [44].

2.2. Soil Characteristics and Specimen Preparation for Flume Tests

The soil selected for this study has characteristics that are typical of soils from the state of Goiás, Brazil. Disturbed samples were collected in the city of Goiânia at the geographical coordinates 16.7289° S and 49.2958° W. Samples were collected between 50 cm and 150 cm below the ground surface, reaching the B horizon. The soil has a clay content above 20%. The soil presents liquid and plastic limits of 33% and 22%, respectively, and a pH of 5.6, which is typical for lateritic soils from the region. The in situ void ratio varies from 1.0 to 1.8, and the saturated hydraulic conductivity ranges from 1×10^{-6} m s^{-1} to 2×10^{-5} m s^{-1}. Finally, the soil has a cation exchange capacity of 2.29 cmol dm^{-3} and an amount of organic matter of 9 g dm^{-3}.

Laboratory specimens were reconstituted under highly controlled conditions. A quasistatic compaction procedure was adopted, using the Emic DL2000 automated loading frame manufactured by Instron (Norwood, MA, USA). The specimen enclosure shown in Figure 2 was used as a mold. Specimen compaction was achieved using four soil layers of 50 mm in height each, each one compressed at a constant rate of 0.017 mm/s. A 50 mm extension collar was used to hold the loose soil before compression. After compaction, the collar was gently removed, exposing 20 mm of excess compacted material. Finally, the excess material was scraped using a beveled ruler so that the specimen would reach the final 180 mm in height. This scraping procedure ensured that the specimen surface roughness was not influenced by the compaction piston, which often produces undesired particle orientation. Specimens were sealed and stored to reach hydraulic equilibrium for at least two days before testing.

Specimens were reconstituted using air-dried samples that were later wetted to a gravimetric water content of 18%. The selected remolding void ratios were equal to 1.0 and 1.5. The mean value and coefficient of variation (CV) obtained for the gravimetric water content, considering all reconstituted specimens, was 17.8% and 0.4%, respectively. The mean value obtained for the void ratio of remolded specimens was 1.02 and 1.51. The corresponding values for CV of the void ratios were 0.6% and 0.4%, respectively. The obtained mean and CV values indicate that the compaction procedure was adequate, producing specimens that correspond closely to the specified conditions. It is important to note that the remolding process does not necessarily produce specimens with the same soil structure as the one found in natural undisturbed soil. However, flume test repeatability benefits greatly from the use of the relatively homogeneous specimens obtained from the compaction procedure.

2.3. Flume Testing Procedure

Before the execution of each test, the specimens were unsealed and weighed to confirm the remolding water content after the equilibrium period. The placement of the specimens in the flume was achieved to ensure the strict alignment between the specimen surface and the hydraulic flume. The relatively narrow gap (smaller than 1 mm) located along the edges of the specimen surface was sealed using silicone glue, removing excess glue, and leveling it. Next, the specimen surface was covered with a polyethylene plastic film. Water was allowed to flow over the covered specimen until the desired flow rate and flow regime uniformity was observed. The start of the test proper was established by carefully removing the protection film.

The eroded material was collected using sieves covered with non-woven polyester geotextile with a mass of 140 g m^{-2} and an apparent opening equal to 0.094 mm, which was calculated based on the retention criteria presented by Christopher and Holtz [47]. The geotextile was oven-dried at approximately 104 °C for at least 24 h and were weighed on a precision scale before use in the test.

To investigate the evolution of erosion over time, from its initial seconds to prolonged periods of flow, the eroded material was collected after 0.25, 0.5, 1, 2, 3, 4, 5, 10, 15, 30, 45, and 60 min. The eroded mass between each time step was determined by placing a new sieve-geotextile set on top of the existing one and by removing the latter. Each sediment collection geotextile was placed in an oven that was set at approximately 105 °C for at least 24 h, and the geotextiles were then weighed to obtain the eroded dry soil mass.

2.4. Flume Testing Program

The experimental program presented herein involved a total of 31 specimens with hydraulic conditions that were selected to allow the determination of the erosion parameters established by the WEPP model framework. The test conditions were combined to produce the following sets:

- Out of the 31 specimens, a total of 16 specimens were compacted at a void ratio of 1.5. The remaining 15 specimens were compacted with a void ratio of 1.0;
- Out of the 31 specimens, 12 were used to assess the repeatability of the tests performed with the new apparatus. This was accomplished by testing six pairs of specimens under identical flow conditions;
- Five different flow conditions were used to analyze the influence of the hydraulic shear stress on the erosion curves.

Hydraulic stresses of 0.98 Pa, 1.26 Pa, and 1.93 Pa were applied to specimens with void ratios of 1.0. Hydraulic stresses of 0.82 Pa, 0.98 Pa, and 1.26 Pa were selected for specimens with void ratios of 1.5. To obtain the aforementioned hydraulic stresses, a 10-degree tilt was adopted and was combined with flow rates of 1.5×10^{-3} m^3 s^{-1}, 2.0×10^{-3} m^3 s^{-1}, 2.5×10^{-3} m^3 s^{-1}, 3.0×10^{-3} m^3 s^{-1}, and 6.0×10^{-3} m^3 s^{-1}. These conditions were based on their recurrence in previous studies.

3. Results and Discussion

As shown in Figure 1b, flume tests executed under multiple hydraulic shear stresses, τ_h, are required to define pairs of values of τ_h and D_c, thereby allowing the determination of the soil erosion properties, τ_h and K_r. Unfortunately, the erosion rate, D_c, may vary during testing. This is particularly true when fixed specimens are employed because the erosion curves under these testing conditions often present a hyperbolic behavior. As a result, several authors recommend that D_c be defined using a secant to the soil loss curve [15]. The secant value of D_c is obviously dependent on the time interval that is adopted. However, there is a lack of consensus in the literature regarding the elapsed time to be used for the computation of a secant erosion rate. Such a relatively imprecise definition of D_c prompted an investigation of the effect of the time interval.

Figure 3 presents the results of 12 flume tests. Accumulated soil loss (E) curves are shown over time for three pairs of hydraulic shear stress values and for two void ratio values. The erosion curves obtained from the tests performed on the specimens with void ratio of 1.5 (Figure 3b) and under hydraulic shear stresses of 0.82 Pa and 1.26 Pa indicated superior repeatability during the beginning of the erosion process, particularly within the first five minutes of testing. On the other hand, curves for the same hydraulic stress diverge for times longer than five minutes. Erosion curves obtained for specimens prepared with a void ratio of 1.0 presented significantly higher variability, which indicates some degree of randomness, even during the first five minutes. This fact can be explained by the low erodibility of the specimen surface with eroded material that is likely to only corresponded to the loosened soil particles present on the specimen surface before the test started.

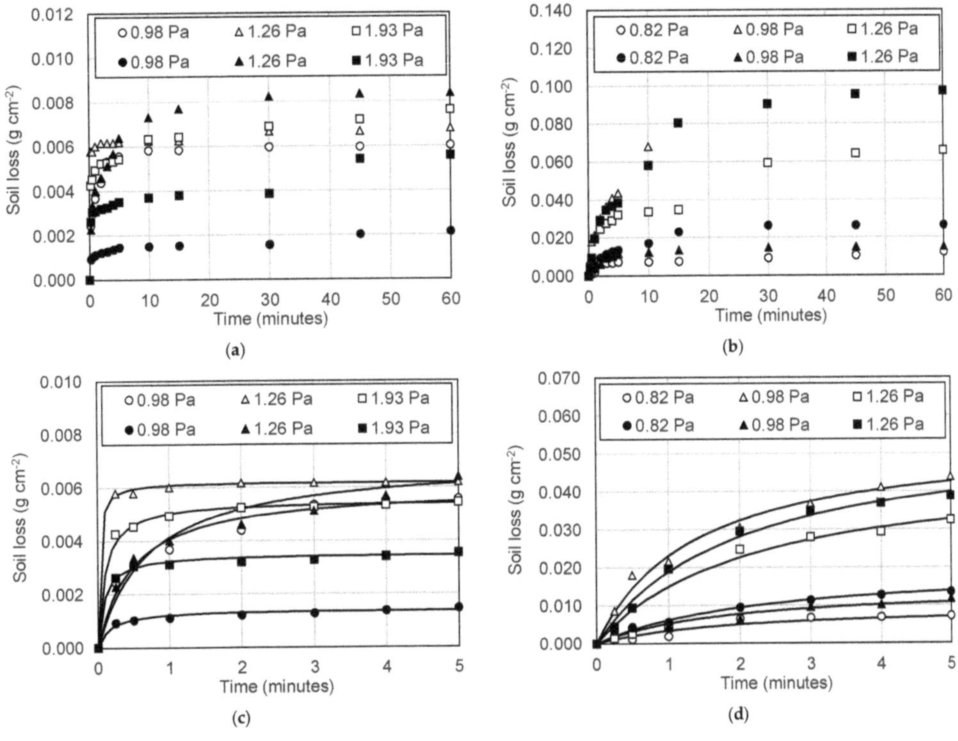

Figure 3. Flume test results for void ratios of: (**a**) 1.0; (**b**) 1.5; (**c**) 1.0 modelled curves; and (**d**) 1.5 modelled curves. Filled and empty markers of the same shape indicate test pairs that were conducted under identical conditions.

The data presented in Figure 3 were modelled using a nonlinear fitting procedure aimed at defining the initial rate of soil loss and the maximum accumulate erosion value in a rigorous unbiased manner. The soil loss accumulated over time has a behavior that may be approximated by a hyperbolic function, as is apparent from Figure 3. The hyperbolic behavior may be mathematically represented as follows:

$$E = \frac{t}{\frac{1}{E_{ult}}t + \frac{1}{a_0}} \quad (6)$$

where E is the soil loss by unit area accumulated at any given elapsed time (g cm^{-2}); t is the elapsed time (min); E_{ult} is the maximum asymptotic value of E, which is somewhat similar to the main erosion parameter defined in [15,42,48]; and a_0 is a fitting parameter (g cm^{-2} min^{-1}).

The fitting parameter a_0 is the rate of erosion taken at the origin of the erosion curve and may be considered as the erosion rate defined in the WEPP model [5], D_c:

$$\left.\frac{dE}{dt}\right|_{\Delta t \to 0} = a_0 = D_c \tag{7}$$

Nonlinear regression was conducted for the data presented in Figure 3 and only considered the first five minutes of testing. The analysis time was limited to the first five minutes because of the random nature of the erosion curves at later stages. Table 1 presents the results of the regression analysis. The continuous curves shown in Figure 3 present the corresponding best-fit curves. The results shown in Table 1 indicate RMSE values that range between 0.66 × 10^{-4} and 17.6 × 10^{-4} and R^2 values that range between 0.912 and 0.999. The quality of the fitting results was considered adequate for the purposes of this analysis.

Table 1. Results of nonlinear regression modelling using the hyperbolic equation.

Void Ratio	Test	τ_h (Pa)	D_c(×10^{-2}) (g cm^{-2} mim^{-1})	E_{ult} (×10^{-2}) (g cm^{-2})	RMSE (×10^{-4})	R^2
1.000	1	0.98	0.81	0.14	0.73	0.931
	2	0.98	1.25	0.60	2.52	0.950
	3	1.26	1.00	0.70	4.46	0.940
	4	1.26	2.85	0.62	0.66	0.937
	5	1.93	4.03	0.35	1.14	0.972
	6	1.93	5.55	0.55	1.20	0.984
1.500	7	0.82	0.89	2.01	6.27	0.993
	8	0.82	0.52	0.97	8.78	0.912
	9	0.98	0.83	1.48	9.24	0.942
	10	0.98	4.00	5.43	15.9	0.999
	11	1.26	2.85	5.54	12.5	0.998
	12	1.26	2.20	4.65	17.6	0.997

The values of D_c shown in Table 1 present a clear increasing trend for larger values of hydraulic shear stress, τ_h. The relationship between τ_h and E_{ult}, however, did not indicate a clear positive correlation for tests 1–6. Some variability is observed in the repeatability analyses comparing the values of tests under the same shear stress. The relative difference of D_c between identical tests ranged between 30% and 380%. The relative difference of E_{ult} between identical tests range between 13% and 326%. The higher relative differences can be partially attributed to the fact that the mass loss of specimens compacted with void ratios of 1.0 was relatively low and only corresponded to loosened soil particles. Unfortunately, previous studies using open channels and fixed specimens have not presented similar systematic evaluations of repeatability that could be compared to those presented herein.

Figure 4a,b show the variation of τ_c and K_r as a function of the adopted analysis time interval for the tests conducted with a void ratio of 1.5. There is some variation in the values of τ_c and K_r according to the adopted time interval, and the values of τ_c reach a constant value after 5 min. Thus, while simultaneously meeting the requirements of repeatability and stabilization of erosion parameters, the analysis time of 5 min is recommended for the soil studied here. The typical behavior of the mass loss curves in the flume trials [41,48] indicate that the analysis time selection criterion proposed here has the potential to be applied to other soils.

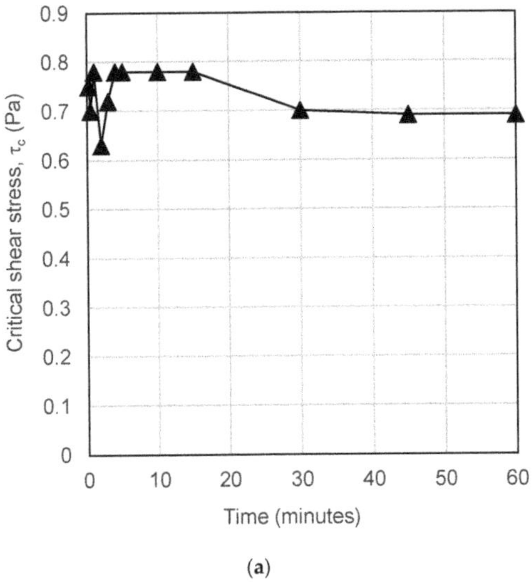

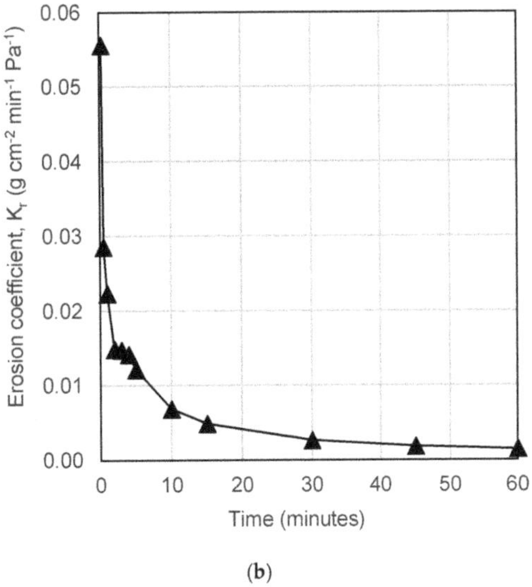

Figure 4. Evaluation of the time interval effect adopted in the definition of the parameters of the WEPP methodology (w_0 = 18%, e = 1.5): (**a**) τ_c; (**b**) K_r.

The criterion for establishing the ideal analysis time seems to be related to the transition between an initial test stage in which erosion occurs uniformly along the specimen surface and a later stage in which cavities start to be randomly formed on the soil surface, as shown in Figure 5. The random character of the formation of the cavities and the loss of uniformity of the hydraulic conditions on the surface of the specimens indicate that the time interval representative of the flume test must be delimited for each soil based on the examination of the accumulated mass loss curve over time. Such an ideal time interval would correspond to the initial erosion stage, which can be identified as an approximately linear relationship between accumulated erosion and time. The curve transition and the maximum mass loss tend to have a random character that depends on the manner in which the specimen surface develops grooves and cavities.

The lateritic soil studied herein presented a τ_c from 0.69 to 0.78 Pa and a K_r from 0.0014 to 0.012 g cm^{-2} min^{-1} Pa^{-1}. The values of τ_c and K_r for the lateritic soils vary greatly and depend on the soil texture, soil weathering, and stability of particle aggregations. Some of the values of τ_c reported in the literature for lateritic soils range between 0.01 and 1.33 Pa [15,41,49]. The range of K_r for the lateritic soils reported by the same authors varies between 0.01 and 36.5 g cm^{-2} min^{-1} Pa^{-1} [15,41,49]. Therefore, the results presented in Figure 4 fall within these typical ranges but also indicate that the soil studied herein has a relatively low erodibility when considering the values for other lateritic soils.

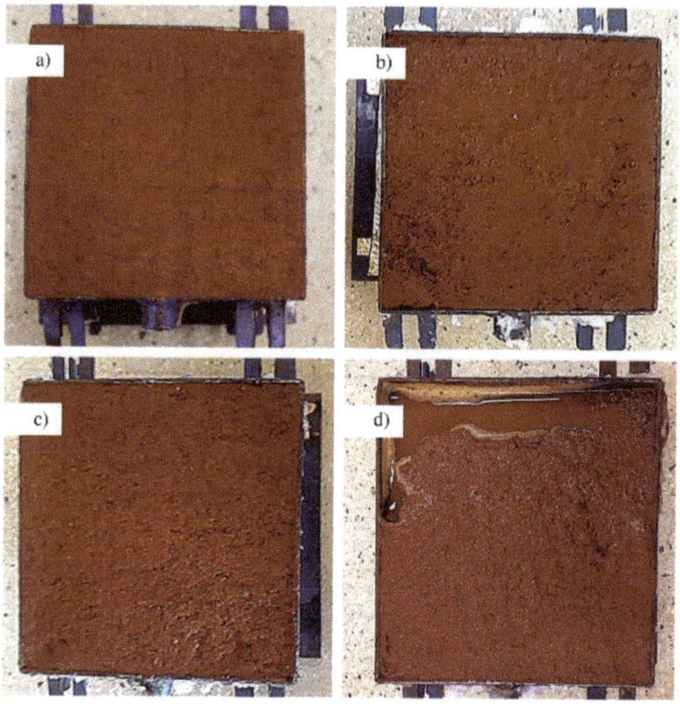

Figure 5. Specimen surface condition after flume tests for various void ratios and hydraulic shear stresses: (**a**) test #5, at e = 1.0 and τ_h = 1.93 Pa; (**b**) test #6, at e = 1.0 and τ_h = 1.93 Pa; (**c**) test #7, at e = 1.5 and τ_h = 0.82 Pa; and (**d**) test #8, at e = 1.5 and τ_h = 0.82 Pa.

4. Final Remarks

The apparatus presented herein met the established requirements, allowing the control of imposed hydraulic conditions, which were defined by the flume's slope and flow rate. The effluent collection method proved to be adequate, allowing the evaluation of the erodibility and of the conditions of the specimen during the test. The use of uniform flow in a supercritical regime allowed the strict control of the water flow height and of the hydraulic shear stresses that were imposed. In this sense, the cost-effective design characteristics of the apparatus (i.e., the use of an open channel and of a fixed specimen) proved to be a viable alternative for studying the erosion characteristics of soils subjected to water flow.

It was possible to perceive a relatively low test repeatability when comparing the results for identical conditions. However, this variability is expected given the complex nature of water erosion processes in soils. Further studies are recommended regarding ideal testing methodologies that would ensure improvements in test result repeatability. Future comparisons of the test results obtained using fixed and moving specimens (i.e., extruded at a speed defined by the erosion rate) would be of great value.

The results from the experimental program presented herein indicate that the slope of the initial stage of the mass loss curve give more consistent results when compared against the final maximum mass loss. Thus, it is proposed that only the initial section of the mass loss curves be considered during the analysis of erosion tests. The interpretation of the mass loss curves must also consider the observation of the physical characteristics of the specimen surface since the randomness of the maximum mass loss value can be attributed to the cavities that form after the linear rate erosion phase. Finally, the definition of the erodibility parameters of the WEPP methodology must consider time intervals that are within the initial erosion phase and that correspond to the linear portion of the mass loss curves.

This study was specifically focused on establishing adequate techniques for the evaluation of soil erodibility parameters. In this sense, the determination of representative field hydraulic conditions is of paramount importance for the adequate planning and specification of flume testing parameters. The hydraulic shear stresses and corresponding flow velocities selected for soil testing must be based on expected field values, which may be obtained based on an adequate hydrological model [50].

Finally, it is important to note that field testing devices and approaches based on the monitoring of field plots offer important alternatives for the assessment of soil erosion, in particular for bioengineering studies, the evaluation of the effect of vegetation, and the study of soil management practices [51,52]. Field testing also offers unique alternatives for considering real atmospheric events, including realistic rainfall patterns [53], which are difficult to reproduce under laboratory conditions [54,55]. Improving laboratory techniques, such as the device presented herein, allows inexpensive and relatively fast evaluations of soil erosion. However, the use of small-scale laboratory apparatuses and techniques have not been widely employed in the study for the effect of vegetation and bioengineering applications, and further studies are welcomed.

Author Contributions: Conceptualization, G.d.F.N.G.J. and V.N.d.O.; methodology, G.d.F.N.G.J., M.M.d.A.M. and V.N.d.O.; formal analysis, V.N.d.O., G.d.F.N.G.J. and L.F.R.V.; investigation, V.N.d.O.; writing—original draft preparation, V.N.d.O. and G.d.F.N.G.J.; writing—review and editing, M.M.d.A.M., M.M.S., L.F.R.V. and M.P.d.L.; supervision, G.d.F.N.G.J. and M.M.d.A.M.; project administration, M.M.S. and M.P.d.L.; funding acquisition, M.M.S. and M.P.d.L. All authors have read and agreed to the published version of the manuscript.

Funding: This research was financed by the Eletrobras Furnas and Agência Nacional de Energia Elétrica (Project PD-ANEEL 0394-1705/2017) and by the Coordenação de Aperfeiçoamento de Pessoal de Nível Superior—Brazil (CAPES)—Finance Code 001.

Institutional Review Board Statement: Not applicable.

Informed Consent Statement: Not applicable.

Data Availability Statement: The data presented in this study are available on request from the corresponding author.

Conflicts of Interest: The authors declare that they have no conflicts of interest/competing interests that could have influenced this research and paper.

References

1. Julian, J.P.; Torres, R. Hydraulic Erosion of Cohesive Riverbanks. *Geomorphology* **2006**, *76*, 193–206. [CrossRef]
2. Meng, X.M.; Jia, Y.G.; Shan, H.X.; Yang, Z.N.; Zheng, J.W. An Experimental Study on Erodibility of Intertidal Sediments in the Yellow River Delta. *Int. J. Sediment Res.* **2012**, *27*, 240–249. [CrossRef]
3. Rinaldi, M.; Mengoni, B.; Luppi, L.; Darby, S.E.; Mosselman, E. Numerical Simulation of Hydrodynamics and Bank Erosion in a River Bend. *Water Resour. Res.* **2008**, *44*. [CrossRef]
4. Kimiaghalam, N.; Goharrokhi, M.; Clark, S.P.; Ahmari, H. A Comprehensive Fluvial Geomorphology Study of Riverbank Erosion on the Red River in Winnipeg, Manitoba, Canada. *J. Hydrol.* **2015**, *529*, 1488–1498. [CrossRef]
5. Foster, G.R. Modeling the Erosion Process. In *Hydrologic Modeling of Small Watersheds*, 1st ed.; Haan, C.T., Johnson, H.P., Brakensiek, D.L., Eds.; American Society of Agricultural Engineers: St. Joseph, MI, USA, 1982; pp. 297–380.
6. Baigorria, G.A.; Romero, C.C. Assessment of Erosion Hotspots in a Watershed: Integrating the WEPP Model and GIS in a Case Study in the Peruvian Andes. *Environ. Model. Softw.* **2007**, *22*, 1175–1183. [CrossRef]
7. Pandey, A.; Chowdary, V.M.; Mal, B.C.; Billib, M. Runoff and Sediment Yield Modeling from a Small Agricultural Watershed in India Using the WEPP Model. *J. Hydrol.* **2008**, *348*, 305–319. [CrossRef]
8. Dey, S. *Fluvial Hydrodynamics—Hydrodynamic and Sediment Transport Phenomena*, 1st ed.; Springer: Berlin/Heidelberg, Germany, 2014; 687p.
9. Shen, N.; Wang, Z.; Wang, S. Flume Experiment to Verify WEPP Rill Erosion Equation Performances Using Loess Material. *J. Soils Sediments* **2016**, *16*, 2275–2285. [CrossRef]
10. Pandey, A.; Himanshu, S.K.; Mishra, S.K.; Singh, V.P. Physically Based Soil Erosion and Sediment Yield Models Revisited. *Catena* **2016**, *147*, 595–620. [CrossRef]

11. Du Boys, M.P. Le rhone et les rivieres a lit affouillable. *Ann. Ponts Chaussés* **1879**, *18*, 141–195.
12. Partheniades, E. Erosion and deposition of cohesive soils. *J. Hydraul. Hydraul. Eng.* **1965**, *91*, 105–138.
13. Nearing, M.A.; Deer-Ascough, L.; Laflen, J.M. Sensitivity Analysis of the WEPP Hillslope Profile Erosion Model. *Trans. ASAE* **1990**, *33*, 839–849. [CrossRef]
14. Zhang, X.C.; Nearing, M.A.; Risse, L.M.; McGregor, K.C. Evaluation of WEPP Runoff and Soil Loss Predictions Using Natural Runoff Plot Data. *Trans. ASAE* **1996**, *39*, 855–863. [CrossRef]
15. Bastos, C.A.B. Estudo Geotécnico Sobre a Erodibilidade de Solos Residuais não Saturados. Ph.D. Thesis, Universidade Federal do Rio Grande do Sul, Porto Alegre, Brazil, 1999.
16. Kimiaghalam, N.; Clark, S.P.; Ahmari, H. An Experimental Study on the Effects of Physical, Mechanical, and Electrochemical Properties of Natural Cohesive Soils on Critical Shear Stress and Erosion Rate. *Int. J. Sediment Res.* **2016**, *31*, 1–15. [CrossRef]
17. Nearing, M.A.; West, L.T.; Brown, L.C. A Consolidation Model for Estimating Changes in Rill Erodibility. *Trans. ASAE* **1988**, *31*, 0696–0700. [CrossRef]
18. Sanford, L.P.; Maa, J.P.Y. A Unified Erosion Formulation for Fine Sediments. *Mar. Geol.* **2001**, *179*, 9–23. [CrossRef]
19. Khanal, A.; Klavon, K.R.; Fox, G.A.; Daly, E.R. Comparison of Linear and Nonlinear Models for Cohesive Sediment Detachment: Rill Erosion, Hole Erosion Test, and Streambank Erosion Studies. *J. Hydraul. Eng.* **2016**, *142*, 04016026. [CrossRef]
20. Wardinski, K.M.; Guertault, L.; Fox, G.A.; Castro-Bolinaga, C.F. Suitability of a Linear Model for Predicting Cohesive Soil Detachment during Jet Erosion Tests. *J. Hydrol. Eng.* **2018**, *23*, 06018004. [CrossRef]
21. Gao, X.; Wang, Q.; Ma, G. Experimental Investigation on the Erosion Threshold and Rate of Gravel and Silty Clay Mixtures. *Trans. ASABE* **2019**, *62*, 867–875. [CrossRef]
22. Khanal, A.; Fox, G.A.; Guertault, L. Soil Moisture Impacts Linear and Nonlinear Erodibility Parameters from Jet Erosion Tests. *Trans. ASABE* **2020**, *63*, 1123–1131. [CrossRef]
23. Lane, L.J.; Foster, G.R.; Nicks, A.D. Use of fundamental erosion mechanics in erosion prediction. *Am. Soc. Agric. Eng.* **1987**, *11*. Available online: https://agris.fao.org/agris-search/search.do?recordID=US8843443 (accessed on 20 September 2021).
24. Lim, S.S.; Khalili, N. An improved rotating cylinder test design for laboratory measurement of erosion in clayey soils. *Geotech. Test. J.* **2018**, *32*, 1–7. [CrossRef]
25. Lyle, W.M.; Smerdon, E.T. Relation of Compaction and Other Soil Properties to Erosion Resistance of Soils. *Trans. ASAE* **1965**, *8*, 419–422. [CrossRef]
26. Zhang, H.Y.; Li, M.; Wells, R.R.; Liu, Q.J. Effect of Soil Water Content on Soil Detachment Capacity for Coarse- and Fine-Grained Soils. *Soil Sci. Soc. Am. J.* **2019**, *83*, 697–706. [CrossRef]
27. Nearing, M.A.; Parker, S.C. Detachment of Soil by Flowing Water Under Turbulent and Laminar Conditions. *Soil Sci. Soc. Am. J.* **1994**, *58*, 1612. [CrossRef]
28. Debnath, K.; Nikora, V.; Aberle, J.; Westrich, B.; Muste, M. Erosion of Cohesive Sediments: Resuspension, Bed Load, and Erosion Patterns from Field Experiments. *J. Hydraul. Eng.* **2007**, *133*, 508–520. [CrossRef]
29. Tolhurst, T.J.; Black, K.S.; Paterson, D.M.; Mitchener, H.J.; Termaat, G.R.; Shayler, S.A. A Comparison and Measurement Standardisation of Four in Situ Devices for Determining the Erosion Shear Stress of Intertidal Sediments. *Cont. Shelf Res.* **2000**, *20*, 1397–1418. [CrossRef]
30. Ghebreiyessus, Y.T.; Gantzer, C.J.; Alberts, E.E.; Lentz, R.W. Soil Erosion by Concentrated Flow: Shear Stress and Bulk Density. *Trans. ASAE* **1994**, *37*, 1791–1797. [CrossRef]
31. Kamphuis, J.W.; Gaskin, P.N.; Hoogendoorn, E. Erosion Tests on Four Intact Ontario Clays. *Can. Geotech. J.* **1990**, *27*, 692–696. [CrossRef]
32. Zhu, J.C.; Gantzer, C.J.; Peyton, R.L.; Alberts, E.E.; Anderson, S.H. Simulated Small-Channel Bed Scour and Head Cut Erosion Rates Compared. *Soil Sci. Soc. Am. J.* **1995**, *59*, 211–218. [CrossRef]
33. Cantalice, J.R.B.; Cassol, E.A.; Reichert, J.M.; Borges, A.L.O. Hidráulica do escoamento e transporte de sedimentos em sulcos em solo franco-argilo-arenoso. *Rev. Bras. Ciênc. Solo* **2005**, *29*, 597–607. [CrossRef]
34. Shepard, D.M.; Bloomquist, D.; Henderson, M.; Kerr, K.; Trammell, M.; Marin, J.; Slagle, P. *Design and Construction of Apparatus for Measuring Rate of Water Erosion of Sediments*, 1st ed.; The National Academies of Science, Engineering, Medicine: Washington, DC, USA, 2005; 66p.
35. Ni, S.; Zhang, D.; Wen, H.; Cai, C.; Wilson, G.V.; Wang, J. Erosion Processes and Features for a Coarse-Textured Soil with Different Horizons: A Laboratory Simulation. *J. Soils Sediments* **2020**, *20*, 2997–3012. [CrossRef]
36. Smerdon, E.T.; Beasley, R.P. Critical Tractive Forces in Cohesive Soils. *J. Agric. Eng.* **1961**, *42*, 26–29.
37. Shaikh, A.; Abt, S.R. Erosion Rate of Compacted NA-Montmorillonite Soils. *J. Geotech. Eng.* **1998**, *114*, 10. [CrossRef]
38. Ting, F.C.K.; Briaud, J.-L.; Chen, H.C.; Gudavalli, R.; Perugu, S.; Wei, G. Flume Tests for Scour in Clay at Circular Piers. *J. Hydraul. Eng.* **2001**, *127*, 969–978. [CrossRef]
39. Ravisangar, V.; Dennett, K.E.; Sturm, T.W.; Amirtharajah, A. Effect of Sediment PH on Resuspension of Kaolinite Sediments. *J. Environ. Eng.* **2001**, *127*, 531–538. [CrossRef]
40. Barry, K.M.; Thieke, R.J.; Mehta, A.J. Quasi-Hydrodynamic Lubrication Effect of Clay Particles on Sand Grain Erosion. *Estuar. Coast. Shelf Sci.* **2006**, *67*, 161–169. [CrossRef]
41. Koetz, M.; Pruski, F.F.; Mehl, H.U.; Silva, D.D.; Marques, E.A.G. Métodos Para a Determinação da Erodibilidade e da Tensão Crítica de Cisalhamento do Solo em Estradas não Pavimentadas. *Rev. Eng. Agric.* **2009**, *17*, 110–119. [CrossRef]

42. Almeida, J.G.R.; Romão, P.A.; Mascarenha, M.M.A.; Sales, M.M. Erodibilidade de solos tropicais não saturados nos municípios de Senador Canedo e Bonfinópolis (GO). *Geociências* **2015**, *34*, 441–451.
43. Wang, Y.-C.; Sturm, T.W. Effects of Physical Properties on Erosional and Yield Strengths of Fine-Grained Sediments. *J. Hydraul. Eng.* **2016**, *142*, 04016049. [CrossRef]
44. Munson, B.R.; Rothmayer, A.P.; Okiishi, T.H.; Huebsch, W.W. *Fundamentals of Fluid Mechanics*, 7th ed.; Wiley: Hoboken, NJ, USA, 2012; p. 792.
45. Vatankhah, A.R. Explicit Solutions for Critical and Normal Depths in Trapezoidal and Parabolic Open Channels. *Ain Shams Eng. J.* **2013**, *4*, 17–23. [CrossRef]
46. Martins, J.R.R.A.; Sturdza, P.; Alonso, J.J. The Complex-Step Derivative Approximation. ACM Trans. *Math. Softw.* **2003**, *29*, 245–262. [CrossRef]
47. Christopher, B.R.; Holtz, R.D. *Geotextile Engineering Manual—Report FHWA-TS-86/203*, 1st ed.; The National Academies of Science, Engineering, Medicine: Washington, DC, USA, 1985; p. 1044.
48. Nascimento, R.O.; Oliveira, L.M.; Mascarenha, M.M.A.; Angelim, R.R.; Oliveira, R.B.; Sales, M.M.; Luz, M.P. Uso de solução de cal para mitigação de processos erosivos em um solo da UHE de Itumbiara. *Geociências* **2019**, *38*, 279–295. [CrossRef]
49. Basso, L. Estudo da Erodibilidade de Solos e Rochas Sedimentares de Uma Voçoroca na Cidade de São Francisco de Assis—RS. Master's Thesis, Universidade Federal de Santa Maria, Santa Maria, Brazil, 2013. (In Portuguese).
50. Grimaldi, S.; Nardi, F.; Piscopia, R.; Petroselli, A.; Apollonio, C. Continuous Hydrologic Modelling for Design Simulation in Small and Ungauged Basins: A Step Forward and Some Tests for its Practical Use. *J. Hydrol.* **2021**, *595*, 125664. [CrossRef]
51. Apollonio, C.; Petroselli, A.; Tauro, F.; Cecconi, M.; Biscarini, C.; Zarotti, C.; Grimaldi, S. Hillslope Erosion Mitigation: An Experimental Proof of a Nature-Based Solution. *Sustainability* **2021**, *13*, 6058. [CrossRef]
52. Stanchi, S.; Zecca, O.; Hudek, C.; Pintaldi, E.; Viglietti, D.; D'Amico, M.E.; Colombo, N.; Goslino, D.; Letey, M.; Freppaz, M. Effect of Soil Management on Erosion in Mountain Vineyards (N-W Italy). *Sustainability* **2021**, *13*, 1991. [CrossRef]
53. Mohamadi, M.A.; Kavian, A. Effects of Rainfall Patterns on Runoff and Soil Erosion in Field Plots. *Int. Soil Water Conserv. Res.* **2015**, *3*, 273–281. [CrossRef]
54. Mendes, T.A.; Pereira, S.A.d.S.; Rebolledo, J.F.R.; Gitirana, G.d.F.N., Jr.; Melo, M.T.d.S.; Luz, M.P.d. Development of a Rainfall and Runoff Simulator for Performing Hydrological and Geotechnical Tests. *Sustainability* **2021**, *13*, 3060. [CrossRef]
55. Mendes, T.A.; Alves, R.D.; Gitirana, G.d.F.N., Jr.; Pereira, S.A.d.S.; Rebolledo, J.F.R.; da Luz, M.P. Evaluation of Rainfall Interception by Vegetation Using a Rainfall Simulator. *Sustainability* **2021**, *13*, 5082. [CrossRef]

MDPI
St. Alban-Anlage 66
4052 Basel
Switzerland
www.mdpi.com

Water Editorial Office
E-mail: water@mdpi.com
www.mdpi.com/journal/water

www.ingramcontent.com/pod-product-compliance
Lightning Source LLC
Chambersburg PA
CBHW061458210926
51850CB00008B/1754